#  History and the Disciplines

# History and the Disciplines
## The Reclassification of Knowledge in Early Modern Europe

*edited by*
Donald R. Kelley

THE UNIVERSITY OF ROCHESTER PRESS

Copyright © 1997 Contributors

All Rights Reserved. Except as permitted under current legislation, no part of this work may be photocopied, stored in a retrieval system, published, performed in public, adapted, broadcast, transmitted, recorded or reproduced in any form or by any means, without the prior permission of the copyright owner.

First published 1997
1–878822–85–3

University of Rochester Press
668 Mt. Hope Avenue
Rochester, New York, 14620, USA
and at P.O. Box 9, Woodbridge, Suffolk IP12 3DF, UK

Library of Congress Cataloging-in-Publication Data

History and the disciplines : the reclassification of knowledge in
  early modern Europe / edited by Donald R. Kelley.
      p.  cm.
    Includes index.
    ISBN 1–878822–85–3 (alk. paper)
    1. Europe—Civilization—Philosophy.  2. Europe—Intellectual
life—17th century.  3. Learning and scholarship—Europe—
History—17th century.—4. Historiography.  5. Knowledge, Theory
of.  I. Kelley, Donald R., 1931–
CB203.H57  1997
940'.01—dc21                                                97–24434
                                                               CIP

British Library Cataloguing-in-Publication Data

A catalogue record for this book is available from the British Library

This publication is printed on acid-free paper
Printed in the United States of America
Designed and Typeset by Cornerstone Composition Services

# Contents

Introduction .................................................. 1

## PART I. THE PROBLEM OF KNOWLEDGE

DONALD R. KELLEY
　The Problem of Knowledge and the Concept of Discipline ............ 13
ANN BLAIR
　Bodin, Montaigne, and the Role of Disciplinary Boundaries .......... 29
PAUL NELLES
　The Library as an Instrument of Discovery: Gabriel Naudé
　and the Uses of History ...................................... 41

## PART II. PHILOSOPHY AND HISTORY

CONSTANCE BLACKWELL
　Thales Philosophus: The Beginning of Philosophy as a Discipline ....... 61
ULRICH SCHNEIDER
　Eclecticism and the History of Philosophy ....................... 83
MARTIN MULSOW
　Gundling vs. Buddeus: Competing Models of the
　History of Philosophy ...................................... 103
J. B. SCHNEEWIND
　No Discipline, No History: The Case of Moral Philosophy .......... 127
DONALD PHILLIP VERENE
　Vico and the Barbarism of Reflection .......................... 143

## PART III. HUMAN SCIENCES

PETER MILLER
　An Antiquary between Philology and History: Peiresc
　and the Samaritans ......................................... 163
ANN MOYER
　Musical Scholarship in Italy at the End of the Renaissance,
　1500–1650: From Veritas to Verisimilitude .................... 185
MICHAEL SEIDLER
　Natural Law and History: Pufendorf's Philosophical
　Historiography ............................................ 203
ANTHONY PAGDEN
　Eighteenth-Century Anthropology and the "History of Mankind" ..... 223

## PART IV. NATURAL SCIENCES

PAULA FINDLEN
   *Francis Bacon and the Reform of Natural History*
   *in the Seventeenth Century* .............................. 239

ANTHONY GRAFTON
   *From Apotheosis to Analysis: Some Late Renaissance*
   *Histories of Classical Astronomy* ......................... 261

HEIKKI MIKKELI
   *Legitimizing a Discipline: James Mackenzie's*
   *History of Health (1758)* ................................ 277

NICHOLAS JARDINE
   *The Mantle of Müller and the Ghost of Goethe: Interactions*
   *between the Sciences and Their Histories* ................. 297

LONDA SCHIEBINGER
   *Gender in Early Modern Science* .......................... 319

*Index* ........................................................ 335

# Contributors

ANN BLAIR, Assistant Professor of History and Literature at Harvard University, is editor (with Anthony Grafton) of *The Transmission of Culture in Early Modern Europe* (1990) and author of *The Theater of Nature: Jean Bodin and Renaissance Science* (1997) as well as several articles in intellectual history.

CONSTANCE BLACKWELL, founding Director of the Foundation for Intellectual History and the International Society of Intellectual History, is editor of *Models of the History of Philosophy* (1993) and author of several articles on the history of philosophy.

PAULA FINDLEN. Professor of Italian History and History of Science at Stanford University, is the author of *Possessing Nature* (1994), various articles in the cultural history of early science, and *When Science Became Serious*.

ANTHONY GRAFTON, Dodge Professor of History at Princeton University, is author of *Joseph Scaliger: A Study in the History of Classical Scholarship* (1983-1993), *Defenders of the Text* (1991), *Die tragischen Ursprünge der deutschen Fussnote* (1995), etc.

NICHOLAS JARDINE, Professor of the History of Science (Cambridge University), has written both on the history of science, e.g., *The Birth of History and the Philosophy of Science*, and the philosophy of science (*The Fortunes of Inquiry* and *The Scenes of Inquiry*), and is author of many articles in these areas.

DONALD R. KELLEY is James Westfall Thompson Professor of History at Rutgers University, editor of the *Journal of the History of Ideas*, and author of *Foundations of Modern Historical Scholarship* (1970), *The Human Measure* (1990), *Renaissance Humanism* (1991), two volumes of studies of history, law, and the human sciences (1984, 1997), etc.

HEIKKI MIKKELI, Junior Research Fellow at the Academy of Finland, University of Helsinki, is author of *An Aristotelian Response to Renaissance Humanism: Jacopo Zabarella on the Nature of Arts and Sciences* (1992) and forthcoming work on European encyclopedias.

PETER N. MILLER has been a Fellow of the Wissenschaftskolleg zu Berlin, Clare Hall (Cambridge), and Mellon Instructor at the University of Chicago, and is author of *Defining the Common Good: Empire, Religion and Philosophy in Eighteenth-Century Britain* (1994) and (with Iain Fenlon) *The Song of the Soul: Understanding Poppea* (1992).

ANN E. MOYER, Assistant Professor of History at the University of Pennsylvania, is author of *Musica Scientia: Musical Scholarship in the Italian Renaissance* (1992) and other studies in this field.

MARTIN MULSOW teaches philosophy at the University of Munich, is the author of a commentary on Bruno's *De Monade* and editor of a volume on Johann Lorenz Mosheim, and has forthcoming books on Telesio, Mathurin Veyssière La Croze, and Georg Shades.

PAUL NELLES, trained at the Warburg Institute in London as well as Johns Hopkins University (Ph.D. 1994), has written on Gabriel Naudé, the history of learning, and the history of libraries in the seventeenth century.

ANTHONY PAGDEN, Harry C. Black Professor of History at John Hopkins University, has published *The Fall of Natural Man* (1982), *Spanish Imperialism and the Political Imagination* (1990), *European Encounters with the New World* (1993), *Lords of all the World* (1995), etc.

LONDA SCHIEBINGER, Professor of History and Women's Studies at Pennsylvania State University, is author of *The Mind Has No Sex? Women in the Origins of Modern Science* (1989) and *Nature's Body: Gender in the Making of Modern Science* (1993).

J. B. SCHNEEWIND is Professor of Philosophy at John Hopkins University, President of the Board of Editors of the *Journal of the History of Ideas*, and author of *Sidgwick's Ethics and Victorian Moral Philosophy*, *Moral Philosophy from Montaigne to Kant* (1990), etc.

ULRICH SCHNEIDER, Professor of Philosophy at the University of Leipzig, is author of *Die Vergangenheit des Geistes* (1990) and many articles in German and French as well as English on the history of philosophy.

MICHAEL J. SEIDLER, Associate Professor of Philosophy at Western Kentucky University, is author of several articles on early modern political and moral philosophy and translator or Samuel Pufendorf's *Political Writings* (1994) and *On the Natural State of Men* (1990).

DONALD PHILLIP VERENE is Charles Howard Candler Professor of Metaphysics and Moral Philosophy; Director of the Institute for Vico Studies at Emory University; and author of *Vico's Science of the Imagination* (1981), *Hegel's Recollection* (1985), and *The New Art of Autobiography* (1991); and coeditor of Ernst Cassirer, *The Philosophy of Symbolic Forms*, IV (1996).

# Introduction

## Donald R. Kelley

These essays are all addressed, in specific historical ways and from particular disciplinary standpoints, to the problem of knowledge and to what used to be called the classification of the sciences. What is or what passes for knowledge? What are its divisions, and how should they be related? Who possesses this knowledge, and to what uses has it been put? How is it transmitted, and how can its history be understood and written? In this volume these questions are limited to the period between the Renaissance and the Enlightenment. Ranging across the epistemological barrier formed by the revolution of modern science, these contributions inquire into the changing disciplinary and interdisciplinary patterns of those parlous and tumultuous times which saw the fragmentation of old ideals and the creation of European modernity.

What gives concreteness, continuity, and intelligibility to the history of Western knowledge is the concept of *discipline*, defined originally as the relationship between disciple and master and so possessing religious as well as pedagogical and perhaps political connotations. In general, disciplines constitute the nations, provinces, and smaller units on the map of learning at any given point in cultural time. A discipline may be seen as a science, art, or a craft, depending on the procedures it has formulated, the following it has collected, and the perspective it has formed on its own activities. What may be regarded as marks of disciplinarity include a characteristic method, specialized terminology, a community of practitioners, a canon of authorities, an agenda of problems to be addressed, and perhaps more formal signs of a professional condition, such as journals, textbooks, courses of study, libraries, rituals, and social gatherings—not to speak of revealing calls for more "interdisciplinary" approaches to knowledge.

The ordering of disciplines has usually been hierarchical, from the Aristotelian system of practical and theoretical science to the "new Organon" of Francis Bacon, for whom "knowledges are as pyramids," with the more rational sorts of knowing placed at the top. In medieval (and indeed modern) universities learning is a ladder, homologous with the classes of society and those of the great chain of being itself—the liberal arts preparing the way for higher sciences such as medicine, law, and theology; and libraries relied on similar distinctions in their classification. The humanist movement of the Renaissance subverted this invidious arrangement, reverting to the classical conceit of a circle of learning (*encyclopaedeia*); but the New Science of the seventeenth century restored the principle of hierarchy, making philosophy again the highest form of secular knowledge. From Descartes to the French Ideologues and Positivists the disciplines were arranged in a "chain," with the more rational sciences given priority; and universities today have in many ways (financially as well as intellectually, institutionally, and socially) preserved these priorities.

The patterns formed by particular disciplinary trajectories and interplay, within the encyclopedia of available knowledge, have been kaleidoscopic and cacophonous. Some disciplines have enjoyed great longevity and extraordinary fortunes. Over many centuries the seven liberal arts of the trivium (grammar, rhetoric, and dialectic) and quadrivium (mathematics, geometry, music, and astronomy) have risen to high and even scientific status. Other disciplines, such as the sciences of medicine and law, after having fallen on hard times, have recovered spectacularly in the modern age. Still others, such as anthropology, hygiene, and many special branches of the physical, social, and life sciences, are of more recent vintage, though in no way less ambitious to achieve professional recognition. The picture is much more complicated by shifting alliances among older and newer fields of knowledge—natural philosophy joining with mathematics and the "mechanical arts" to form the new empirical and mathematical science, music forging links with poetry as well as mathematics, the study of religion seeking assistance from philology (the "scientific" form of grammar), and philosophy making connections with many other fields, especially with the Eclectic school of the seventeenth and eighteenth centuries, which turned to history as the foundational discipline.

This is not to say that the idea of discipline should be taken for granted; for interdisciplinary criticisms and crossovers have always operated to question, to weaken, or to deny intellectual barriers; and so, from the time of Ramus and Descartes down to the present century, have notions of the unity of science, or of scientific method. Still more fundamental critiques were launched by skeptics such as Henry Cornelius Agrippa of Nettesheim and Francesco Sanches, who cast doubt on the value and even possibility of positive knowledge. The stability and legitimacy of traditional, disciplinary knowledge were further undermined by human factors of race, class, and gender, which even for philosophers were occasionally acknowledged to be the conditions of learning and teaching. One may see the ideological and epistemological critiques of Foucault and Derrida as continuations of such skeptical lines of argument, as also postmodern attitudes toward history and intellectual tradition; but even here the concept retains at least a negative value, as a condition to be overcome. Indeed "discipline" is a locus of many of the modernist assumptions—the "founding function of the subject," intellectual "influence" and continuity, and ideas of progress—which have been the targets of such poststructuralist criticism.

Disciplines have often been the objects of study, in theoretical ways as well as in the history, analysis, or celebration of particularly disciplinary constructs. Both approaches are in different ways artificial and inadequate, theory tending to accept disciplines as more or less equivalent cultural units (commensurable paradigms, to adopt the ubiquitous Kuhnian terminology), and "disciplinary histories" being bound in many ways to a presentist, internalist, and perhaps Whiggish approach. The papers assembled here are much more concerned with the cultural, contextual, institutional and historically problematic aspects of disciplines and with the investigation of particular texts, authors, intellectual projects, and what Nicholas Jardine calls the "scenes" and "fortunes" of inquiry.

The problem of knowledge takes many forms in the early modern period, especially during the search for a proper method of understanding and discovery. Ann Blair addresses the question through a comparative study of the work of Jean Bodin and Michel de Montaigne. Despite similar backgrounds and experiences in an age of religious conflict, Bodin and Montaigne took fundamentally different views of the problem of knowledge and the conditions of thought. Bodin's approach was encyclopedic but uncritical; for in his study of natural history as well as history and politics he plucked the "flowers"—accepted unproblematically as facts—from his vast reading, gathering his data indiscriminately into commonplace books, which he then shaped magisterially into useful and conventionally disciplined knowledge. By contrast Montaigne, in the pursuit of self-knowledge, questioned the philosophical and literary works that attracted his interest and focused not on prerecorded knowledge but on whatever insight into the human condition he could attain. If Bodin's construction was a grand theater, a replica perhaps of God's own creation, that of Montaigne was a literary self—a new and more profound Michel—created in an unprecedented book representing a critical journey across many intellectual boundaries in a search for a personal wisdom transcending disciplinary knowledge.

Bodin and Montaigne both worked within the tradition of Renaissance humanism, an intellectual project aspiring to encompass not only the liberal arts but the whole heritage of learning which print culture made available, and which Bacon called *historia literaria* ("literary history" in the sense of everything that had been written down). In this explosion of knowledge another new discipline came into play—that is, "library science," represented especially in the bibliographical works of Konrad Gesner, Antonio Possevino, and other pioneers in the effort to accommodate the old encyclopedia to the new learning. Drawing on these precedents, as Paul Nelles shows, Gabriel Naudé helped to define the structure and role of the public library, which Nelles called the "defining institution" of the seventeenth-century Republic of Letters. In the modern *Bibliotheca* Naudé envisioned not only an institutional embodiment of the *encyclopedia* and the old art of memory but also, in Baconian style, an instrument of research and the transformation of culture. His plan, following the precedent of humanists like Polydore Vergil and anticipating the modern faculty system of higher learning, made history, and the history of disciplines in particular, the organizational principle of modern, fast-accumulating knowledge. In the institution of the modern library all the disciplines of early modern European culture could find a place, continue to compete for elevated and even foundational status, and make contributions to the kaleidoscopic problem of knowledge that still occupies historians as well as philosophers.

In the library, as in the university and the Republic of Letters at large, the foundational discipline was philosophy, and philosophy was also establishing ties with history. In its largest extent philosophy continued to be virtually coterminous with human knowledge; in a more restricted sense it is an academic discipline which, while identified in medieval universities as the teaching of Aristotle, took on new and more "eclectic" forms in the early modern period and was separated from particularly doctrinal schools. This became clear through the work of scholars and critics who, out of the writings of philosophers drawn from many centuries, created a new self-proclaimed discipline called the history of philosophy (*historia philosophica*). From the seventeenth century this new "science," which had itself recently emerged from the larger field of the "history of literature," gave both legitimacy and an illustrious genealogy to the field which, following the work of Diogenes Laertius, it surveyed, and thereby, in a historical way, defined. Constance Blackwell's paper shows how philosophy, through efforts to reconstruct its origins, established its intellectual identity and separated itself from myth and religion. Renaissance scholars often confused philosophy with "wisdom" (*sapientia*) in general, including near eastern occult traditions; and it was the task of self-proclaimed "critical" historians like J. J. Brucker to extricate true philosophy from the ignoble and irrational antecedents of *prisca theologia*. So, following Diogenes Laertius, Brucker restricted the discipline of philosophy—terms, concept, and practice—to the Greeks, beginning with Thales, who has ever since headed the Western philosophical canon.

The philosophical movement which gave a conceptual basis to this alliance with history was that of Eclectic Philosophy. Ulrich Schneider traced the transition of Eclecticism from a "sect" to a general method (*methodus Eclectica*) which—upholding the Ancients in the philosophical version of their quarrel with Moderns such as the skeptical Cartesians—at once liberated philosophy from the authority of the schools and at the same time joined it to its ancient heritage in a positive and hermeneutical way. According to Christian Thomasius, Johann Sturm, J. F. Budde, and other Eclectics, the best procedure for philosophy was to select, on the basis of critical scholarship, the best from this inheritance and thus, while avoiding the fallacies of syncretism, to advance from error to enlightened truth. The motto of Eclecticism was that there could be no valid philosophy without the history thereof, so that philosophy was in effect historicized, and the history of philosophy legitimized as a discipline in its own right.

A more problematic disciplinary formation was moral philosophy, which was situated, paradoxically, both in the circle of liberal arts (being one of the five *studia humanitatis*) and among the practical sciences of Aristotelian philosophy. J. B. Schneewind shows the mixed heritage of this inchoate discipline, defined alternatively as art and as science and torn as it were between Athens and Jerusalem—or (in Matthew Arnold's famous formula) between Hebraism and Hellenism—in its efforts to find a satisfactory genealogy and a usable past. Was moral philosophy a legacy of Socrates, who gave philosophy a practical turn in the famous characterization of Cicero, or of an older, pre-Greek tradition indicated clumsily by the (later discredited) hypothesis that Pythagoras was Jewish? In other words, was moral philosophy primarily rational or religious in inspiration? Was it tied to assumptions about the essentially rational or essentially sinful nature of humankind? Later thinkers, such as Barbeyrac, were of two minds about this historical-philosophical issue; but in any case each of these "grand narratives" demanded a "single-aim" premise which, Schneewind argues, is untenable in the light of the history of moral philosophy—which leaves as questionable the disciplinary status of moral philosophy.

With the work of Giambattista Vico we remain in the tradition of Eclectic philosophy in a sense, but we also reach, in a sense, the end of disciplines. What Vico envisaged, likewise on the basis of historical and philological scholarship, was a totalizing metaphysical "new science" which, while drawing on the entire heritage of Western arts and sciences, especially rhetoric, philology, and law, aspired to bring them all together into one historicist encyclopedia. Donald Verene tries to capture an important aspect of Vico's metahistorical creation by examining a paradoxical coinage which joins one term associated with the primitive stages of human experience (parallel to the contemporary "savagery" discussed by Lafitau) with another associated with its maturity—or senility. The "barbarism of reflection" is a characterization of and a lament for the abstract and hyperrational line of argument, traceable usually to Descartes and Locke, which has alienated human thinking from experience, diverted it from virtue, and divorced it from piety. According to Verene, Vico's New Science is a grand oration on moral philosophy. Yet it transcends this discipline (if discipline it was) in its inter- and trans-disciplinary aspiration—through a critical and cyclical return to unified wisdom—to be at once philology, law, philosophy, theology, and a universal "history of ideas."

It is characteristic of lowly liberal arts to aspire to the status of science; and this is just what happened to the *ars grammatica* as it was transformed into the science of philology. This discipline—likewise claiming an ancient pedigree—had risen above its trivial origins in the work of scholars such as Lorenzo Valla, whose discipline was rhetoric, and Angelo Poliziano, who (like Erasmus) still called himself a "grammarian." Philology expanded interdisciplinary contacts without the whole encyclopedia of arts and sciences, which itself was transformed, of course, during the century that separated Erasmus (d. 1536) and Nicholas-Claude Fabri de Peiresc (d. 1637), whom Peter Miller takes as illustrating the condition of the Republic of Letters in the wake of discoveries, geographical as well as scientific. Peiresc carried on the tradition of Biblical studies associated especially with Erasmus; but unlike that pioneer of Biblical humanism Peiresc could draw upon a vastly increased knowledge of oriental languages (thanks to the labors of the younger Scaliger and others), a wider scholarly network, extended by expeditions to the Levant, and grammars and other reference tools unavailable to earlier generation. The Peirescan project examined by Miller is the study of the newly discovered manuscript of the Samaritan Pentateuch, which introduced the (still debated) question of variants in the Hebrew text of the first five books of the Bible. In keeping with humanist scholarship and in contrast to the critical disciplinary efforts of Kepler and Brucker, Peiresc continued to be fascinated with prehistorical, indeed pretextual, origins. Regarding the Samaritan language as giving potential access to an original text of Hebrew scriptures and perhaps to the lost language of ancient Egypt, Peiresc anticipated the later work not only of the "oriental renaissance" and the Higher Criticism of the Bible but also of the new disciplinary formations, notably modern comparative and historical linguistics.

If philology asserted claims to interdisciplinary relevance as well as disciplinary status and a certain conceptual autonomy, so did other members of the old circle of liberal arts, though not always with the same success. Music, for example, was situated in the quadrivium (locus of the mathematical subjects) and yet claimed links as well with the art of poetry, drama, and other disciplines within the humanities. In the Renaissance Pythagorean constructions of music as a transcendent expression of universal harmonies at once elevated and confused its intellectual status. Music displayed an ambiguous character, too, in being divided not only between theory and practice but also between composition and performance—and of course introducing aesthetic considerations as well. Ann Moyer traces the career of this discipline, especially in terms of humanist treatises on the *ars musica*, from the early Renaissance to the period when, as in the work of Vincenzo Galilei, physical and scientific (e.g., acoustical) questions came more fully into play. In the later period, too, psychological and medical questions overshadowed the old focus on Pythagorean interpretation. In the musical counterpart to the quarrel between ancients and moderns, music history and antiquarian research became important, although, in contrast to visual art as well as literature, the recovery of the object of this study—music itself—was in large part speculative. By the seventeenth century music improved its disciplinary status and historical perspective, but at the cost of losing its claims to be a universal key to understanding the mysteries and harmonies of the universe.

History also played a role in practical—or what we would now call human—philosophy, including law, ethics, and anthropology in a modern sense. In the sixteenth century history had already formed alliances with what was already called jurisprudence and "political science"; in the seventeenth century it established ties also with the, as Otto Gierke called it, "antique-modern" philosophy of modern natural law. Best known as one of the two or three most influential members of this school, Samuel Pufendorf, like his friend Leibniz, wrote and though extensively about history, especially in terms of contemporary political issues. As Michael Seidler shows, Pufendorf interpreted historical action within a secular moral framework, defined by the modern *ius naturale* and *ius gentium*, which liberated history from transcendent and theodicean assumptions and represented it as a field of free human agency, diverse conditions, and undetermined form. History has practical value but had to be viewed not *sub specie aeternitate* but according to the multiple—and as it were monadic—perspectives of modern European political life.

In the transfigurations of the old encyclopedia and its divisions, older disciplines have, in order to survive, been renewed or modernized; but occasionally a wholly new form of knowledge has emerged from unorganized practices and concepts within traditional structures of knowledge. A notable case in point is modern "anthropology"—a term which was equivalent originally to "psychology" but which came to designate the ethnographic study of human behavior based on travel (field work), systematic observation, and theoretical interpretation. Historical scholarship had long supplemented philosophical speculation in the effort to understand human nature in its widest sense; the exploration of the New World provided no less exotic examples of human diversity; and it was out of this growing mass of ethnographic data and its increasing sophisticated analysis that this new discipline developed and that the human science more generally expanded. Anthony Pagden illustrates this extraordinary process with special attention to the observations on the customs and culture of the "American savages," which was a major contribution to the Enlightenment project of "the history of humanity" in a global sense and to anthropology in a contemporary sense.

The natural sciences, too, looked to history to tell their disciplinary stories and give themselves honorable pedigrees. Paula Findlen shows how Francis Bacon attempted to reform the field of natural history, essential to natural philosophy, and so to transform this classical pursuit into a modern scientific discipline and profession. For Bacon—very much in the spirit of Kepler and Brucker—"disciplining" natural history meant extricating it from its "trivial" humanist (and one might say Bodinian) condition and setting in on a methodical and scientific course. Clearly, this disciplinary reform was directed not only to more reliable (that is, more direct and less literary or historical knowledge of the natural world) but also to the formulation of a proper method by which to give shape and direction to the serious projects of modern philosophy. Although Bacon's efforts bore little immediate fruit, they did serve as a model for later research; and they set a widely accepted story line for the history of science and of the many subdisciplines which it was to spawn.

Anthony Grafton illustrates the phenomenon of disciplinary self-definition with reference to astronomy, an art which was elevated from its quadrivial condition to the level of philosophy by Ptolemy and his descendants and correctors, Copernicus, and especially Johann Kepler. Before Kepler Renaissance scholars, notably the mathematician Georg Joachim Rheticus and the nature philosopher Girolamo Cardano, told the story of the heavens—an area shared by astronomy and astrology—as a search parallel to that of philosophy in general; and they also traced it back to pre-Greek practices, especially those of the Egyptians and Chaldaeans, based on belief in a lost astronomical wisdom analogous to the pre-philosophical *prisca sapientia* detectable in the Hermetic texts and other occult sources. In his new astronomy Kepler, while he was sympathetic to astrology and a certain sort of number mysticism, rejected speculation about a lost science of the stars. As Brucker restricted the history of philosophy as a modern discipline to the Greek wise men, so Kepler, seeking "astronomy without hypotheses," designated the Greeks, in the person of Ptolemy, as the founders of the discipline which he devoted his life to renewing.

Heiki Mikkeli shows the emergence of another new scientific discipline out of one of the oldest of human sciences. From antiquity hygiene, or diatetics, was recognized as part of medicine; but in the early modern period, within the framework of medical humanism, it established itself as an independent field of study. Practical as well as theoretical, popular as well as academic in character, health science looked to "non-naturals" (diet, exercise, and other physical actions) to induce physiological change and so to operate not only as remedies but also as "preventive medicine" and, according to the agenda of Cardano, Bacon, and others, to furnish a means of pursuing longevity. By way of illustration Mikkeli looks in particular at James Mackenzie's *History of Health* (1758), which not only sets down rules of health but also gave legitimacy to this new discipline within the framework of old-fashioned humoral (and astrological) medicine by showing its progress over many generations.

Nicholas Jardine carries the question of the disciplinary history of science, via medicine in particular, into the nineteenth century by examining the contrasting views of two German colleagues and rivals, Emile Du Bois-Reymond and Rudolf Virchow. In terms not only of their conceptions of their discipline, its past and future, but also of its relations with society and culture more generally, these two scientists stood poles apart. Virchow took a traditionalist view of medicine (looking sympathetically upon speculative German *Naturphilosophie* and the poetic influence of Goethe's controversial natural philosophy); and yet socially he was a radical, committed to the reformation of society as well as clinical practice. By contrast Du Bois-Reymond, no activist, was a champion of the new experimental methods, and his career line followed this privileging of laboratory medicine. The interpretations of the history of medicine by the rivals displays a similar contrast, with Du Bois-Reymond adopting a Baconian notion of progress through experimental science, and Virchow taking a wider but softer focus that set the history of medicine in a social and cultural context. As Jardine concludes, the history of science is still in a sense divided between what would now be called triumphalist and socio-constructionist interpretations.

That natural sciences, in the early modern period, in their drive to achieve high disciplinary status, including official support and a place in the university curriculum (and library catalogues), tended to be exclusive as well as elitist; and one consequence was the barring of women from participation in professional scientific activities. Londa Schiebinger, in an essay that is a plea as well as an analysis, shows how the value-free pretensions of early modern science and the notion that the operations of the mind were above sexual influences, instead of leading to the sort of objectivity that these beliefs seemed to entail, were in fact turned against women scholars. Instead of rising above the human condition, male scientists had a skewed view of gender which distorted their perceptions of nature in general, as illustrated most conspicuously by the efforts of biological classification—the sexual parts of plants, for example, and especially the Linnaean class of "mammals." What, Sciebinger asks would science have been like without these invidious attitudes toward gender?

These are just a few soundings on the changing chart of knowledge in the Republic of Letters in early modern Europe, illustrating the problematic role of disciplines and the expansion, traversing, and obliteration of disciplinary frontiers by a variety of scholars—humanists as well as scientists, theoreticians as well as practitioners, critics as well as historians. One common pattern is the employment of history to give identity, antiquity, and (therefore) legitimacy to disciplinary formations such as music, astronomy, and philosophy itself—processes that may also, as in the case of natural history, be accompanied by projects of reformation and modernization. These historical efforts were generally cast in the form of stories of progress and a critical awareness that—even in the case of traditional disciplines such as philology—entailed separation from earlier, often mythical stages of traditional disciplinary histories. The result was modernized stories of the rise of particular disciplines and the formation of new perspectives, canons, methods, and authorities—notions of mastership and discipleship—for the further advancement of particular arts and sciences.

At the same time, the notion of disciplinary knowledge came into question, as is evident by the divergent conceptions of science formed by Bodin and Bacon—the shift from a largely literary to a large experiential, or experimental, method. Some fields, such as moral philosophy and perhaps anthropology, transcended academic convention and seem difficult to situate within the framework of disciplinary knowledge at all, as attempts to write their history revealed large cultural disparities that impeded simple definition. Still other scholars, from Montaigne to Vico and Lafitau, carried on their unconventional enterprises beyond the disciplinary grid of academic knowledge in search of deeper and more human forms of understanding. Their inter-, meta-, or antidisciplinary ventures are also part of the problem of knowledge and the concept of discipline in this early period of Western modernity and pre-postmodernity.

*Part I*
*The Problem of Knowledge*

# THE PROBLEM OF KNOWLEDGE
# AND THE CONCEPT OF DISCIPLINE

## Donald R. Kelley

Since the Renaissance, Western philosophy has been dominated by the problem of knowledge (*episteme, scientia, cognitio, science, connaissance, scienza, Wissenschaft, Erkenntnis*, etc.). For philosophers this has meant above all epistemology or gnosiology, that is, the debates excited in the wake of the Cartesian *cogito*, the Lockean "way of ideas," and the Kantian critique of pure reason.[1] Can we know at all, and if so, how? Through the senses, reason, or a combination of the two? Can knowledge be certain? How can we achieve such certainty? Such questions have led academic philosophy down an abstract road toward a vision of "rigorous science" (*strenge Wissenschaft*, in Husserl's phrase)[2] and hegemony over other fields, but—such is the price that must be paid for such rigor—they have also encouraged a break between the form and content, the acquisition and transmission, and the nature and history of knowledge.

Attendant on these issues is a second major problem, which is that of the unity of knowledge. Is science an integrated whole, or should it be so integrated, or is it, and should it be, as diversified as the human condition itself? Plato described the question in the *Sophist* in this way:

> Knowledge is one, but each separate part of it which applies to some particular subject has a name of its own; hence there are many arts [*technai*] . . . and kinds of knowledge, or science [*epistemai*].[3]

Philosophers have usually sought to transcend the particular (often regarded as the "lower") arts and disciplines and their languages through a unified method or metalanguage, which has usually been demonstrative, whether logical, mathematical, or metaphysical. In the mid-sixteenth century, for example, Peter Ramus argued "that there should be but one method for establishing science" (*quod sit una doctrinae instituendis methodus*),[4] and this neo-scholastic line of argument was pursued in various ways by Bacon, Descartes, Leibniz, and other major figures in the philosophical canon.

Related to these questions, too, is the problem of the proper scope of philosophy. In the medieval scheme of learning philosophy, or science, held a subordinate position. According to the Venerable Hildebert,

> *Scientia* is the lower part of reason by which men can correctly manage their earthly affairs and secular business, and arrange to live rightly in this depraved world which everyone knows to be mortal.[5]

The possibility of true philosophy and perfect knowledge had been eliminated by the Fall, which itself, ironically, had been produced by an excess of curiosity.

In modern times philosophy regained some of its original prestige, but there continued to be disagreement about its scope. On the one hand there is the belief that philosophy consists in first principles—knowledge through causes (*scientia est per causas cognitio*) is one common formulation, which usually implies metaphysics; on the other hand there is the polyhistorical or encyclopedic idea that philosophy is the sum of the intellectual practices which have been produced by human culture and which have been communicated through methods and institutions of formalized learning. "Philosophia est omnis scibilis cognitio" is the way the encyclopedist Johann Alsted put it in the early seventeenth century: "Philosophy is the understanding of everything intelligible."[6]

Now this encyclopedic and eclectic view is obviously a world apart from the revolution of modern science celebrated in more recent times, and yet it has never entirely disappeared from the agenda of philosophical inquiry. In any case what is at issue here is this second option—not the high road to science opened up by Galileo and Descartes or even Bacon and Locke but the low road travelled by the old-fashioned philologists and the polyhistors. The purpose, in other words, is to explore the problem of knowledge not in terms of formal philosophy—not as the abstract construction of ideas out of reason or experience—but as a process of what I will call *mathesis*. By this old term, "mathesis," I mean not the study of mathematical quantity (as seventeenth-century scholars like Gerard Vossius employed the word) or the "universal mathesis" of Descartes or Leibniz or certainly the "mad mathesis" of Pope ("mathesis too mad," he says, "for mere material chains to bind").[7] Rather, reverting to an earlier usage, I mean that system of study which Renaissance scholars called "discipline" but which, as Juan Luis Vives noted, the ancients called "mathesis."[8] That is, the connection between the master (*didaskalos*) and the disciple (*mathetes*), the process of intellectual transmission or succession which ostensibly gave structure and continuity to Greek philosophical tradition from the very beginning.

The most notorious example of this authoritarian pattern can be seen in the legendary career of Pythagoras, the very inventor of the term *philosophos*. The "godlike one" as his disciples called him, bound himself to no master. According to Iamblichus,

> He neglected no doctrine valued in his time, no man for understanding, no rite honored in any region, no place where he expected to find some wonder. So he visited all the priests, profiting from each one's particular wisdom.[9]

Yet despite his own eclectic background, the pedagogical method of Pythagoras was, legend has it, masterly in every sense of the term; and according to a Cartesian critic, his disciples (*discipuli*) resolved questions not by reason but by authority (*non rationis sed auctoritatis causa*), following the famous Pythagoreran formula, cited also by Vives, *ipse dixit* (*autos epha*).[10] And as founder of the Italic sect, Pythagoras passed on his magisterial habits (learned, it was believed, from the Egyptian priests he had visited) to Socrates, Plato, Aristotle, and the "sects" and academies in which philosophy came to be taught, with structure and continuity being provided by the master (*scholarch*) and a regularized succession (*diadoche*).

In the history of philosophy this "diadochism" is given classical expression in the foundational work of Diogenes Laertius, the *Lives and Opinions of the Philosophers*. Assembled in the third century A.D., this uncritical collection of lives, opinions, documents, anecdotes, memorable sayings, and poems (including his own) was based not only on oral tradition but also on a mass of previous writing, including historical treatises *On the Succession of Philosophers* by Sophion and Alexander Polyhistor and philosophical works and comments of philosophers on their predecessors. Literary fortune, rather than scholarly or conceptual quality, made Diogenes Laertius in effect the founder of philosophical doxography and a treasury of information about the human aspects of ancient philosophical tradition.[11]

The pattern established by Greek diadochism was continued by medieval scholars, who likewise received doctrine on the basis of the principle of *magister dixit*. Initially at least, (discipular) truth had its foundation in (doctrinal) authority—learning (*discere*) and teaching (*docere*) being, for Isidore of Seville, the respective etymological roots of "discipline" (*discere* and *plena*) and "doctrine."[12] According to a lecturer at the University of Paris in the thirteenth century, "*discipline* expresses the spiritual link between the master and the pupil."[13] *Disciplina in discipulo, doctrina in magistro* was the often-repeated formula—or, according to the entry in Thomas Cooper's Latin dictionary of 1565, "learnyng as it is perceyved of the scholar, doctrine as it is taught of the mayster."

However remote from epistemology, the concept of "discipline" is essential to the problem of knowledge in human terms, representing as it does the categories of learning and vehicles of intellectual transmission over many centuries; and in this connection *disciplina* has been essential to pedagogical theory and practice, which has its own sort of rudimentary epistemology. Medieval scholars such as Boethius and Gilbert de la Porrée regularly distinguished between the doctrinal and the disciplinary (knowledge transmitted *disciplinaliter* and *doctrinaliter*) aspects of learning as a human process aimed at, though not identical with, an intellectual and pedagogical ideal.[14]

"Discipline" has had a special relationship with philosophy, which according to Cassiodorus was the "discipline of disciplines" (*disciplina disciplinarum*)[15] and, in the still more hyperbolic phrase of the great humanist Guillaume Budé, the "inventor of disciplines" (*inventrix disciplinarum*).[16] The term "discipline" has regularly been regularly represented in philosophical dictionaries. In Goclenius's *Lexicon Philosophicum* (1613) discipline could be either an art or a science ("disciplina interdum accipitur pro arte vel scientia") and constituted for him what, in a restricted sense, the Greeks called mathesis. And according to the *Lexicon Philosophicum* of Etienne Chauvin (1692),

Discipline is a conception accepted from a master, such that disciples follow the master's example through his teaching. It is in this way that philosophy is formed, so that [discipline] is properly compared to the light of natural reason.

If the idea of discipline has not figured prominently in speculative philosophy, it was certainly central to the work of encyclopedic authors such as Bartholomew Keckermann and Johannes Alsted. "Method is the soul and form of disciplines," Keckermann wrote in 1612, "and without them there is coherence neither in things nor in the human understanding of things."[17] In his *Philosophia Restituta* of the same year (and again, in more detail, in his *Encyclopedia*) Alsted offered an elaborate theory of discipline in connection with his polyhistorical vision, describing the principles, nature, difference, and various ways of teaching the disciplines making up his encyclopedia.

For Alsted disciplines could be considered in three ways—as universal, if their principles are general (such as physics or ethics), as common, if their principles are similar (such as geography or astrology), or as singular if they are unique (such as a language).[18] Internal discipline (*disciplina interna*) was an intellectual *habitus*, while external discipline was a methodical system organized for the instruction and "perfection" of individuals—an encyclopedia being a system of such disciplinary subsystems. In its disciplinary deployment, according to Alsted, knowledge ranged from the most certain *scientia*, which is demonstrable and necessary, down to the lowest *opinio*, which is probable rather than certain and dependent on local consensus rather than universal reason.

In philosophical tradition "opinion" (*doxa*) has been a derogatory term at least since Plato, and modern philosophy has preserved this pejorative usage.[19] Ian Hacking associates opinion with the seventeenth-century "emergence of probability," but for many philosophers it represented an even more degraded form of knowledge, being identified with unreflective custom, partisanship or material interest, and arbitrary authority.[20] As Hume observed, "though men be much governed by interest; yet even interest itself, and all human affairs, are entirely governed by *opinion*."[21] For Hegel the conventional history of philosophy is a mere "gallery of opinions" (*Meinungen*), which "becomes a thing of idle curiosity, of interest only to the merely erudite; for erudition consists mainly in knowing a mass of useless things."[22]

But useless for what, one may ask, and for whom? And is it really legitimate to reserve the word "opinion" just for the views of others, as philosophers from Plato to Popper have commonly done? For clearly, the term "opinion" may have positive connotations. In his *Treatise on Opinion* of 1733 Gilbert-Charles Le Gendre reminded his readers that all human sciences are grounded in opinion, and in this connection he cited the famous aphorism of Pindar, invoked long before by Montaigne, Charron, and other skeptics, which recognized that custom, or opinion, was ruler in human affairs: "opinion, sovereign over gods and men." "Pindar appelle l'Opinion la souveraine des dieux et des hommes" is Le Gendre's rendering of Pindar's *nomos basileus*.[23]

The "new philosophy that placed all in doubt" affected to be free of "prejudice" (a pejorative buzz-word of the dawning Enlightenment) and indeed any subjection to previous authority.[24] Pierre Sylvan Régis, who was one of Descartes's most faithful disciples, portrayed Pythagoras as a paradigm of dogmatic philosophy; and he repeated the old charge that the opinions of Pythagoras were cited by his discipuli to resolve questions not by reason but by authority (*non rationis sed auctoritatis causa*). Not until the sixteenth century, Regis concluded, was philosophy finally "liberated" from such servile habits. In the seventeenth century such declarations of independence from magisterial authority were also expressed by the common slogan, "liberty of philosophizing" (*libertas philosophandi*), and by the Horatian formula, *nulla in verbis magistri jurare*.[25]

Yet to some critics such claims sounded hollow—Regis himself being a prime example of idolatrous Cartesianism. In any case, how was it possible to escape one's education, as Descartes affected to do? To devotees of the older learning this was a dream based on an uncritical psychology. As Bartholomew Keckermann insists (invoking Aristotle), "all learning and all disciplines proceed from previous knowledge."[26] These *praecognita*, as he and Alsted called them, stemmed both from nature and from philosophy, and they were essential for disciplinary knowledge. They are equivalent to that essential "forestructure of the understanding," as Gadamer (with reference to Heidegger) calls it; and they constitute precisely that sort of "prejudice" which the *novatores* (especially the Cartesians) were trying to purge from philosophy—that "prejudice against prejudice," as Gadamer has called it, which has always haunted the projects of rationalism.

On the threshhold of the Enlightenment the history of philosophy was thus situated between two intellectual poles, one the attitude associated with Descartes which relegated this field to the history of error and untrustworthy "opinion." Descartes's own experience with philosophy was most disillusioning—"seeing that it has been cultivated for many centuries by the best minds . . . and that nevertheless so single thing is to be found in it which is not the subject of dispute." His own solution was in effect to divorce reason from memory. In the first of his "rules for the direction of the mind" Descartes warned against "certain special investigations" and the distractions which they presented to the achievement of "universal wisdom"; and in the third rule he advised directing attention "not to what others have thought, nor to what we ourselves conjecture, but to what we can clearly and perspicuously behold and with certainty deduce. . . ."[27] The Cartesian model of wisdom is the lone conceptualizer working with the simplest analytical concepts and ostensibly unencumbered by the erudition of the past.

At the opposite extreme from Descartes was his nemesis, Pierre-Daniel Huet, who likewise prefixed an autobiographical declaration (written in 1692) to his profession of philosophical faith:

> From my school reading so ardent a love of ancient philosophy had seized me, that from that period classical literature seemed to me the handmaid of this science. By this passion I was led to a knowledge of the sects of those ancient

high priests of philosophy which are treated of in Diogenes Laertius . . . , [which], compared with the observations of others gave me an intimate acquaintance with the history of philosophy.[28]

For Huet understanding was inseparable from this "boundless" science established by Diogenes Laertius, whom (he confessed) he took as his constant companion in all leisure times and travels.

Here indeed is an opposition that suggests a meeting of "two cultures," one devoted to ideals of certainty and logical closure, the other to accommodation and encyclopedic coverage. In some cases the opposition appears in a single author. In his philosophical work, for example, Condillac rejected "opinion" as "frivolous" and less than useless, while in his voluminous historical writings he took a very different view. He turned in particular to the famous prize question posed by the Academy of Sciences of Berlin in 1757: "What is the reciprocal influence of the opinions of a people on its language and of its language on its opinions?" (*opinions*, *Meinungen*, being the operative term).[29] Historically, he concluded, "opinion" and the "revolution of opinions" followed the "revolutions of empires" and formed the very substance of history. Opinion represented the first stages of "the progresses of the human mind" ("les progrès de l'esprit humain"—note the plural form) leading to higher levels of culture which made possible Condillac's own enlightened and quite unsystematic philosophy.

Viewed in historical and what I have called mathetic terms, then, the problem of knowledge concerns not philosophically whether or psychologically how the thinking subject knows, but rather the extent, structure, and divisions of knowledge as acquired, communicated, recalled, criticized, transformed, and advanced or retarded by human effort in the course of time. In this sense philosophy is identified with the whole encyclopedia of arts and sciences, not only that of Vincent of Beauvais and of the medieval university but also that of Diderot and his colleagues. According to Diderot's *Encyclopédie*,

> "Science," as a philosophical term, is the clear and certain understanding [*connoissance*] of something through principles either self-evident or demonstrable. "Science" in this sense is opposed to "doubt," and "opinion" is the mean between the two.

It is in this middle area between the polar extremes of certainty and doubt that the problem of knowledge is situated in its most practical and historical guises.

It is in this area, too, that the history of philosophy has for the most part been cultivated. The history of philosophy has had its own disciplinary history, beginning as a department of humanist scholarship and ending up as a "science" in its own right and an essential preparation for the study of philosophy itself.[30] The literary tradition of *historia philosophiae* may be traced back to the foundational work of Diogenes Laertius and, in modern times, those of Georg Horn and Thomas Stanley (both in 1655) and Gerard Vossius (1658); but these were labors of humanist scholarship and criticism and conceptually unsatisfying to philosophers who wanted what Leibniz called (in a distinction made famous more recently by Thomas Kuhn) internal, as distinguished from external, history. Leibniz saw this turn away from Laertian doxography to the history of ideas in the teachings of Jakob Thomasius and (in a famous letter of 20/30 April 1669) wrote to his former mentor, "Most of the others are skilled rather in antiquity than in science and have given us lives rather than doctrines. You will give us the history of philosophy, not of philosophers" (*non philosophorum, sed philosophiae historia*).[31]

The new discipline of the history of philosophy, with roots in humanist scholarship and Protestant theology, reached its first high point in the modern school of Eclecticism, which was associated especially with Jakob Thomasius's son Christian, J. F. Buddeus, N. N. Gundling, and other colleagues, especially at the University of Halle. The classic work of Eclectic history was J. J. Brucker's *Historia critica philosophiae* (1742–44), which not only summed up the story of philosophy from prehistorical times but also fixed the canon of the new discipline for generations to come. The scientific status of *historia philosophica* at this time was signalled in a number of ways (including studying the history of the history of philosophy) characteristic of the emergence of academic disciplines—in celebratory orations, university courses and professorships, scholarly correspondence, monographs, textbooks, and scholarly journals, and in variety of fascinating iconographic representations. The drift of these organs and media of communication, however, was generally away from the conceptual philosophical focus championed by Leibniz and Thomasius and back to doxographical interest in the personalities and the "literature" of philosophy and with the human conditions of philosophizing.[32]

In 1715 C. A. Heumann, another Eclectic philosopher at the Halle, founded a journal called *Acta Philosophorum*, modelled on the *Acta Eruditorum* (founded a generation earlier) and featuring not only articles but also bibliographies and reviews of recent literature.[33] The journal was devoted to the proposition that Eclecticism was (as Heumann declared in his editorial statement) "the best method of philosophizing and, furthermore, that no one deserved the name 'philosopher' who was not an Eclectic" (*dass keiner den Namen eines Philosophi verdiene der nicht ein Eclecticus ist*). In successive issues of the *Acta* Heumann gave a comprehensive, up-to-date survey of the history of philosophy—word, concept, and discipline—from its origin (*Ursprung*) through all stages of succession and growth (*Wachsthum*).

In this connection Heumann published an extensive dissertation on the scholarly theme and ancient topos of the credibility of history (*fides historiae*), especially as it applied to the myth-ridden canon of philosophy.[34] According to Heumann,

> There are two methods of writing history; for one can represent either it in terms of disciplines, such as the history of logic, of metaphysics, etc., or in terms of geography and chronology and show how philosophy has been treated, according to each discipline, in different places, from the beginning of the world down to our own time.[35]

For the history of philosophy self-knowledge required not merely inward-looking speculation but inquiry into the cultural and historical conditions of philosophical practice.

In significant contrast to poets, Heumann wrote, "Philosophers are made, not born" (*Philosophi fiunt, non nascuntur*), and this leads to another major theme of the *Acta*, the character of the philosopher, or "philosophical genius" (*Von dem Ingenio Philosophico*).[36] Heumann recognized geniuses of the first and second magnitude, depending on their powers of judgment and criticism, freedom from prejudice and superstition, and temperamental balance—between doubt and enthusiasm (illustrated by men like Bruno and Cardano, who believed everything he read). He also wondered (following Augustine) if bastards had a special talent and whether women or *castrati* were capable of philosophy. Perhaps the mind had some sexual elements after all.

Beyond psychological factors, Heuman inquired into the influence of environment, climate, the stars, and historical periods: was there a *genius locorum* or a *genius seculi*? "All ages have their genius," he remarked, adding that this was also something for historians to consider.[37] All nations seemed to have claimed precedence in this connection, but all these claims, including those by the Germans, were exaggerated. Nevertheless, historians needed to take into account factors of nationality, and to examine the special quality of philosophizing in England, France, and other modern nations. Philosophy was thus not only the desire for truth but also a human creation; and proverbially, Heumann did not omit to note, to be human is to err (*errore humanum est*). Philosophy was carried on by all sorts of people, in all sorts of times and places, under all sorts of historical conditions, and in all sorts of languages. In short and in more familiar terms, philosophy was a cultural construction, and it was accessible only through literary texts (except perhaps for secret and "esoteric" oral traditions) that had to be understood in historical context.

This was not a lesson that many philosophers wanted to learn; and what emerged from these disciplinary discussions was the great issue which, from the eighteenth century down to the present, has been called "the philosophy of the history of philosophy" and which centers on Thomas Kuhn's famous question about the role of history in science—in this case philosophy.[38] On one side of this question were the enthusiasts for an eclectic and historical approach; on the other side were the philosophers who either despised or were becoming weary of or intimidated by the huge accumulation of learning chronicled and criticized in such publications as Johann Jonsius's bibliography of 1659,[39] the *Acta Philosophorum*, and Brucker's history.

In his Latin dissertation on the *Method for Writing the History of Philosophy* of 1768, for example, Kant's younger colleague, Christian Garve, describes his early enthusiasm for the history of philosophy and then its sad aftermath.[40] "But how can I describe how much my hope was disappointed?" he asks. "Out of that great and splendid apparatus nothing has issued but the lives of philosophers and the listing of dry opinions." Thus Garve turned to the new transcendental mode of philosophizing, which saw history at best as a rehearsal of the arguments which led to Enlightenment—something which Kant redefined as "the a priori history of philosophy" and which Hegel fashioned into a metanarrative of reason.[41]

With Hegel the theoretical separation between erudition and philosophy seems complete. His "philosophical encyclopedia" was distinguished from earlier encyclopedias (and especially that of Diderot) "by the fact that the ordinary one is an assemblage of sciences, taken up in a contingent and empirical manner," such as philology, and other disciplines of a merely "positive" type.[42] For Hegel true philosophy transcended and encompassed all such inferior sciences and pseudo-sciences. Similarly, the history of philosophy was "internal" and had be "freed from that historical outwardness" which had obsessed scholars like Brucker.[43] True science was systematic, and so the history of philosophy was the story of this system in logical development—culminating, needless to say, in the Hegelian Absolute. "Positive" knowledge and modern "positivism" were the very antithesis of the Idea pursued by Hegel.

Here we return to our point of departure, the divergence of views taken toward the problem of knowledge—one rational but humanly impoverished, the other dogmatic but historically rich. In his last work, *The Conflict of Faculties*, Kant gives his somewhat invidious view of the status of the major disciplines. While theology, law, and the other traditional academic fields were bound to texts and canons, philosophy, because of its special relationship with reason, was not only independent of prejudice, dogma, and authority but also free of the burdens of historical erudition (*historische Gelehrsamkeit*).[44] In this way the disciplines of the old encyclopedia are in effect placed under the surveillance of the sovereign and imperialist metadiscipline of philosophy, and as a result the problem of knowledge is detached from—or set above—the disciplines with which it was once associated. In the Kantian critique philosophy reasserted its old claim to be, in the words of Macrobius, "the discipline of disciplines."

If Kantian philosophy enthroned pure reason, however, it was at the expense of traditional learning, and the effect of Hegelianism was much the same. As Trendelenberg wrote in the 1830s, "the production of pure thought demands that it be purified of all content."[45] Modern philosophy has retained a rhetoric of content and concreteness, of course, as Kantian and Hegelian philosophy retained a rhetoric of practice (or rather *praxis*), but there was an alienation from learning and history itself, except in the formalized doctrine of "historicism," which itself became estranged from its original scholarly context and which was transformed—and in the process "dehistoricized"—from a mode of interpretation into a philosophical doctrine that allowed philosophers to appropriate as well as to purify "history."[46]

The problem of knowledge has continued since the days of the Kantianer and the Hegelianer, but so have critiques (and "metacritiques") of their unitary and totalizing view of reason and the imperially thinking subject on which their views relied—and Kant's notion, at least as vulgarly interpreted, that "reason has the sources of knowledge in itself, not in the objects and their observation."[47] Objections to idealism's neglect of "positive" knowledge were not only theoretical but also practical, taking the form of various declarations of disciplinary independence, to which philosophers themselves, including some Neo-Kantians, bowed. Thus in the last, much delayed volume of his *Erkenntnisproblem* Cassirer shifted from the familiar canon of philosophy to the epistemological ideals of particular modes of knowledge, especially mathematical, biological, and historical.

> For philosophy is gradually losing its leadership in this domain that it had held and treasured for centuries. The individual sciences will no longer delegate their authority but mean to see and judge for themselves.[48]

Wilhelm Dilthey, too, while he continued the systematic drive of idealism, had a generous appreciation of the role of positive knowledge, which he conceived in a dualist rather than unitary form (based on his famous distinction between *Geisteswissenschaften* and *Naturwissenschaften*); and he insisted that "we need analysis of special domains" (*Einzelanalysen*).[49]

These are significant concessions for disciples of Kant, but in fact, as I have been suggesting, the individual disciplines had been seeing and judging for themselves for a long time. Like philosophy all sorts of "sciences," human as well as natural, have sought identity and legitimacy by constructing their own separate histories, forming their traditions and authorized canons, devising their own terminology, formulating a common set of questions (if not answers), and a common methodology, defining an intellectual community, finding an institutional base, especially in the form of the modern university endorsed by public support, and so achieving disciplinary status. In this still "encyclopedic" arena habits of mastery and discipleship still thrive, as do other elements of "discipline"; and it is these circumstances that justify the employment of the concept as a long-term historical category.

It is also true that "discipline," with its connotations of rigidity and (thanks to Foucault, in English anyway) associations with "punish," has lost some of its attraction in modern times. We no longer share the authoritarian notions of the structure of knowledge linked with the old tradition of humanist learning. Many scholars prefer to work in the interstices of the disciplines or to reject the notion of discipline altogether; and some, with Paul Feyerabend, may rage against the imperialism of capital "M" Method that subordinates disciplinary practices to a unitary reason. Yet we cannot entirely escape the premises of this disciplinary perspective, if only because of the conventions of academic language and argument. Knowledge is a culturally constructed phenomenon, and particular disciplines represent the cultural forms in which this knowledge has been preserved, transmitted, and transformed throughout the history which is accessible to us. The fashionable concept of interdisciplinarity only reinforces the centrality of disciplines to the modern problem of knowledge.[50]

The most familiar recent discussions of the concept of discipline and disciplinary history centers on Thomas Kuhn's much used and much abused notion of intellectual paradigms, their "revolutions," and "normal science."[51] In violation of Kuhn's own warnings his ideas have been extended beyond their original application to the hard sciences to other, more amorphous disciplinary contexts and "multiparadigm" disciplines, including political theory, historical thought, linguistics, sociology, anthropology, archeology, rhetoric, and literary theory.[52] Stephen Toulmin has tried to refine Kuhn's ideas specifically by distinguishing between theoretical and disciplinary—universal and local—principles of a science; and this surely applies to the softer disciplines. As Toulmin argues,

> Within any particular culture and epoch, men's intellectual enterprises do not form an unordered continuum. Instead they fall into more or less separate and well-defined 'disciplines,' each characterized by its own body of concepts, methods, and fundamental aims.[53]

Except for its gender and evolutionist bias, this line of argument also supports a disciplinary approach to the problem of knowledge.

The radically different (but hardly less influential) critique of knowledge offered by Foucault also deserves notice in this connection. Foucault's notion of *episteme* is a sort of historical a priori or (in Heideggerian terms) forestructure of understanding embedded in the practices and discourse of an age but marked by fundamental discontinuities. An episteme is a "grill" (a term Foucault borrows from Levi-Strauss); and the disciplines which are part of this grill shape and indeed make possible perception and expression of matters both of nature and of culture. "Discipline," a concept which has been associated famously with "punish," can suggest restriction and subjection to authority and its rules and penalties; and in any case it is bound to a larger and more problematic set of assumptions about the world of practice and power. As Foucault explains,

> The sciences always carry within themselves the project, however remote it may be, of an exhaustive ordering of the world; they are always directed, too, toward the discovery of simple elements and their progressive combinations; and at their center they form a tableau in which knowledge is displayed as a system contemporary with itself. . . . As for the great controversies that occupied men's minds, they are accommodated quite naturally in the folds of the organization.[54]

Foucault treats only the disciplines, or "classical sciences," of of biology, linguistics, and political economy; but these go far beyond the horizons of the concerns of philosophy, encompassing as they do life, language, and labor.

What Foucault offers is a metadisciplinary analysis, or method of decipherment, that (going beyond Kuhn and Toulmin) would give entry to the historical reality behind disciplinary structures, and resolve the problem of knowledge through a politics of knowledge. There have been many other recent attempts to reveal the hidden forces underlying knowledge and the relationships, in effect, between disciplinary and power structures. For Foucault's episteme Pierre Bourdieu substitutes a Marxoid concept of the "intellectual field," which is a structure of orthodoxies and heterodoxies that determine, or limit, the issues, choices, and language fixing the range of possibilities of disciplinary activity—normal science, in Kuhn's usage—within the broader "cultural field," which is Bourdieu's term for the material and political conditions of intellectual exchange.[55] Habermas, beginning also in Marxist mode, originally looked to "interests" as a way of understanding the ideological dimension of knowledge, though he has since joined this, following a remarkably eclectic strategy, with a theory of "communication action," which mitigates the reductionism of his earlier position while preserving the old universalist projects of the critique of reason and of the sociology of knowledge.[56]

These efforts of "unmasking" the claims of reason, as Habermas has called them, are important contributions to the problem of knowledge; but they have been criticized from the standpoints both of poststructuralism and of philosophical hermeneutics—which agreed at least in this, that there is no Archimedean point from which to judge, let alone move, the pursuits of knowledge, interested or disinterested, and therefore its history. It is impossible to get "behind the back of" language, Gadamer has argued; and more specifically Derrida has protested that Foucault has no language in which to carry on his "archeological" critique of knowledge—that is, he has only the language of the reason which he ostensibly rejects.[57] In different ways both hermeneutics and deconstruction look to language as a way of resolving, or of bypassing, epistemological questions.

These are a few of the modern, and one of the postmodern, efforts to deal with the problem of knowledge; but they all seem to fall short, or else to exceed the aims of historical inquiry. With Foucault and Habermas positive and disciplinary knowledge is subjected to reductionist analysis; with Gadamer and Derrida it is marginalized in favor of universalizing philosophical goals. Either way they contribute only indirectly to the purpose of this paper, which is to consider the problem of knowledge in what I have called mathetic and disciplinary terms.

Of more relevance here, perhaps, are the much-discussed views of J. G. A. Pocock and Quentin Skinner, especially since they in fact operate for the most part in a disciplinary mode (although they make broader claims for intellectual history). Both focus, too, on the question of the interpretation of language, and both are intentionally "intentionalist," being concerned largely with human agency—authorially created and readerly derived meaning.[58] It seems to me that not only "idioms of political discourse" but also disciplines in a more general way represent language communities, and perhaps "paradigmatic structures" (in another of Pocock's phrases, derived ultimately from Kuhn), whose history can most directly be followed through its distinctive discourse.

It may be concluded that mine is a very Whiggish agenda, accepting as it does current definitions of particular disciplines for the purposes of writing their history, and such indeed is the case, but I do not take this as a criticism. Intellectual Whiggery is an unavoidable condition not only of writing the history of particular modern constructs but also, as hermeneutics has made clear, of history in general. We look back from the standpoint of our own intellectual and linguistic—or disciplinary—condition. Writing the history of "anthropology," "sociology," and of even "Literature," "Art," and "Science" are anachronistic undertakings which introduce a terminology and a large freight of assumptions alien to earlier historical contexts; and it is within this context that we tell our disciplinary stories. Writing the history of science, Marcello Pera concludes, "one can only be a whig."[59]

Such are the conditions, it seems to me, of mathetic learning and of the problem of knowledge from the particular standpoint which I have chosen, leading to the concept of discipline, of which I conclude with a brief summary. Within Western tradition the major disciplines are rooted in the arts and sciences of classical education (even if they have been torn away from these roots); they were reshaped in the universities of the later Middle Ages; and they have undergone extraordinary transformations since their deployment in the medieval *studium* and rearrangement in modern classifications of knowledge. Mathematics, for example, emerged from the depths of the quadrivium to join with philosophy in the work of Galileo, while the trivial art of grammar was transformed by humanist scholarship into a critical science—"philology," as it was called—which (in Vico's "new science") likewise formed an unconventional union with philosophy. The history of philosophy was another interdisciplinary hybrid that sought legitimacy between two divergent approaches to knowledge—one rationalist and abstractive and the other literary and mathetic.

Every discipline has its own story, but there are some common patterns. One is the impulse to achieve parity or superiority to rival fields of knowledge—and of course in the universities this parity has always been proportionate not only to social status but also to fiscal remuneration. In times past law and medicine sought such eminence; in the eighteenth century academic philosophy, having emerged from the faculty of arts, aspired to the status of a profession and of what, from Kant to Husserl, philosophers called a "rigorous science" (*reine Wissenschaft*). One of the major reasons why thinkers like Garve, Kant, and Hegel turned against the learned discipline of the history of philosophy was precisely because it approached philosophy as a human creation that had to be understood in the lowly terms of geography, environment, nationality, anthropology (that is, psychology), and (from at least the early eighteenth century) gender. Today, in my interdisciplinary experience, this is still a taboo among many academic philosophers and scientists.

In general disciplines may pretend to be necessary and logical structures, but they are also cultural constructions which non-disciples may well see as unstable opinion or expressions of interest. Disciplines serve particular needs; adopt formal language, methods, agendas, and organization; find a consensus, a social and an institutional base, and ways of preserving and extending intellectual continuity through a more or less mathetic relationship; and finally fashion their self-image and legitimacy through construction of their histories. The construction of a disciplinary history not only defines the subject matter and method but also provides the most convincing justification of a field, its claims and aspirations, compares it with other disciplines, determines its classic authors and canon, its major questions and answers, its institutional and social tradition, its place in the larger intellectual and cultural fields, and perhaps its role in the hierarchies and networks of power which Marxian, Freudian, and Foucauldian analysis and other forms of the "hermeneutics of suspicion" have tried to uncover.

There is much more to be said on the subject; suffice it here to conclude by suggesting that the virtue of focusing on disciplinary structures is that they represent a social, economic, and political as well as intellectual presence in our own world, and that they offer a way of pursuing historical inquiry through intellectual and cultural continua which preserve a certain linguistic coherence across many cultural and chronological divides and political and social transformations. And finally, they suggest concretely historical ways of posing—if not finally of resolving—the problem of knowledge.

## Notes

1. The classic discussion is Ernst Cassirer, *Das Erkenntnisproblem in der neueren Geschichte*, 3 vols. (Berlin, 1906–20).

2. Edmund Husserl, "Philosophie als strenge Wissenschaft," *Logos* 1 (1910–11): 289–344.

3. Plato, *Sophist* 357c; and see Robert McRae, *The Problem of the Unity of the Sciences: Bacon to Kant* (Toronto, 1961).

4. Peter Ramus, "That There is but One Method of Establishing a Science," in *Renaissance Philosophy*, ed. Leonard A. Kennedy (The Hague, 1973), 153.

5. Cited in George Boas, *Essays in Primitivism and Related Ideas in the Middle Ages* (Baltimore, 1948), 127.

6. Johann Alsted, *Philosophia digne restituta* (Herborn, 1612), 10.

7. Alexander Pope, *Dunciad* iv, 31. René Descartes, *Rules for the Direction of the Mind*, in *The Philosophical Writings*, trans. John Cottingham, Robert Stoothof, and Dugald Murdoch, 3 vols. (Cambridge, 1984–91), I, 19 (Fourth Rule); and on Vossius see Wilhelm Schmitt-Bigemann, *Topica Universalis: Eine Modellgeschichte humanistischer und barocker Wissenschaft* (Hamburg, 1983), 258. Nor do I follow the usage of Foucault, who identifies "mathesis" with "a general science of order" in *The Order of Things: An Archeology of the Human Sciences* (New York, 1970), 71.

8. Juan Luis Vives, *De Initiis sectis et laudibus philosophiae*, in *Early Writings*, ed. and trans. C. Martheeussen, C. Fantazzi, and E. George (Leiden, 1987), 20.

9. Iamblichus, *On the Pythagorean Life*, trans. Gilliam Clark (Liverpool, 1989), 8.

10. Vives, *Discursus philosophicus in quo Historiae Philosophiae Antiquae et Recentioris recensetur* (n.p., 1705), 25; cf. Vives, ibid., 36.

11. See Richard Hope, *The Book of Diogenes Laertius: Its Spirit and Method* (New York, 1930).

12. Isidore of Seville, *Etymologiae*, I, 1: "Disciplina a *discendo* nomen accipit; unde et scientia dici potest. Nam *scire* dictum a discire, quia nemo mostrum scit, nisisquod discit. Aliter dicta disciplina, quia discitur plena."

13. Etienne Gilson, *History of Christian Philosophy in the Middle Ages* (New York, 1955), 316.

14. Otto Mauchg, *Der lateinische Begriff "Disciplina": Eine Woruntersuchung* (Freiburg, 1941), and H.-I. Marrou, "'Doctrina' et 'disciplina' dans la langue des pères de l'église," *Bulletin Du Cange* 9 (1934): 5–25.
15. Cassiodorus, *Institutiones* II, 3.
16. Guillaume Budé, *De Studio literarum* (1528), *L'Etude des lettres*, trans. M. M. de la Garanderie (Paris, 1988), 91.
17. Bartholomew Keckermann, *Praecognitorum philosophicorum libri duo* (Hanover, 1612), vol. 1.
18. Alsted, *Philosophia digne restituta*, 48, and *Encyclopaedia* (Herborn, 1630), 49.
19. D. R. Kelley, "Philodoxy: Mere Opinion and the Question of History," *Journal of the History of Philosophy* 34 (1997): 117–32; and "Second Nature: The Idea of Custom in European Law, Society, and Culture," in *The Transmission of Culture in Early Modern Europe*, ed. Anthony Grafton and Ann Blair (Philadelphia, 1990), 131–72.
20. Ian Hacking, *The Emergence of Probability* (New York, 1975); and see Edmund F. Byrne, *Probability and Opinion: A Study in the Medieval Presuppositions of Post-Medieval Theories of Probability* (The Hague, 1968).
21. David Hume, *Essays: Moral, Political, and Literary*, ed. Eugene F. Miller (Indianapolis, 1985), 51.
22. Georg Wilhelm Friedrich Hegel, *Introduction to the Lectures on the History of Philosophy*, trans. T. M. Knox (Oxford, 1985), 15–16 (1820).
23. Gilbert-Charles LeGendre, *Traité de l'opinion: Memoires pour servir à l'histoire de l'esprit humain* (Paris, 1733), I, 9.
24. Werner Schneiders, *Aufklärung und Verurteilskritik: Studien zur Geschichte der Vorurteilstheorie* (Stuttgart, 1983).
25. See especially Michael Albrecht, *Eklektik* (Stuttgart, 1994).
26. Keckermann, *Praecognita philosophica*, vol. 2.
27. Descartes, *Rules*, 9, 13.
28. *Memoirs of the Life of Pierre-Daniel Huet*, trans. John Aikin (London, 1810), II, 203.
29. Etienne Bonnot de Condillac, *Oeuvres complètes* (Paris, 1821–22), XIV, 564; cf. *Treatise on Sensations*, III, ii, 3, in *Philosophical Writings*, trans. Franklin Philip (Hillsdale, NJ, 1982), 393.
30. In general see Giorgio Santinello, ed., *Storia della storie generali della filosofia*, 5 vols. (1981), and vol. I, updated and translated as *Models of the History of Philosophy*, trans. C. W. T. Blackwell and Phillip Weller (Dordrecht, 1993).
31. Preface to Mario Nizolio, *De Veris Principis* (Frankfurt, 1670), fol. 2v, and in *Philosophical Papers and Letters*, trans. Leroy E. Loemker (Dordrecht, 1969), 93.
32. See Gottlieb Stolles, *Anleitung zur Historie der Gelahrheit* (Jena, 1718).
33. *Acta Philosophorum, Gründl. Nachrichten aus der Historia Philosophica*, ed. C. A. Heumann (Halle, 1715–21).
34. *Acta Philosophorum*, I, 381.
35. *Acta Philosophorum*, I, 2.
36. *Acta Philosophorum*, I, 567.
37. *Acta Philosophorum*, I, 622.
38. Including Richard Rorty, J. B. Schneewind, and Quentin Skinner, eds., *Philosophy in History* (Cambridge, 1984); A. J. Holland, ed., *Philosophy, Its History and Historiography* (Dordrecht, 1985); Bernard P. Dannenhauer, *At the Nexus of Philosophy and History* (Athens, GA, 1987); Peter H. Hare, ed., *Doing Philosophy Historically* (Buffalo, NY, 1988); T. Z. Lavine and V. Tejera, eds., *History and Anti-History in Philosophy* (Dordrecht, 1989); Jorge J. E. Gracia, *Philosophy and Its History: Issues in Philosophical Historiography* (Albany, 1992).
39. Johann Jonsius, *De Scriptoribus historiae philosophiae libri*, ed. J. C. Dorn (Jena, 1716).
40. Christian Garve, *De ratione scribendi historiam philosophiae* (Leipzig, 1768); and cf. G. G. Fülleborn, ed., *Beiträge zur Geschichte der Philosophie* (Züllichau, 1791), XI, 88.
41. Immanuel Kant, *Critique of Pure Reason*, "Transcendental Doctrine of Method," ch. 4.
42. *Encyclopedia of the Philosophical Sciences in Outline*, trans. Steven A. Taubveneck (New York, 1990), 53.

43. *The Encyclopedic Logic*, trans. T. F. Geraets, W. A. Suchting, and H. S. Harris (Indianapolis, 1991), 38.

44. Kant, *Der Streit der Fakultäten*, in *The Conflict of the Faculties*, trans. Mary J. Gregor (Lincoln, NE, 1979), 38.

45. Klaus Christian Köhnke, *The Rise of Neo-Kantianism*, trans. R. Hollingdale (Cambridge, 1991), 29.

46. A good example of this is Michael Allen Gillespie, *Hegel, Heidegger, and the Ground of History* (Chicago, 1984), who finds historians (mentioning in particular Arnaldo Momigliano, J. G. A. Pocock, Julian Franklin, George Huppert, and myself) going "crucially astray" in their efforts to seek the origins of historicism in the history of scholarship.

47. *Prolegomena to Any Future Metaphysics*, trans. L. W. Beck (New York, 1950), 110.

48. Ernst Cassirer, *The Problem of Knowledge: Philosophy, Science, and History since Hegel*, trans. William H. Woglom and Charles W. Hendel (New Haven, 1950), 10.

49. Wilhelm Dilthey, *Introduction to the Human Sciences*, ed. Rudolph Makreel and Frithjof Rodi (Princeton, 1989), 117.

50. For Jean Piaget, *Main Trends in Inter-Disciplinary Research* (New York, 1970), this approach involves collaborations between the disciplines, scientific and humanistic. See also, for example, Jürgen Kocha, ed., *Interdiziplinarität* (Frankfurt, 1987).

51. Some of these questions have been taken up in Loren Graham and Wolf Lepenies, eds., *Functions and Uses of Disciplinary Histories* (Cambridge, Mass., 1983); and see also *Knowledges: Historical and Critical Studies in Disciplinarity*, ed. Ellen Messer-Davidow, David R. Shumway, and David J Sylvan (Charlottesville, 1993).

52. It is a rare theoretical work in any of these areas that does *not* cite Kuhn.

53. Stephen Toulmin, *Human Understanding: The Collective Use and Evolution of Concepts* (Princeton, 1972), 139.

54. Foucault, *The Order of Things* 74–75; and see Paul Bové, "The Rationality of Disciplines: The Abstract Understanding of Stephen Toulmin," in *After Foucault: Humanistic Knowledge, Postmodern Challenges*, ed. Jonathan Arc (New Brunswick, NJ, 1988), 43–70.

55. Pierre Bourdieu, *The Field of Cultural Production: Essays on Art and Literature*, trans. Claude Du Verlie (New York, 1993).

56. Jürgen Habermas, *Knowledge and Human Interests*, trans. Jeremy Shapiro (Boston, 1971), and *The Theory of Communicative Action*, trans. Thomas McCarthy (Boston, 1981–85), vol. 1, *Reason and the Rationalization of Society*, and vol. 2, *Lifeworld and System: Critique of Functionalist Reason*.

57. Jacques Derrida, "Cogito and the History of Madness," *Writing and Difference*, trans. Alan Bass (Chicago, 1978), 78, and Hans Gadamer, *Philosophical Hermeneutics*, trans. David E. Linge (Berkeley, 1975), 35.

58. See especially J.G.A. Pocock, *Virtue, Commerce, and History* (Cambridge, 1985), "Introduction: The State of the Art," and Quentin Skinner, "A Reply to My Critics," in *Meaning and Context: Quentin Skinner and His Critics* (Princeton, 1988).

59. *The Discourses of Science* (Chicago, 1994), 179.

# BODIN, MONTAIGNE, AND THE ROLE OF DISCIPLINARY BOUNDARIES

## Ann Blair

Michel de Montaigne (1533–92) and Jean Bodin (1529–96) were contemporaries, compatriots and colleagues. Both served as officers in the Parlements, Montaigne as a magistrate in Bordeaux and Bodin as a barrister in Paris. They both ended up dissatisfied with their lives at the center of public activity and withdrew from the political fray.[1] Montaigne retired to his family estate, while Bodin earned a living as a royal officer in the provincial town of Laon.[2] Both owe their considerable fame, at the time and since, to their prolific writing during this period. But, at least at first view, their writings seem so strikingly different in tone as to belie any similarities of cultural context.

Montaigne's *Essays*, first published in 1580 and constantly revised and expanded in later versions (identified as "A," "B" and "C" texts), have delighted readers and literary critics for centuries with their digressive mix of anecdotes, pointed criticism of contemporary practices and excruciating self-reflection. Montaigne is considered modern, even post-modern, for his questioning, challenging, often ironic, always hyper-self-conscious tone. In emphasizing the endless diversity and inconstancy of all things, including the impressions and passions that constitute the self, Montaigne erodes not only traditional beliefs and customs (which he often in the end advocates keeping, for lack of anything with which to replace them), but also the possibility of knowledge, and the very existence of a self-consistent and identifiable subject.

Bodin on the other hand writes on a similarly wide range of topics, in separate and supposedly systematic treatises, with an unquestioning assurance of truth. In deadly earnest Bodin describes the heinous crime of witchcraft, for example, and the need for tougher persecution to eradicate it, in his *Démonomanie* of 1580. In other works on the method of writing history and on political theory, Bodin's views—notably his call for a critical appraisal of sources and his new concept of sovereignty—have been praised for their modernity.[3] But he is never a self-conscious author, guilty in the *Method for the Easy Comprehension of History*, for example, of the same kinds of hasty judgments and nationalist biases against which he warns his readers. In his last work, the *Universae naturae theatrum* or *Theater of All of Nature* (1596), which I will discuss in more detail here, Bodin treats of "universal nature," starting with the first principles of physics and rising through the stages of the chain of being from elements, to minerals and metals, plants and animals and finally the soul and the heavenly bodies. But this "theater," purportedly a true and complete representation of nature, is little more than a succession of unquestioned natural "facts" for which Bodin offers new causal explanations. Bodin questions neither the truth of the myriad "facts" which he has garnered from various sources, largely from ancient and modern historians, nor his method of representing all of nature through this unsystematic and idiosyncratic accumulation of specific details and explanations.[4]

Whereas Montaigne's voice is heard constantly throughout the *Essays* in first-person accounts and most strikingly in direct self-reflection about the purpose, nature and method of his work, Bodin reserves his very few methodological comments to an introduction and dedication and constructs the *Theatrum* as a dialogue between an inquiring pupil Theorus and his knowledgeable master Mystagogus. Bodin's voice is thus technically absent from the work, although there is hardly ever any ambiguity about Mystagogus' authority in presenting Bodin's own views.[5] But, faithful to Bodin's general strategy, Mystagogus himself speaks with treatise-like authority rather than self-reflexivity, and uses the first person only for occasional eye-witness reports.

This difference in tone and authorial strategy is a most conspicuous point of divergence between the two authors. Yet, as I will try to show, Montaigne and Bodin also have in common a number of cultural practices and presuppositions, which can be detected, notably, in the methods of composition and structure of their works, as well as in a number of their precepts about nature and about knowledge. In attempting to explain how these shared characteristics are overlaid with strikingly different voices, and to do so somewhat more systemically than by invoking differences of "temperament" or personality, I argue that the attitudes of Montaigne and Bodin toward disciplinary boundaries play a crucial part in shaping their authorial strategies.

## Methods of Composition and Structure

Montaigne and Bodin both draw on a vast store of material collected from bookish (primarily historical) sources, using a method of reading and notetaking widely taught in Renaissance schools, which involved a commonplace book. In this personal notebook students and scholars would copy out under appropriate headings material of interest which they had encountered in their reading, selected for example for a turn of phrase, an argument, a piece of factual information or an authoritative quotation which might some day prove useful in composing prose of their own. I have presented elsewhere the indirect but substantial evidence for Bodin's having composed his *Theatrum* from a book of natural commonplaces (or commonplaces about the natural world).[6] Montaigne's use of a commonplace book is widely acknowledged (even by Montaigne himself) and is the indispensable source of the countless explicit quotations and tacit borrowings which pepper the thousand-plus pages of the *Essays*. But Bodin and Montaigne use this same method of composition from a book of commonplaces, attractive precisely because of its great flexibility, to quite different effects.

For Bodin, as I have argued in more detail elsewhere, the commonplace book serves as a storehouse of tidbits of knowledge about the natural world garnered from sources of many kinds, primarily from books, but also from personal observation, hearsay and second-hand reports, from which Bodin draws material to motivate and corroborate his new causal explanations. In the *Theatrum* Bodin never stops to discuss or establish these "facts" which surface as self-evident truths in Theorus's questions as well his master's answers. Why do snakes bite women more than men? Theorus asks, tacitly taking the "fact" that they do for granted (316/452); or, the effectiveness of cabbage juice as an antidote against the excessive consumption of wine is used to demonstrate the natural antipathy that Mystagogus asserts between the two plants involved (294/419). For Bodin the commonplace book is a treasury of "pre-approved" facts, so that he is guilty of exactly the behavior that Montaigne finds so reprehensible among his witch-hunting contemporaries (and he may well have Bodin's *Démonomanie* in mind in writing these lines):

> I see ordinarily that men, when facts are put before them, are more ready to amuse themselves by inquiring into their reasons than by inquiring into their truth.... They ordinarily begin thus: 'How does this happen?' What they should say is: 'But does it happen?' . . . We know the foundations and causes of a thousand things that never were. (III, 11; 1026–27/785)[7]

Montaigne uses the method of commonplaces to store not facts to be used uncritically as Bodin does, but snippets of poetry and prose and anecdotal examples, with which he illustrates or embellishes his accounts and, occasionally, supports his own skeptical conclusions. Like Bodin, Montaigne especially likes historians, both ancient and modern, which he considers "his thing."[8] Thus Montaigne's unrevised early essays (e.g. book I, ch. 2–18, probably written as early as 1572) are composed as "lessons," a fairly traditional literary and moral exercise in which one compiles examples taken from one's readings, interspersing them with a few personal reflections. Montaigne's chapters on various passions, precepts of political actions or the virtues and vices (of idleness, constancy, lying and so on) are typical of this genre: cases from ancient and modern history including his direct experience of the French civil wars are gathered to illustrate the point, for example, that "parley time is dangerous" (I, 6)—a specific case of the dictum that no one can be trusted. Over the years, in revisions and new compositions, Montaigne increasingly shifts the balance away from compilations on standard topics toward more personal reflection and an idiosyncratic association of ideas. He still builds his reflections around a plentiful supply of explicit quotations and tacit borrowings from his wide-ranging reading, "other people's flowers" as he calls them, but Montaigne increasingly turns the method of commonplaces into a process of self-discovery.

Thus, in a late essay, Montaigne defends his work against any resemblance with contemporary "concoctions of commonplaces," books which are mere feats of learning, or worse yet, "made out of things never either studied or understood, the author entrusting to various of his learned friends the search for this and that material to build it, contenting himself for his part with having planned the project and piled up by his industry this stack of unfamiliar provisions." On the contrary, Montaigne explains: "Indeed I have yielded to public opinion in carrying these borrowed ornaments about on me. But I do not intend that they should cover and hide me; that is the opposite of my design, I who wish to make a show only of what is my own" (III, 12; 1055–56/808). Francis Goyet has studied Montaigne's method of composition and reflection from the marginal annotations and cross-references that Montaigne left in the books he read as well as in his own copy of the *Essays*, where he kept track of new ideas and additions for future versions. Goyet shows how Montaigne subjected the material he selected from his reading to a complex and very personal process of "digestion." After marking a passage as interesting Montaigne would explore alternative directions in which to take it instead of just compiling it under standard headings in the manner of the "concoctors" he scorns; returning to the passage in a second reading, for example, Montaigne might correct a heading that he had proposed earlier or create a new and non-traditional one under which he would finally use the passage. The result is a very personalized journey (as Goyet concludes) in which "Montaigne is searching less for what his sources have to say, than to understand what he is seeking himself in his reading of them."[9]

Rather than the stock of truisms which it generated for Bodin and most contemporaries, the commonplace method for Montaigne constituted a process of self-discovery, which he shares with his readers by leading them down the idiosyncratic path of his reflective reading and borrowing. Thus it is not a failing, but rather an essential part of Montaigne's intention that the issues treated in each chapter stray so notoriously from the heading (often characteristic of a commonplace book) which he has chosen for the title. "Of the lame" (III, 11), for example, famous for Montaigne's denunciation of the witchcraze, opens with a discussion of calendar reform and ends with a discussion of the Italian saying that lame women are more fulfilling in bed, interspersed with quotations from Persius, Vergil, Tacitus, and Erasmus among other favorites. Montaigne's point is to show, as he says somewhere in the middle of this ramble, how "our reasons often anticipate the fact, and extend their jurisdiction so infinitely that they exercise their judgment even in inanity and non-being" (III, 11; 1034/791). Our imagination is so powerful as to "anticipate the fact," to attach great significance to events that are paltry or even non-existent. Thus the adoption of the Gregorian calendar in France in 1582, which involved dropping ten days from the month of December, caused much more disruption in people's mental anticipation of it, Montaigne notes, than in the actual event itself: "my neighbors find . . . the harmful and the propitious days exactly at the same point to which they had always assigned them" (III, 11; 1026/784). Similarly, he implies, witchcraft looms larger in the imagination than in reality, as do the greater sexual charms of lame partners.[10]

Montaigne's message that, subject as we are to such delusions, we can never trust ourselves nor the supposed powers of our reason, is conveyed appropriately enough, he indicates, through a meandering and diverse collection of material. In both the *Theatrum* and the *Essays* the method of commonplaces produces texts constructed loosely by association of ideas and juxtaposition of borrowed tidbits in an open-ended structure to which new material could be added indefinitely. Indeed Montaigne constantly added to his essays new quotations and reflections.[11] Bodin's succession of questions and answers would have lent itself easily to expansion too, if Bodin had been so inclined (which I tend to doubt, since, as I will argue, he perceived his unsystematic work as complete); in any case, Bodin died a few months after the *Theatrum* was published. But although both texts are equally "mobile" in Terence Cave's characterization of much French Renaissance literature,[12] Bodin and Montaigne differ absolutely in the ways they present and perceive their work. Montaigne is happy to speak "in disjointed parts" (III, 13; 1076/824); he calls his work at various points a "fricassee" (III, 13; 1079/826) or a "bundle of disparate pieces" ("fagotage" II, 37; 758/574). By choosing the unusual title of *Essays* Montaigne suggests tentativeness, experimentation and exploration. Bodin, on the other hand, announces in both his introductory remarks and his title that his treatment of "all of nature" will be orderly and systematic, even though the text itself is fragmented and loosely organized.

Bodin's text moves idiosyncratically (like Montaigne's) through topics and themes as they are called up by the discussion immediately preceding. He considers the horns of cattle and deer, for example, but the teeth of hares and horses (which are flat), and operates a seamless transition between the two clusters of questions with a discussion of sexual differentiation in deer (creating the link to the first cluster) and hares (leading on to the second one) (355–60/509–16). By and large Bodin does follow the broad topical sections announced in the table of contents at the beginning of the work (such as "snakes," "quadrupeds" and "fish"), and the order which he proudly explains in his introduction, rising up the chain of being from simplest to most complex. But the text itself proceeds in chunks of some one hundred pages of continuous prose with no indication of these subsections, meandering along a loosely associative chain of thought which is carefully prepared by thematic or rhetorical transitions and yet entirely unpredictable and unsystematic.

Contrary to Montaigne, who revels explicitly in his inconstancy, Bodin assumes that his collection of natural facts constitutes a theater or an orderly and distributive table of "all of nature." Bodin explains that "the Theater of Nature is nothing other than a sort of table of the things created by the immortal God placed for all to see, so that we may contemplate and love the majesty ... and admirable providence of the author himself in things great, middling and small."[13] With characteristic ambiguity Bodin has phrased his sentence so that the "Theater of Nature" could designate either nature itself or the book that he has written to describe nature, or both. I see this as evidence of a conflation of the object (the theater of nature) with its description or representation (the book entitled *Theater of Nature*) that runs throughout Bodin's work and is both cause and consequence of his unself-conscious strategy. One gets a better sense of Bodin's ideal of "the theater which is nothing other than a sort of table" from a later passage in which the pupil asks Mystagogus to "unfold, if you will, the table of the universe, as in a theater, so that, when the distribution of all things is displayed for the eyes to see, the essence and faculty of each thing can be more clearly understood" (129/170).[14] What exactly "the table . . . , as in a theater" designates is still unclear: possibly a stage backdrop, a poster or program,[15] or a theater like Camillo Delminio's (as described by Frances Yates[16]), a free-standing structure in which one could contemplate the elements of philosophy brought together under one roof. In any case, it is clear from Mystagogus' answer what Bodin thinks a table in natural philosophy should be: in response to this request, the master outlines a kind of verbal dichotomous diagram (of the type associated with Peter Ramus, although it was by no means exclusively Ramist[17]), in which plants and animals are sorted into groups and subgroups—ruminants, for example, according to whether they have horns or not, are covered with hair or wool, are domestic or wild, and so on. This systematic overview is a far cry from what Bodin actually provides in the rest of the text—by modern standards at least: Bodin never mentions ruminants as a category in the text and applies such diverse criteria to each species he discusses that there is no hierarchical subdivision possible. Yet Bodin does not comment on any discrepancy between theory and practice; on the contrary he concludes in his introduction that "there is nothing we have searched for more diligently than the succession of all things and the indissoluble coherence of nature, its interrelations and agreements, and how the first things correspond to the last, the middle ones to both extremities and everything to everything else" (6/[7]).[18] I can only conclude that Bodin feels satisfied that his accumulation of particulars *is* the theater of nature, as if his discussion constituted a transparent and immediate representation of nature itself. Similarly, in Bodin's prefatory praise of natural philosophy his proclamations of the nobility of the discipline insensibly become a paean to the beauties of nature itself. Nature and its representation in natural philosophy are unself-consciously conflated.

## Philosophical Principles

In an attempt to explain why Montaigne and Bodin turn similar textual methods and structures to such different uses, I will briefly compare some of their philosophical principles—in particular, their attitudes toward knowledge and doubt, which are remarkably harder to distinguish than one might expect, and, more conclusively, their divergent stances on the role of disciplinary boundaries.

Stated bluntly, a comparison of the two authors' epistemological positions would seem to pit skeptic against dogmatist, but I will suggest that the picture is actually more complex, to the extent that it cannot suffice to explain their different authorial strategies. To Bodin, nature speaks loud and clear of the greatness, goodness and providence of God. Human reason is thus an effective tool against the impious. Applied to the particulars of nature it produces causal explanations that reveal the divine plan. Applied to the principles of natural philosophy, reason successfully refutes the eternity of the world and proves the immortality of the soul, as Bodin claims to do in books I and IV of the *Theatrum*. With the confidence of a rationalist Bodin hails natural philosophy as the ultimate weapon with which to assail the impious:

> How valuable it is that those who cannot be dragged by any precepts of divine laws or oracles of the prophets from their ingrained folly or led to the worship of the true deity, are forced by the most certain demonstrations of this science, as if under the application of torture and questioning, to reject all impiety and to adore one and the same eternal deity![19]

Bodin also uses this striking torture metaphor later in the text, to argue against philosophical skeptics who would assert that one can know nothing: "although an infinite number of mathematical demonstrations forces them to confess willy-nilly, as if under torture, the truth which they would not acknowledge" (474/ 684).[20] By upholding the possibility of certain knowledge, Bodin grounds a natural theological project designed to show both philosophical and religious skeptics that there can be no doubt about the sacred attributes of the godhead.

On the other hand, Bodin does not dogmatically resolve all the questions he poses. Some causal explanations are not asserted but suggested as questions themselves: Why do snakes bite women more than men? Perhaps because snakes are wily and choose as victims those who are weaker? Or because divine providence wisely preserves men, who are most useful, from their attacks? Or is it because of a natural antipathy which stems from that fateful encounter in the Garden of Eden? (316/542). And, more strikingly, the *Theatrum* also includes questions to which Mystagogus declines to offer a solution. Thus, after ridiculing the theories of Aristotelian meteorology, which would explain comets, earthquakes, storms or the formation of metals from the condensation or combustion of earthly vapors,[21] Mystagogus declares that it is enough to refute false reasons and offers no alternative explanations of his own. Instead of overreaching our abilities in the search for rational explanations, he concludes, we should modestly confess our ignorance.[22] At most, we can attribute these inexplicable phenomena to the action of demons; even this solution serves less for its explanatory value itself than because it places the topic beyond the purview of the physicist, who deals only in natural causes: at least this strategy, as he explains in applying it to the cause of violent storms, "refutes the opinions of those who prefer to repeat silly causes of things rather than referring them to demons or confessing ignorance" (178/243–44).[23] Nonetheless, even as it fails, reason teaches the greatness of God. Bodin concludes for example his discussion of the inexplicable phenomena of magnetic attraction: "It is better to admire in silence the majesty of the greatest Workman than to want rashly to go insane with reasoning" (249/349).[24] Similarly, in giving hoarfrost the power to burn plants although it is not even as cold as snow, the "Creator wanted to make something as if against the laws of nature, so that he would draw men to a greater admiration and love for him" (206/284).[25] Bodin's sense of the weakness of reason combined with the omnipotence of God can lead to prudent confessions of ignorance, as in these cases, but in other cases can serve to justify his acceptance of "facts" that cannot be understood, such as the travel of witches to the sabbath. Who are mere mortals to decide what is and is not possible? The fact that we cannot explain something does not invalidate its existence.

For Montaigne, to read the book of nature and the signs of divine providence in nature is virtually impossible. But in his perversely argued "apology" for the natural theology of Ramon Sebon, Montaigne allows that Sebon is successful in demonstrating the existence of God through nature because the faith guiding his arguments makes them "firm and solid" (II, 12; 448/327). Montaigne then goes on to show at great length how human reason, unaided, can establish nothing certain, not against religion (to that extent Montaigne "defends" Sebon's arguments), but not for it either. Montaigne believes that divine providence is present in things great and small (II, 12; 529/394), but questions our ability to uncover it: "[God] has left the stamp of his divinity on these lofty works, and it is only because of our imbecility that we cannot discover it" (II, 12; 446–47/326). Ironically, though, his conclusion at times rejoins that of Bodin: the infinite power of nature and of God combined with the awareness of human ignorance and weakness should make us guard against the presumption of disdaining what we do not comprehend (I, 27; 179/134). So that, at one point, Montaigne actually criticizes Bodin, whom he considers in general to be of good judgment (words probably written before the *Démonomanie*), for being excessively incredulous. Notably when Bodin dismisses as fabulous Plutarch's report of the extraordinary stoicism of a Lacedaemonian boy who let a fox hidden under his shirt rip out his bowels rather than disclose his theft. Montaigne complains that the story should not be considered unbelievable simply because we could not imagine ourselves capable of such a feat. He acknowledges that Bodin is not in fact generally guilty of this kind of arrogance common to many "who balk at believing of others what they themselves could not do" (II, 32; 722–25/546–48). But clearly Montaigne too would have us guard against hastily rejecting seemingly unbelievable facts—like the travel to the witches' sabbath? No—on that issue he questions the belief rather than the disbelief.

The difference between Bodin and Montaigne seems to be less one of theory or epistemology, but more a matter of practice and emphasis: each acknowledges and makes use at various points of both natural theology and doubt. But Montaigne extends his doubt everywhere, systematically attacking facile certainties wherever they are, whereas Bodin uses it only here and there to tear down a traditional construction or two. On balance, Bodin thinks he can represent and explain nature, whereas Montaigne retreats from the investigation of the world out there to an investigation of the self, and even there refuses to find certainty. "It is only personal weakness that makes us content with what others or we ourselves have found out in this hunt for knowledge" (III, 13; 1068/817). But we are left trying to explain why Montaigne pushes in one direction and Bodin in another within a similar theoretical framework. In the face of the inevitable uncertainties of human judgment Bodin and Montaigne take different stances on the same question. Is it better to be inclusive, even if it means accepting some false facts, for fear of losing some true ones (Bodin's emphasis), or is it preferable to be cautious, perhaps at the cost of losing some truths (Montaigne's choice)? One might compare this epistemological decision to the legal one that both magistrates faced in the case of witchcraft: Bodin preferred to condemn a few innocent witches rather than let any guilty ones go. Montaigne, on the other hand, would rather let the guilty go than condemn someone to death falsely.

What distinguishes the two thinkers more decisively than these divergent emphases in the application of doubt, is an apparently minor philosophical precept to which Bodin is nonetheless forcefully attached, as he explains in his introduction:

> We consider that there is nothing more ugly nor more depraved than confusion and disorder . . . , which blame Aristotle himself incurred . . . when in the books about nature he tried to explain the way of teaching it, a topic which is appropriate rather to the dialecticians, and nonetheless left it untreated. (1–2/[2])[26]

Bodin is only being consistent with this principle, therefore, when he forgoes any metalevel discussion in the text itself. A self-conscious tone, such as we might like to see him take (like Montaigne's), would be a major breach of methodological propriety as far as Bodin is concerned. Instead Bodin relegates his few reflections on his work (of which in fact this is the principal one) to a separate introduction; in the body of the text he only reiterates the absurdity of wanting "to treat a science at the same time as the way of knowing it" (13/7).[27] Bodin is deeply attached to the traditional separation of the disciplines: dialectic, mathematics, physics, metaphysics—each has its own separate subject matter, principles, and place, not only in the university curriculum but also beyond it (indeed the *Theatrum* is certainly not a university textbook). Thus physics must not treat of the first principles on which it rests (which are proper to metaphysics) nor with the order in which it should be presented (a topic appropriate for dialectic): as a result it can never legitimately question the parameters of its practices.

For Montaigne, on the contrary, these disciplinary distinctions are part of a blindered scholastic system that he would like to see brought down:

> In this trade and business of knowledge, we have taken for ready money the statement of Pythagoras, that each expert is to be believed in his craft. . . . For

each science has its presupposed principles, by which human judgment is bridled on all sides. If you happen to crash this barrier, in which lies the principal error, immediately they have this maxim in their mouth, that there is no arguing against people who deny first principles (II, 12; 540/404).

Montaigne's self-reflexivity is thus an explicit attack on the standard classification and separation of the disciplines, which Bodin (who never mentions Montaigne) would have found utterly shocking. Montaigne's willingness to create disorder among the disciplines is a crucial prerequisite to his radical challenge of received knowledge and methods and to his stunningly "modern" authorial strategy. Conversely, Bodin's insistence on respecting strictly the traditional boundaries between the disciplines, compounded by the format of the *Theatrum* which treats natural philosophy as a long series of separate questions, perpetuates the scholastic tendency to atomize and subdivide knowledge. As Edward Grant has argued, this strategy insured the long survival of the medieval system of knowledge, through the sixteenth century and even beyond, by easily accommodating new opinions and observations (including Bodin's humanist interests and anti-Aristotelian explanations) and by keeping in separate categories, and thus invisible, many potential contradictions and logical inconsistencies.[28]

Although I may not have finally explained why Montaigne and Bodin are so different despite their similar backgrounds and methods, I hope I have indicated at least how further study of the diversity of positions on the classification of knowledge in the early modern period holds the promise of interesting results.

## Notes

1. Nannerl Keohane, *Philosophy and the State in France: The Renaissance to the Enlightenment* (Princeton: Princeton University Press, 1980), 79ff., 99ff.

2. For biographies of the two, see Donald Frame, *Montaigne: A Biography* (New York: Harcourt, Brace and World, 1965) and Jean Bodin, *The Six Bookes of the Commonweale*, ed. Kenneth McRae (Cambridge, MA: Harvard University Press, 1962), introduction. For comparisons, see Maryanne Cline Horowitz, "Montaigne versus Bodin on Ancient Tales of Demonology," *Proceedings of the Annual Meeting of the Western Society for French History* 16 (1989): 103–14, and Raymond Esclapez, "Deux Magistrats humanistes du XVI$^e$ siècle face à l'irrationnel: Montaigne et Bodin," *Bulletin de la Société des Amis de Montaigne*, 7e série 7–8 (1987): 47–74.

3. See Julian H. Franklin, *Jean Bodin and the Sixteenth-Century Revolution in the Methodology of Law and History* (New York and London: Columbia University Press, 1963), ch. 9.

4. For more on this point, and in general, see Ann Blair, *The Theater of Nature: Jean Bodin and Renaissance Science* (Princeton: Princeton University Press, 1997), chs. 2 and 5, here esp. pp. 72–74, 159–65.

5. For the one exception, see Jean Bodin, *Universae naturae theatrum* (Frankfurt: Wechel, 1597), 221; François de Fougerolles, trans., *Le Théâtre de la nature universelle* (Lyon: Pillehotte, 1597), 309. All further citations will refer by page number to these two editions respectively. English translations of this work are my own.

6. Ann Blair, "Humanist Methods in Natural Philosophy: The Commonplace Book," *Journal of the History of Ideas* 53 (1992): 541–51.

7. References to the Essays are made by book and chapter, then by page number, to Pierre Villey, ed., *Les Essais* (Paris: Presses Universitaires de France, 1965), and to the English translation which I have used: Donald Frame, trans., *The Complete Essays of Montaigne* (Stanford: Stanford University Press, 1965). For a helpful introduction to the *Essays*, see I. D. McFarlane and Ian Maclean, eds., *Montaigne: Essays in Memory of Richard Sayce* (Oxford: Clarendon Press, 1982).

8. "Les historiens sont ma droite bale" (II, 10; 416/303).

9. Francis Goyet, "A propos de 'ces pastissages de lieux communs' (le rôle des notes de lecture dans la genèse des *Essais*)," *Bulletin de la Société des Amis de Montaigne* 5-6 (1986): 11-26 and 7-8 (1987): 9-30; quotation from (1987): 25.

10. I am indebted to Jean Céard for his interpretation of this essay, which he presented in a paper delivered at Princeton University in April 1990.

11. See, most recently, Michel Jeanneret, "Montaigne et l'oeuvre mobile," in *Carrefour Montaigne, Quaderni del Seminario di Filologia Francese 2* (Pisa-Geneva, 1994).

12. Terence Cave, *The Cornucopian Text: Problems of Writing in the French Renaissance* (Oxford: Clarendon Press, 1979), 22.

13. "Et quidem Naturae Theatrum aliud nihil est quam rerum ab immortali Deo conditarum quasi tabula quaedam sub uniuscuiusque oculos subiecta, ut ipsius auctoris maiestatem, potentiam, bonitatem, sapientiam, atque etiam in rebus maximis, mediocribus, minimis admirabilem procurationem contemplemur et amemus." Bodin, sig. 3v (not in the French translation).

14. "Explica, si placet, universitatis tabulam, velut in theatro, ut quasi ob oculos rerum omnium distributione ad intuendum proposita, essentia cuiusque ac facultas planius intelligatur."

15. In introducing his dichotomous "tables" summarizing the *Theatrum*, the French translator Fougerolles refers to them as a "door or *planche* with which to enter the theater." Fougerolles, "Au lecteur," 917 (not in Latin edition).

16. Frances Yates, *The Art of Memory* (Chicago: the University of Chicago Press, 1966, ch. 6.)

17. See Charles Schmitt, *Aristotle and the Renaissance* (Cambridge, MA: Harvard University Press, 1983), 56-59.

18. "Nihil autem curiosius consectati sumus, quam rerum omnium seriem atque indissolubilem naturae cohaerentiam, contagionem et consensum, et quemadmodum responderent prima extremis, media utrisque, omnia omnibus."

19. "At illud quanti est, quod qui nullis divinarum legum perceptis, nullis Prophetarum oraculis ab inveterata amentia deduci, aut ad veri numinis cultum perduci possunt, certissimis huius scientiae demonstrationibus, quasi tormentis et quaestionibus admotis, omnem impietatem exuere, atque unum eundemque aeternum numen adorare cogantur!" Bodin, sigs. 3r–v (not in the French translation).

20. "Quanquam innumerabiles Mathematicorum quaestiones, veritati vim inferendo, ab invitis quasi adhibita quaestione, assensionem extorquere videmus."

21. See Friedrich Solmsen, *Aristotle's System of the Physical World* (Ithaca, NY: Cornell University Press, 1960), 404ff.

22. See his conclusions on the causes of earthquakes (179/245), violent storms (177/243-44), comets (222/310), magnets (243-44/342), on the origin of metals (259/363-64) and the nature of hoarfrost (205/283).

23. "Eatenus pertinet [disputatio de natura geniorum] quatenus refellendae sunt opiniones eorum, qui malunt ineptissimas rerum causas ingerere, quam ad genios, referre, vel de ignorantia confiteri."

24. "Sed praestat silentio summi Opificis maiestatem admirari, quam temere cum ratione insanire velle." "Cum ratione insanire" appears in Cicero, *Tusculan Disputations*, IV. 35. 76 in a quotation from Terence, *Eunuchus*, I. 1. 14.

25. "Illud tantum subiiciam, quamplurima mundi Conditorem, quasi contra naturae leges facere voluisse, ut homines in maiorem sui raperet admirationem et amorem: nec aliter a nobis naturae princeps esse putaretur."

26. "Confusione vero, ac perturbatione nihil aspectu foedius, aut deformius esse iudicamus. . . . Quam reprehensionem ne ipse quidem Aristoteles effugit, cum libris de Natura quaestionem de ratione docendi, quae Dialecticorum propria sit, explicare conatur, quam tamen indiscussam reliquit."

27. "Denique absurdum est scientiam simul et sciendi modum tradere velle."

28. Edward Grant, "Aristotelianism and the Longevity of the Medieval World View," *History of Science* 16 (1978): 93–106.

# THE LIBRARY AS AN INSTRUMENT OF DISCOVERY
## Gabriel Naudé and the Uses of History

Paul Nelles

### Cognitio Librorum: Philology and Curiosity

One of the most valuable guides to the early modern preoccupation with the cultural role of the library in all its forms is the 1627 *Advis pour dresser une bibliotheque* of Gabriel Naudé. In his *Advis*, Naudé combined a revised Renaissance universalism with an insistence on the unimpassioned judgment of the textual tradition in terms which would continue to animate discussion of the nature and purpose of the library into the eighteenth century. Naudé articulated a series of historical criteria for determining universality within the library, and argued that these criteria would be best implemented in what was quickly becoming the defining institution of the *res publica litteraria*, the public library: "une Bibliotheque dressee pour l'usage du public doit estre universelle, et... elle ne peut pas estre telle si elle ne contient tous les principaux Autheurs qui ont escrit sur la grande diversité des sujets particuliers, et principalement sur tous les Arts et Sciences."[1] It comes as no surprise that universality entailed the representation of the encyclopedia of particular disciplines: predecessors such as Conrad Gesner and Antonio Possevino had made similar claims. The significance of the *Advis* lies in its premise that the function of the library was to guarantee sound foundations for knowledge through the provision of means for the historical criticism of the disciplines. In formulating this view of the library, Naudé was able to draw upon his experience of contemporary Parisian libraries, in particular the collections of René Moreau and Henri de Mesmes. He gave this experience methodological form through a confrontation with Francis Bacon, advancing principles for the organization of scholarship and the library which countered contemporary sceptical criticism of the state of learning.[2]

The *Advis* registers an important shift in contemporary experience of the library. It initiated a methodological discussion which supplanted the dominant bibliographical conception of the library as a static repository of existing knowledge with a recognition of the library as an institution actively engaged in the production of new knowledge. The *Advis* bears witness to the ongoing early modern redefinition of the central purpose of the library from one of determining the authority of texts to one of evaluating the validity of sources. In addition, it confronted the most pressing dilemma facing the universalization of the library in this period: the need to establish a nonhierarchical classification of knowledge that avoided the charge of arbitrariness but which could yet claim to accurately serve all disciplines of knowledge and provide a stable basis for investigation. Drawing on Renaissance historiographical initiatives launched by Polydore Vergil, Angelo Poliziano, Jean Bodin and others, Naudé articulated a method of library organization which provided historical foundations for the encyclopedia of disciplines.[3]

As the student of the Parisian physician René Moreau, who established one of the best medical libraries in France, and as an intimate of one of the most prominent *robe* libraries in Paris, that of Henri de Mesmes, Naudé had gained direct experience of a vigorous institutional model of the library as a locus of erudite investigation.[4] Some years later, he would describe the kind of knowledge to be gained in the libraries of de Mesmes, Moreau, the Dupuys, and Jean Descordes as *cognitio librorum*, knowledge of the substantive content of books as well as of bibliographical minutiae—*nomenclatura aut potius oeconomia*. It was in fact Naudé who coined the term *bibliographia* in 1633, in an attempt to characterize the early modern bibliographical enterprise as more of a cognitive than an enumerative undertaking.[5] The *Advis*, though addressed to Naudé's patron, Henri de Mesmes, unlike earlier library treatises is not concerned with ordering a single library, but rather advances a universal method for establishing a library of this type, through recourse to a unique figure, Francis Bacon.[6] Bacon's presence in the *Advis* is complex; while Bacon blatantly rejected *cognitio librorum* as a valid intellectual instrument, Naudé was nonetheless much indebted to Bacon in formulating his views. With Mersenne, Naudé furnishes early evidence for Bacon's reception in France, and from Mersenne's comments it is evident that Bacon was read critically by the Parisian medical community of which Naudé was a member.[7] Bacon's influence on Naudé has been unduly neglected, and has had a particularly distorting effect: many of Naudé's much vaunted libertine views turn out to be reformulations of Baconian ideas.

Leaving more substantive issues for below, Naudé's relationship with Bacon can be characterized in the following terms. Naudé accepted Bacon's dissection of the world of learning, and he avidly appropriated Bacon's inclusion of the mechanical arts and low sciences in the encyclopedia of disciplines. He shared Bacon's conviction that progress in learning was to be had only cautiously, requiring constant vigilance over the projections of the intellect and the supply of helps to the senses. He shared too Bacon's methodological scepticism as well as Bacon's final rejection of radical sceptical conclusions. Both viewed doubt as the temporary suspension of judgment, and held that judgment was valid only when exercised independently of the authority of the object of judgment, whether a book or a philosophical system; Naudé's concern was to establish viable criteria for judging the intellectual validity of texts.[8] Naudé never accepted Bacon's thorough anti-Aristotelianism, and his own approach has more in common with the evenhanded survey of the state of learning of *De augmentis* than with the vitriolic campaign of the first book of the *Novum organum*. Importantly, despite Naudé's abiding allegiance to natural philosophy, he did not accord it the foundational status envisioned in the *Novum organum*, nor did he embrace Bacon's radical division of investigative labor between natural philosophy and the rest of the arts and sciences.

Naudé sought to render the library an *instrumentum* of learning, in the Baconian sense of an *adminiculum* or *auxilium* of inquiry, an institutional safeguard against worshipfulness and credulity, and an apparatus which would correct the weaknesses of the intellect and senses. In *De augmentis*, Bacon had described three "works or acts" which advance the state of learning: places of learning, books of learning (understood as libraries and new textual editions), and the persons of the learned. While he looked favorably upon books as *imagines ingeniorum*, in general Bacon was more disposed to institutions and scholars than he was to books and libraries; here he compared the library, negatively, to a shrine of relics of the "ancient saints" (AS, I, 483, 486–87). In the *Novum organum*, Bacon further observed that though the great variety of books in a library might by widely admired, eventually admiration would turn to astonishment at the poverty of subjects to which the human mind had been directed.[9] Bacon claimed to have abandoned the textual tradition in natural philosophy in two ways: first, insofar as he was trying a new method, the contributions of ancient and modern predecessors alike had no bearing upon his work;[10] second, he had banished the textual practices—*antiquitates, citationes, omnia denique philologica*—and compilatory method of Renaissance natural history in favor of the aphorism and induction.[11] There was room for only two books in Bacon's library: the book of nature and his own notebook of aphorisms.

Though if Naudé, like Bacon, had abandoned the late Renaissance encyclopedia—"les Panspermies, les cahos et abysmes de confusion" as he alluded to them in the *Advis* (78)—whether occult compendia, the *Theatra* of anthropological and natural phenomena of Zwinger and Bodin, or the didactic *summae* of Keckermann and Alsted, unlike Bacon he did not reject encyclopedism out of hand. Instead, Naudé located the library within the rather amorphous contemporary encyclopedic values of *curiosité*. The "premier poincte" of the *Advis* advises that "on doit estre curieux de dresser des Bibliotheques," and *curieux* and *curiosité* are terms which appear frequently and significantly throughout the *Advis*. Long viewed as an intellectual vice which transgressed the legitimate institutional and divine limits of human knowledge and which indulged rather than guarded against the passions, by the late sixteenth century curiosity had come to provide a positive investigative model of a *libertas inquirendi*.[12] In the early seventeenth-century circles which Naudé frequented, *curiosité* denoted a cluster of cultural and cognitive values which embraced the totality of knowledge, privileged the observational over the rational, and cultivated and redefined mediating institutions such as the laboratory, cabinet, museum and library.[13]

Naudé employs the register of curiosity in at least four different ways. First, closest to the main subject of the *Advis*, *curieux* describes books themselves, denoting rarity or unusualness, "livres grandement rares et curieux," or more specifically, collections of ephemera (pamphlets, placards, theses) which are "quelque fois des plus curieuses pieces d'une Bibliotheque" (61, 39, 100). Thus, *curieux* and *rare* can be evaluations of a book's intellectual worth: contrast Naudé's contempt for those who collect books for their physical characteristics, "enrichis avec toute sorte de mignardise, de luxe et de superfluité."[14] Rarity in this sense is a cognitive rather than a material assessment. In collecting commentaries on classical authors, for example, Naudé advises that those which are the least common, and consequently *plus curieux*, should not be neglected (*Advis*, 43). Second, *curieux* designates a rarified or neglected subject matter, its synonyms being *peu commun* and *non vulgaire*. Curiosity is thus associated with specialized knowledge and arcana, particularly in natural philosophy, and *communs et triviaux* subjects are distinguished from those which are *difficiles et peu connus* (ibid., 96). Curiosity also constitutes an alternative to both dogmatic learned tradition and common opinion (ibid., 40). The third usage of *curieux* in the *Advis* distinguishes certain material objects of a collection, as were held in the cabinet of the de Mesmes library itself. Naudé notes an acquaintance who was "curieux de Medailles, Peintures, Statuës, Camayeux, et autres pieces et jolivetez de Cabinet;" sounding a high Senecan note, he advised that instead of luxurious and superfluous decoration and furniture, money would be better spent on "les instruments de Mathematiques, Globes, Mappemonde, Spheres, Peintures, animaux, pierres, et autres curiositez tant de l'Art que de la Nature" (103–4, 150). This sense of *curiosité* grants books parity with non-textual instruments of investigation.

The final sense of curieux employed in the *Advis* designates the constituency of the public library. From Naudé's usage it is clear that *curieux* does not yet carry connotations it would acquire later in the century of *un amateur*, and that *docte* and *curieux* are almost interchangeable terms (103). For Naudé, the principal benefit of a library is that "un homme docte," who like Seneca is *curieux* for books not as ornaments but as *instrumenta studiorum*, might by means of the library become "Cosmopolite ou habitant de tout le monde, qu'il peut tout sçavoir, tout voir, et ne rien ignorer," and that "sans contredict, sans travail et sans peine il se peut instruire, et cognoistre les particularitez plus precises de /Tout ce qui est, qui fut, et qui peut estre / En terre, en mer, au plus caché des Cieux."[15] But what, in Naudé's view, distinguished curiosity from the excesses of Renaissance polymathy which he elsewhere deplored? The *curieux* is critical, disinterested and exercises an independent judgment: these are Naudé's criteria for the librarian in the first pages of the *Advis*. He refers to the trouble and difficulty required

> de s'acquerir une cognoissance superficielle de tous les arts et sciences, de se delivrer de la servitude et esclavage de certaines opinions qui nous font regler et parler de toutes choses à nostre fantaisie, et de juger à propos et sans passion du merite et de la qualité des Autheurs.[16]

Curiosity for Naudé is manifestly *not* wonder. Naudé's allegiance is rather to a Ciceronian, erudite and sceptical model of curiosity, a model which entailed the control rather than the indulgence of the passions, and which posited a cautious encyclopedism as a general intellectual good.[17] Thus curiosity in the *Advis* as the positive identity of *une bibliothèque curieuse* encompasses topics of investigation, books valued for their empirical contributions, the physical objects and instruments of a collection, and the encyclopedic and intellectually independent constituency of the library. Such a library type, though it incorporated strong Baconian elements, was as little inclined to banish antiquities, citations and the philological tradition from natural as it was from historical investigations.[18]

It was thus the library as the ecumenical domain of the *curieux* which held its varied intellectual and cultural denominations together. To a large extent, Naudé confronted Bacon with his own experience of the Parisian library. Together with the libraries of Sirleto and especially Pinelli,[19] it was the great *robe* libraries of de Thou, de Mesmes, Fontenay, Hallé, Jacques Ribier, the Dupuys, and Jean des Cordes which Naudé summoned as models in the *Advis*.[20] In so doing, Naudé was led to develop a practical methodology which sought the universalization of the library as an institution of learning on a Baconian model. Occasional reference to his allegiance to "le progrez et l'augmentation" of both the de Mesmes library and the arts and sciences in general is clearly derived from Bacon.[21] Less obviously Baconian, however, is the "ordre et methode" of the *Advis* as a handbook, a method of often disjointed maxims, "les preceptes et moyens sur lesquels il est à propos de se regler," a distillation or summary of Naudé's observations and experience (*Advis*, 11). The 1627 *Advis* is not divided into formal chapters, but instead progresses by a series of considerations of "points principaux."[22] Like Bacon's use of the aphorism, Naudé's precepts are, he claimed, incapable of being reduced into an art or method on account of their very inexhaustibility (*Advis*, 96–97); more pointedly, they are better put into practice by those "qui ont une grande routine des livres, et qui jugent sainement et sans passion de toutes choses" than they are "deduites et couchées par escrit" (ibid., 42). And upon considering means of acquiring manuscripts, Naudé observed that "j'aime mieux le laisser à la discretion de ceux qui en voudront user, que non pas de le prescrire comme une regle generale et necessaire" (ibid., 122). The constant mediation between precept on the one hand and experience on the other would enable the library to achieve its "usage principal," a procedure which validated both Naudé's model of the librarian and the *Advis* as a technical handbook. Use is the final criterion: objects of inquiry are to be judged not according to the opinions of others, but "eu esgard à leur propre usage et nature" (ibid., 97–98). The "fin" and "usage principal" of Naudé's maxims and principles was to "consacrer l'usage" of the de Mesmes library "au public," while it was the librarian's function to "regler cet usage" (ibid., 151–52, 156). Naudé's elevation of use or custom and his anti-systematic procedure constitutes an overt appeal to scepticism. Naudé advanced neither a theory of the library nor, as we shall see, a system of library classification and organization. At once challenging and incorporating Baconian methods and ideas, Naudé sought, through observation and experience, to determine the nature and purpose of those of Bacon's "works or acts" of the advancement of learning Bacon himself held in lowest regard, in order to render it a valid instrument of discovery.

## Dispositio Librorum: Scepticism and the Disciplines

Naudé's incorporation of sceptical elements into a methodological procedure had direct consequences for the organization of a library. Books in the library should be "rangez et disposez suivant leurs diverses matieres, ou en telle autre façon qu'on les puisse trouver facilement et à poinct nommé." Without such an "ordre et disposition," a heap of books cannot become a library (ibid., 130). The best order to follow is that which is the easiest, the least puzzling, the most natural and the most commonly used, the order of the disciplines themselves: theology, medicine, jurisprudence, history, philosophy, mathematics, the humanities and unspecified "others." Within each discipline, divisions should be made according to existing subdisciplines (ibid., 134). Naudé's proposed categories are not original, and the faculty system was in wide use in Parisian libraries, including the de Mesmes library itself: this is entirely the point. The most arresting feature of Naudé's disciplinary approach is that it makes no taxonomical pretensions to completeness or necessity. He does not, for example, claim that it corresponds to either nature or natural reason, nor does he suggest that it represents a true logical or metaphysical system. Still less is it a system of classification: it does not in its own right generate knowledge of the disciplines. Its only "natural" characteristic is that it already obtains in the scholarly *sensus communis*. Naudé's appeal to "cet ordre qui est le plus usité" (ibid., 139) is an overtly sceptical resolution of the problem. While a traditional hieararchical order of the disciplines thus emerges intact, it is in a much weaker position and cannot itself form the basis of an investigative programme.

The disciplinary scheme advanced by Naudé was common enough that there was no need to provide illustrative examples or a validating intellectual genealogy. However, the sole allusion Naudé makes to any such consideration of the order and relationship of the disciplines—a passing reference to Poliziano's *Panepistemon*—reflects tellingly upon the method advanced in the *Advis* (ibid., 36). Naudé's allusion was to Poliziano's procedure of disciplinary enumeration rather than the order of the disciplines detailed in the *Panepistemon*. Poliziano had proposed a method of reading Aristotle's ethical works which recommended restructuring Aristotle's own epistemological categories:

> My intention in expounding Aristotle's books on ethics is to undertake, insofar as it is possible, to divide them in such a way so that not only (as they are called) the liberal and mechanical disciplines and arts, but also those mean and manual

ones which nonetheless life requires, are comprised within the compass of such an arrangement. I will therefore imitate the dissections of the physicians, known as anatomical; I will imitate the reckonings of the accountants. For I shall divide each into nearly all its parts, and bring the whole to account, whereby each part will be easily grasped as well as accurately retained.[23]

Poliziano's ironic appeal to anatomical observation, his widening of the disciplinary field to include disciplines normally excluded from the encyclopedia, and his quiet challenge to the Aristotelian hierarchy of the liberal and mechanical arts might equally describe Naudé's approach to the disciplines. Poliziano proposed a simple inventory of the disciplines rather than a systematic classification which, he claimed, was both straightforward and easy to remember.

Naudé criticized a variety of systems of library organization on the grounds that they failed to satisfy such empirical criteria. He summarily dismissed both the compilatory, commonplace method of La Croix du Maine's *Bibliothèque Françoise* of 1584, and the image-oriented artificial memory system of Giulio Camillo's *L'Idea del theatro* of 1550. Naudé charged Renaissance commonplace methods and arts of memory with violating Cicero's injunction that orderly arrangement is essential for good memory, an injunction which figures such as Camillo claimed to have fulfilled.[24] For Naudé, the professional faculties of theology, medicine and law, and the disciplines of history, philosophy and mathematics, arranged according to their disciplinary divisions, function as both *lieux de mémoire* and *lieux communs*. He criticized alphabetical or symbolic systems of classification, including those of the Ambrosiana and other libraries where books were arranged in random order, distinguished merely by size and symbols, located only by consulting alphabetical author catalogues (*Advis*, 139–40). Instead Naudé proposed two catalogues, the first a subject catalogue "disposez [s]uivant les diverses matieres et Facultez," where one might apprehend in the twinkling of an eye all who had written on a particular subject. A second, author catalogue was also to be kept, though more to avoid buying books twice and to detect missing items than to aid "beaucoup de personnes qui sont quelquefois curieuses de lire particulierement toutes les oeuvres de certains Autheurs" (ibid., 158–59). The subject catalogue was thus the most important, and Naudé suggested that

> les Traictez particuliers suivent le rang et la disposition que doivent tenir leur matiere et sujets dans les Arts et Sciences [ . . . ] et [ . . . ] tous les livres de pareil sujet et mesme matiere soient precisément reduits et placez au lieu qui leur est destiné, parce qu'en ce faisa[n]t la memoire est tellement soulagee, qu'il seroit facile en un moment de trouver dans une Bibliotheque plus grande que n'estoit celle de Ptolomee, tel livre que l'on en pourroit choisir ou desirer (ibid., 136).

The order of the library should mirror the world of learning itself. This was a universally applicable procedure which could be followed in any discipline and, he argued, would perform essential mnemonic functions. Moreover, it permitted the investigator to follow his own good sense and judgment, and to avoid following "la trace et les opinions des autres" (ibid., 143–44). While such a position is heavily Baconian, la trace et les opinions des autres held considerably more interest for Naudé than for Bacon; he consequently propounded a method for the retrieval of both.

## Inventio Librorum: History and Discovery

The order of the disciplines thus constituted the natural method of library organization. It did not rigidly enforce a disciplinary hierarchy in the library, and it allowed for the expansion of the encyclopedia of disciplines. Naudé also established historical criteria for the selection of books within each discipline, which sought to harness sceptical objections within a positive investigative program. It was through the implementation of these criteria that the library's universality was to be maintained, and they performed at least five functions.[25] First, they reinforced a nonhierarchical arrangement of the disciplines by granting each discipline an autonomous past, validating the utility of even those disciplines whose empirical or theoretical authority had been revoked, such as astrology or alchemy. Second, they allowed for the anti-dogmatic revision of dominant intellectual schools and traditions through provision of unorthodox and noncanonical genealogies. Third, such criteria facilitated the identification and elimination of error as well as the discovery of truth. Already in his 1623 Instruction Naudé had cited Lactantius to this effect: "primus sapientiae gradus est falsa intellegere," and the identification of error would remain an on-going project.[26] Fourth, they allowed for the apprehension of empirical knowledge produced by past practitioners of a discipline, of potential value to contemporary investigations. History thus had an intrinsic investigative role. Fifth, they allowed for the resolution of controversy and debate, both religious and philosophical, through reference to those passages in the sources directly in dispute, or through the investigation of noncontroversial areas of shared technical interest.

The history of the disciplines was thus crucial to the library's goal of rendering the universality of knowledge accessible. As Kristeller has pointed out, Naudé's vision of the history of learning furnishes a valuable key to an intellectual world where the "new science" had yet to replace the "new learning," and the substantive content as well as the method of Naudé's history of the disciplines warrants consideration.[27] It is here that the great differences between Naudé and Bacon emerge in sharp constrast. Though it is difficult to determine the degree to which Naudé took issue with Bacon directly, two divergent positions on the domain of natural philosophy and human learning are nonetheless distinguishable.

While Naudé's disciplinary method provided almost unlimited lateral scope, history penetrated vertically through each discipline. In addition to arranging a discipline in chronological order, Naudé advocated the inclusion of the major disputes in the history of a discipline, an approach which Bacon had actively excluded from natural philosophy. Thus Naudé listed both the good sceptics who had "upset all the sciences" (Sextus Empiricus, Agrippa, Sanches) as well as both successful and unsuccessful ("sans toutesfois rien innover ou changer des principes") attacks on particular disciplines: Pico on astrology, Scaliger *père* on Cardanus, Casaubon on Baronius, Thomas Erastus on Paracelsus, and Charpentier on Ramus. While this documents the sceptical technique of hearing an argument *in utramque partem*, Naudé's purpose was to form rather than suspend judgment: "il y auroit autant de faute à les lire separément, comme à juger et entendre une partie sans l'autre, ou un contraire sans celuy qui luy est opposé." Moreover, this principle effectively reduced scepticism to merely a controversial phase of a discipline's history.[28]

Closely associated with scepticism and controversy were the *novatores*, those who had "doctement examiné ce que les autres avoient coustume de recevoir comme par tradition." Thus Copernicus, Kepler and Galileo were summoned for astronomy, and Paracelsus, Severinus, Joseph du Chesne and Croll for medicine. The founders of disciplines were also important—*la doctrine*, like water, never so clear and clean as at its source. Adapting the theme of Polydore Vergil's *De rerum inventoribus*, Naudé listed Reuchlin on the Hebrew language, Budé on Greek and coins, Cocles on physiognomy, Bodin on politics, and Peter Lombard and Thomas Aquinas on scholastic theology, all of whom, he contested, were better than the many who followed.[29] But Naudé also considered others who had made less distinguished, often less reputable, but nonetheless original contributions. While acknowledging that the books of the "curieux et non vulgaires" such as Cardanus, Pomponazzi, Giordano Bruno, and all those who had written on artificial memory, divination, the cabala, the system of Lull, and the philosopher's stone teach nothing which is not "vain et inutile" and serve only as stumbling blocks for those who read them, Naudé argued that one must be as equally familiar with such *esprits foibles* as with the *esprits forts*. And again, in order to be refuted, texts needed to be read. Bacon, by contrast, considered Pliny, Cardanus, Albertus Magnus and the Avicennists best forgotten.[30]

On the same principle ("cette maxime nous doit faire passer à une autre de pareille consequence"), Naudé advocated including heterodox texts. With official license, the works of the "principaux Heresiarques ou fauteurs de Religions nouvelles et différentes de la nostre" were all to be collected in the library. The early reformers, Naudé argued, had come from among the most learned men of the preceding century, though somehow their religion had gotten in the way of their learning: "par je ne sçais quelle fantaisie et trop grand amour de la nouveauté, [ils] quittoient leur froc et la bannière de l'Eglise Romaine pour s'enroller sous celle de Luther et Calvin" (*Advis*, 56–57). But if sensitive and controversial points were avoided, the vast learning of heterodox authors could be profitably utilized, and Naudé helpfully provided an appropriate heterodox catalogue (ibid., 58–60).

Naudé, as had Bacon, excluded extreme enthusiasm for the moderns as well as extreme veneration of the ancients. But where Bacon had categorically dismissed medieval learning, Naudé sought to re-examine the period of learning extending from late Christian antiquity to the fifteenth-century humanists.[31] Valuable, usable contributions had been made in this interval in the fields of philosophy, theology, law, medicine and astrology. Naudé complained that books of former ages of scholarship had been neglected simply because "leur seule impression noire et Gothique met dans le dégoust des plus delicats Estudians de ce siecle" (Advis, 84). In a direct anti-sceptical charge, Naudé stated that it was "tousjours l'ordinaire des esprits" who made so much out of disciplinary dispute and change but yet failed to derive any benefit from it. The diversity of sects and disciplinary flux had long served as a sceptical "proof" of the vanity of learning, employed by Bacon among others.[32] Naudé contended that the result had been disastrous. The majority of authors thus remained abandoned and neglected on the shoals of scholarship, while the nouveaux Censeurs ou Plagiaires helped themselves liberally to the spoils. He observed that while the philosophy of Suarez and the College of Coimbra was currently in vogue, the works of Albertus Magnus, Agostino Nifo, Pomponazzi and others stood completely neglected, even though contemporary school philosophy pillaged these authors relentlessly.[33] The same, he believed, was true of modern medicine, which had only superficially exploited the contributions of the medieval Latin West and of the Avicennists, who, clumsy and heavy as their Latin might be, had nonetheless "tellement penetré le fonds de la Médecine." Bacon, on the other hand, had ruled out the validity of both Arabic and scholastic traditions in natural philosophy.[34]

The serious consideration Naudé gave non-Galenic medicine (Avicenna, Paracelsus) in the Advis, the interest exhibited in memory systems, natural magic, the occult, alchemy, and "vain" or ineffectual philosophy constitutes a significant difference with Bacon on the value of the history of learning. While Naudé and Bacon both valued sixteenth-century novatores for their double assault on ancient tradition and received opinion, Naudé welcomed "vain" philosophy for exactly the same reason. As we have seen, Bacon's position was quite different, though he did not reject historical procedures from natural philosophy out of hand. In De Augmentis, Bacon had proposed to compile a collection of Placita antiquorum philosophorum, an evaluation of both ancient and modern philosophical systems. The tenor of Bacon's proposal is anti-Aristotelian, and it is strictly an evaluation of first principles; empirical problems were to be inventoried in a nonhistorical survey of Problemata naturalia, or a Kalendarium dubitationum (AS I, 561–62). Among the moderns to be considered were Paracelsus, Severinus, Telesio, Patrizi, and Gilbert. Paracelsus aside, the inclusion of Gilbert is significant, for together with the alchemists, Gilbert was Bacon's favourite example of doomed progression from laudable empirical investigation to unwarranted speculation and ill-founded natural philosophy.[35]

However, Bacon considered both the Kalendarium and the Placita as appendices to natural philosophy: such projects were central for Naudé. Where Bacon turned to the history of philosophy for a survey of philosophical systems, Naudé maintained a revised Renaissance position that a critical examination of the history of philosophy could yield useful empirical knowledge, regardless of theoretical orientation. The priority Bacon accorded theoretical criteria effectively ruled out a role for history in empirical investigations as, unlike in Naudé, false theoretical assumptions discredited the validity of empirical findings. And while Bacon directly denied the possibility of obtaining *veritas aliqua purior* from the *Placita*, Naudé stressed that even unsuccessful efforts could, in the right hands, produce unexpected results:

> il est certain que la cognoissance de ces livres est tellement utile et fructueuse à celuy qui sçait faire reflexion et tirer profit de tout ce qu'il voit, qu'elle luy fournit une milliace d'ouvertures et de nouvelles conceptions, lesquelles estans receues dans un esprit docile, universel et desgagé de tous interests, *Nullius addictus iurare in verba magistri*, elles le font parler à propos de toutes choses, luy ostent l'admiration, qui est le vray signe de nostre foiblesse, et le façonnent à raisonner sur tout ce qui se presente avec beaucoup plus de jugement, prevoyance et resolution, que ne fait pas le commun des autres personnes de lettres et de merite.[36]

The value of considering weak or even spurious contributions in natural philosophy for Naudé is that once properly evaluated, they would potentially be capable of producing new, effective knowledge. But more interesting is Naudé's suggestion that the effort required to achieve such an evaluation is itself propaedeutic, as it requires a constant mental vigilance and instills a routine incredulity. The emphasis on observation, universality, disinterestedness and the specific attack on admiration (wonder), while reminiscent of Bacon, is to be deployed in the sort of assessment of tradition Bacon explicitly excludes from the normal practice of natural philosophy.

The movement of Naudé's argument from vain philosophy, to occultism to heterodoxy is important. In all instances he actively supplanted a notion of the determined authority (or lack of it) of a text with a criterion of its usefulness as a source. While Naudé had nothing but disdain for the magical and chemical world views, he was nonetheless interested in their empirical achievements. He suggested a similar approach to theology, patristic scholarship, and ecclesiastical history, arguing that heterodox challenges, especially in the domains of the textual criticism of the Bible and patristic scholarship, had effected disciplinary renewal and driven scholars on both sides of the confessional divide to investigate hitherto neglected aspects of the Christian tradition (*Advis*, 85–86). A similar process of change and renewal had occurred in all the disciplines (ibid., 87). If anywhere, Naudé's own achievement lies here: history provided a means of prying valid knowledge from discredited and invalid systems of thought and belief. He ultimately stopped short, however, of arguing directly that a methodical scepticism would secure new foundations for knowledge.

All of this had direct consequences for collecting. In addition to acquiring bulky multi-volume works in a given subject, brief, less well known, physically smaller (and, he argued, often more incisive) treatments of subjects should also be purchased (ibid., 76–78, 80). Moreover, the library should play a positive, active role in editorial scholarship by ferreting out unedited manuscripts of neglected authors. Thus Naudé took the common run of textual scholars to task. How unseemly to follow the caprice and fantasies of "ces nouveaux Censeurs et Grammariens" who had spent a lifetime, Naudé quipped, forging conjectures (ibid., 92). Meanwhile, the manuscripts of an infinite number of scholars who had toiled and devoted their entire lives to discovering that which was formerly unknown or to the clarification of some useful subject lie mouldering in the hands of ignorant collectors (ibid., 92–93). And while it was fully within the discretion of the library to amass a great number of manuscripts, only manuscripts of actual cognitive worth should be actively collected, obtained with the help of bibliographical aids such as Gesner's *Bibliotheca universalis* and library catalogues. Manuscripts of works already in print were of little concern (ibid., 94–95), as were manuscripts prized solely for their antiquity, decoration, or "autres foibles considerations" (ibid., 116). Naudé argued straightforwardly for the extension of editorial scholarship into periods and domains of human learning which had hitherto been excluded from this process of historical evaluation, and for which there was an abundance of material. Nothing that was of potential use for the advancement of any discipline could be neglected; this indeed is the dominant concern of the *Advis*—"les Bibliotheques ne sont dressees ny estimees qu'en consideration du service et de l'utilité que l'on en peut recevoir" (ibid., 115–16).

The significance of the *Advis* lies in its articulation of an *institutionalized* practice of the history of scholarship, a concept of the function of the library which would become increasingly important as the library came to be seen as a locus for the public verification of scholarship. At the very end of the *Advis*, Naudé had proposed to write, in pursuit of "le progrez et augmentation" of the de Mesmes library, a *Bibliotheca Memmiana*, which would be

> ce qu'il y a si long-temps que l'on souhaite sçavoir, l'histoire tres-ample et particuliere des Lettres et des Livres, le jugement et censure des Autheurs, le nom des meilleurs et plus necessaires en chaque Faculté, le fleau des Plagiaires, le progrez des Sciences, la diversité des Sectes, la revolution des Arts et Disciplines, la decadence des Anciens, les divers principes des Novateurs, et le bon droict des Pyrrh[o]niens fondé sur l'ignorance de tous les hommes.[37]

As Rudolf Blum reminded us many years ago, Naudé had taken the idea of such a history of learning from Bacon's *De augmentis scientiarum*, where Bacon, finding the historiography of learning in as poor condition as the patient itself, called for a "just and universal history of learning," giving full expression to the concept of *historia litterarum*:

> the argument of which is none other than to recover out of the records what doctrines and arts flourished in which regions and ages of the world. Their an-

tiquities, progress, and peregrinations to the various parts of the earth (for knowledge also migrates, just as do peoples), and their reversals, oblivions and renewals should be commemorated. At the same time the occasion and origin of the invention of each art should be observed; its manner and method of transmission; the method and order of its cultivation and excercise. The sects should also be considered, and the most celebrated controversies conducted by the learned; the calumnies they suffered; the praises and honors with which they have been rewarded. The most distinguished authors and outstanding books should be noted, schools, successions, academies, societies, colleges, the orders, and finally everything else which bears upon the condition of learning.[38]

Although in practice Naudé and Bacon held widely divergent views on the nature and purpose of the history of learning, Naudé's evocation of *historia litteraria* is nonetheless one of the strongest Baconian elements of the *Advis*. The projected *Bibliotheca Memmiana* effectively disarmed sceptical criticisms of the state of learning by proposing to write a critical history of them. Just as for Bacon *historia naturalis* put nature to the test, so for Naudé *historia litteraria* put all of knowledge to the test: Naudé very early recognized the potential of historical pyrrhonism as a methodological safeguard against radical sceptical attacks upon the textual foundations and transmission of knowledge. The library emerged as the institutional domain of *historia litteraria*.

While the common order of the disciplines performed the traditional encyclopedic task of arranging knowledge, the historical principles expounded by Naudé allowed for the discovery of a discipline's potential through the propaedeutic evaluation of its own past and that of related disciplines. More than this, history broke down traditional disciplinary hierarchies and allowed for the expansion of the encyclopedia of disciplines, a problem which encyclopedias with logical, metaphysical or theological foundations did not pretend to address. Though building on Bacon's achievement, in identifying the library with the history of learning Naudé rendered the library the apotheosis of a Renaissance historiographical tradition long intent upon the recovery of the *monumenta litterarum*. Perhaps more importantly, in so doing he secured intellectual and methodological foundations for the library which would last a century and more. The final, resounding joke of the *Advis* is that it itself constitutes a *histoire très-abrégé* of the proposed *Bibliotheca Memmiana*, an "autre Catalogue fort exact" of the arts and sciences to accompany Poliziano's *Panepistemon*. As the library emerged as an autonomous institution of learning over the course of the next century, *historia litteraria* would be refined to fulfil Naudé's double injunction that the method of the library adequately reflect the world of scholarship in addition to ordering the library itself.[39]

## Notes

1. Gabriel Naudé, *Advis pour dresser une bibliotheque* (Paris, 1627), 36, abbreviated as *Advis*. Other abbreviations used: NO = Francis Bacon, *Novum organum*, in *Works of Francis Bacon*, ed.

J. Spedding (London, 1857–58), vol. 1; AS = Francis Bacon, *De augmentis scientiarum*, in *Works*, vol. 1. Original orthography and punctuation are maintained in all quotations, while minor typographical conventions have been modernized.

2. On the development of the bibliographical sciences, see Luigi Balsamo, *La bibliografia: storia di una tradizione* (Florence, 1984); Rudolf Blum, *Bibliographia: An Inquiry into Its Definition and Designations*, trans. Mathilde V. Rovelstad (Folkestone, 1980); and the exhaustive survey and summary of the full range of early modern bibliographical and encyclopedic literature of Alfredo Serrai, *Storia della bibliografia*, 6 vols. (Rome, 1984–94). For institutional overviews, see I.R. Willison, "The Development of the British Library to 1857 in Its European Context: A Tour d'horizon," in *Öffentliche und private Bibliotheken im 17. und 18. Jahrhundert*, ed. P. Raabe (Bremen and Wolfenbüttel, 1977), 33–61, and Luciano Gargan, "Gli umanisti e la biblioteca pubblica," in *Le biblioteche nel mondo antico e medievale*, ed. G. Cavallo (Bari, 1988), 165–86. See also Roger Chartier, *The Order of Books* (Cambridge, 1994), 61 ff., and Helmut Zedelmaier, *Bibliotheca universalis und Bibliotheca selecta. Das Problem der Ordnung des gelehrten Wissens in der frühen Neuzeit* (Cologne, 1992), especially valuable for his analysis of the structural relationship of bibliography and *lectio historiarum*. But the majority of studies do not relate the early modern forms of *bibliotheca* to their scholarly constituencies, technical competencies, and intellectual and confessional commitments, though see still Henri-Jean Martin, *Livre, pouvoirs et société à Paris au XVIIe siècle* (Geneva, 1969), esp. 472–551, 625–34, 922–58. For an exemplary redress, see Pierre Petitmengin, "Deux 'bibliothèques' de la Contre-Réforme: la *Panoplie* du Père Torres et la *Bibliotheca Sanctorum Patrum*," in *The Uses of Greek and Latin*, ed. A.C. Dionisotti et al. (London, 1988), 127–53.

3. Most studies of early modern encyclopedism have focused on Protestant encyclopedias with explicitly theological or metaphysical foundations. See Ulrich Dierse, *Enzyklopädie: zur Geschichte eines philosophischen und wissenschaftstheoretischen Begriffs* (Bonn, 1977); Cesare Vasoli, *L'enciclopedismo del seicento* (Naples, 1978). Essential and fundamental characteristics are drawn out by Charles Lohr, "Metaphysics," in *The Cambridge History of Renaissance Philosophy*, ed. C.B. Schmitt et al. (Cambridge, 1988), 537–638, esp. 632 ff. For an overview, valuable synopses and further bibliography, see Joseph S. Freedman, "Encyclopedic Philosophical Writings in Central Europe During the High and Late Renaissance (ca. 1500–ca. 1700)," *Archiv für Begriffsgeschichte* 37 (1994): 212–56. Erudite and historical encyclopedism has received comparatively little attention. In addition to Zedelmaier, *Bibliotheca universalis* (above, n. 2), see Donald R. Kelley, "History and the Encyclopedia," in *The Shapes of Knowledge from the Renaissance to the Enlightenment*, ed. D.R. Kelley and R.H. Popkin (Dordrecht, Boston, and London, 1991), 1–22; R.J.W. Evans, "Culture and Anarchy in the Empire, 1540–1680," *Central European History* 18 (1985): 14–30. Useful studies of individual figures include Aldo Scaglione, "The Humanist as Scholar and Politian's Conception of the 'Grammaticus'," *Studies in the Renaissance* 8 (1961): 49–70; Brian P. Copenhaver, "The Historiography of Discovery in the Renaissance: The Sources and Composition of Polydore Vergil's 'De inventoribus rerum,' I-III," *Journal of the Warburg and Courtauld Institutes* 41 (1978): 192–214; Françoise Waquet, "Le *Polyhistor* de Daniel Georg Morhof, lieu de mémoire de la République des Lettres," in *Les Lieux de mémoire et la fabrique de l'oeuvre*, ed. V. Kapp (Paris, Seattle, and Tübingen, 1993), 47–60; and on the concept of *polymathia*, Luc Deitz, "Johannes Wower of Hamburg, Philologist and Polymath. A Preliminary Sketch of his Life and Works," *Journal of the Warburg and Courtauld Institutes* 58 (1995): 136–54.

4. The most thorough and useful study of Naudé's career is now Maria Cochetti, "Gabriel Naudé, *Mercurius Philosophorum*," *Il bibliotecario* 22 (1989): 61–106, with full bibliography. Cochetti also provides a valuable list of bibliographical literature from the catalogue of Naudé's personal library. For a comprehensive analysis of the principal parts of the *Advis*, see also Alfredo Serrai, *Storia della bibliografia*, vol. 5, *Trattatistica Biblioteconomica* (Rome, 1993), 295–331. See the interesting and valid thesis of Robert Damien, *La Bibliothèque et l'état* (Paris, 1995); though still pertinent are Paul Oskar Kristeller, "Between the Italian Renaissance and the French Enlightenment: Gabriel Naudé as an Editor," *Renaissance Quarterly* 32 (1979): 41–72, and Rudolf Blum, "*Bibliotheca Memmiana*. Untersuchungen zu Gabriel Naudés *Advis pour dresser une bibliothèque*," in *Bibliotheca docet. Festgabe für Carl Wehmer*, ed. Siegried Joost (Amsterdam, 1963), 209–32. Considerations of

the *Advis* tend to interpret it either somewhat naïvely as a straightforwardly groundbreaking work in library management, or, as in the case of the classic work of René Pintard, *Le Libertinage érudit dans la première moitié du XVIIe siècle* (Paris, 1943), e.g., 249, 451–52, and those who follow him, as a mere vehicle for libertine views. Despite a promising title, Lorenzo Bianchi, "Tradizione scettica e ordinamento dei saperi in Gabriel Naudé," *Studi filosofici* 7 (1984): 117–34, does not come to grips with the central issues, and passes over the *Advis* lightly. Idem, "Per una biblioteca libertina: Gabriel Naudé e Charles Sorel," in *Bibliothecae selectae da Cusano a Leopardi*, ed. E. Canone (Florence, 1993), 170–215, reads the *Advis* for, 177, "alcune affermazioni importanti e sintomatiche di una ampia libertà di pensiero." Bianchi publishes excerpts from Naudé's personal library catalogue which bear on his *libertinismo*, 206–15, though he is apparently unaware of Cochetti's efforts in the same area. See also (cautiously) Louis Desgraves, "Naissance de la 'science' des bibliothèques," *Revue française d'histoire du livre* 60 (1991): 3–30.

5. Gabriel Naudé, *Bibliographia politica* (Venice, 1633), 5–7.

6. Cf. Pintard, *Libertinage* (above, n. 4), 163, 446. Though the *Novum organum* was published in London in 1620, Naudé does not mention it in his *Instruction à la France sur la verité de l'histoire des Freres de la Roze-Croix* (Paris, 1623), and the *Instruction* bears none of the Baconian characteristics of the *Advis*. Pintard's erroneous claim for Bacon's presence in the *Instruction* should be revised to refer to Campanella. *De augmentis* was published in London in 1623, and in Paris in 1624. A French translation of *The Advancement of Learning* was also published in 1624. For a discussion of editions and translations, see Michèle Le Doeuff, "Bacon chez les grands au siècle de Louis XIII," in *Francis Bacon. Terminologia e fortuna nel XVII secolo*, ed. M. Fattori (Rome, 1984), pp. 155–78.

7. See Marin Mersenne, *La Verité des sciences* (Paris, 1625), 209, "il [Bacon] se trompe en plusieurs choses, comme quelques excellens Medecins ont remarqué, lors qu'il parle de la Medecine." On Mersenne's criticisms of Bacon, see Robert Lenoble, *Mersenne ou la naissance du mécanisme* (Paris, 1943), 328–35; Peter Dear, *Mersenne and the Learning of the Schools* (Ithaca and London, 1988), 187–88. Gassendi's critique of Bacon's inductive logic was formulated only in the 1630s.

8. Cf. AS I, 456, 461–62.

9. NO I, 192. Cf. also AS I, 492; NO I, 125; *Filum labyrinthi*, in *Works* III, 497; *Cogitata et visa*, in *Works* III, 591.

10. Cf., for example, NO I, 153. Though in the *Parasceve* Bacon does make allowances for testing empirical observations retrieved textually.

11. NO I, 396–97. Again, Bacon has not abandoned the compilatory method altogether. In natural history he does allow for textual sources, though his position on their use is not as clear as his condemnation of it; in natural philosophy, aphorisms rather than testimonies are to be compiled, cf. NO I, 218–19.

12. *Advis*, 14, "I. On doit estre curieux de dresser des Bibliotheques, et pourquoy." On *curieux* and *curiosité*, cf. Krzysztof Pomian, *Collectionneurs, amateurs et curieux. Paris, Venise: XVIe-XVIIIe siècle* (Paris, 1987), esp. 61–80; Edward Peters, "*Libertas inquirendi* and the *vitium curiositatis* in Medieval Thought," in *La Notion de liberté au Moyen Age. Islam, Byzance, Occident*, ed. G. Makdisi et al. (Paris, 1985), 89–98; and especially William Eamon, *Science and the Secrets of Nature. Books of Secrets in Medieval and Early Modern Culture* (Princeton, 1994), 59–65, 314–18. See also Jean Céard et al., *La Curiosité à la Renaissance* (Paris, 1986). See the excellent lexical discussion available in Gérard Defaux, *Le Curieux, le glorieux et la sagesse du monde dans la première moitié du XVIe siècle* (Lexington, Kentucky, 1982), 69–110, especially for the fortune of a positive Ciceronian model of curiosity. The present discussion of curiosity is intended only to forge an entrée into Naudé's world, not to define it.

13. For other institutional initiatives and milieux of which the ongoing redefinition of the library was a part, see R.J.W. Evans, "Learned Societies in Germany in the Seventeenth Century," *European Studies Review* 7 (1977): 129–51; Owen Hannaway, "Laboratory Design and the Aim of Science: Andreas Libavius versus Tycho Brahe," *Isis* 77 (1986): 585-610; Paula Findlen, *Possessing Nature: Museums, Collecting, and Scientific Culture in Early Modern Italy* (Berkeley, 1994).

14. *Advis*, 108. On this aspect of Naudé, see Jean Viardot, "Livres rares et pratiques bibliophiliques," in *Histoire de l'édition française*, vol. 2, *Le Livre triomphant 1660–1830*, ed. H.-J.

Martin and R. Chartier (Paris, 1984), 448; idem, "Naissance de la bibliophilie: les cabinets de livres rares," in *Histoire des bibliothèques françaises*, vol. 2, ed. Claude Jolly (Paris, 1988), 270, 282. Viardot registers Naudé's anti-connoisseurial stance but equates this with a negative attitude towards the *cabinet* and *curiosité* in general, an attitude which crystallized only much later in the century.

15. *Advis*, 22-23. I have been unable to trace a source for the two lines of verse Naudé quotes.

16. *Advis*, 9–10. Cf. Pomian, *Collectionneurs* (above, n.12), 74.

17. Cf. Defaux, *Le Curieux* (above, n.12), 36–37, 72–73, 100.

18. This was fully consonant with the currents of the "new science" in which Naudé was moving. See Lynn S. Joy, *Gassendi the Atomist: Advocate of History in an Age of Science* (Cambridge, 1987); Dear, *Mersenne* (above, n. 7).

19. Sirleto's library was one of the first great Cardinalate libraries of the Counter-Reformation; see I. Backus and B. Gain, "Le Cardinal Guglielmo Sirleto (1514–1585), sa bibliothèque et ses traductions de saint Basile," *Mélanges de l'Ecole française de Rome: Moyen Age, temps modernes* 98 (1986): 889–955. The library of Gian Vincenzo Pinelli at Padua furnished an important model (for Pereisc, among others) of the working library of a *curieux*; see Maria Grendler, "A Greek Collection in Padua: the Library of G.V. Pinelli (1535–1601)," *Renaissance Quarterly* 33 (1980): 386–416.

20. *Advis*, e.g., 25–26. In general, see Martin, *Livre, pouvoirs* (above, n.2), 481ff., 516–34.

21. *Advis*, e.g. 25–26, stating that the example of other Parisian libraries "vous pourra tousjours fournir quelque nouvelle addresse et lumiere qui ne sera peut estre pas inutile au progrez et à l'avancement de vostre Bibliotheque."

22. Cf. *Advis*, 6, "Table des poincts principaux qui sont traictez en cet Advis." The division of the *Advis* into chapters was introduced only in the second edition of 1644 edited by Louis Jacob. In the 1627 edition, the "poincts principaux" of the table are simply itemized in the margins.

23. Angelo Poliziano, *Panepistemon*, in *Opera* (Basle, 1553), 462.

24. *Advis*, 132–33. See Frances Yates, *The Art of Memory* (London, 1966).

25. Several studies have helped in the clarification of Naudé's turn to the history of the disciplines. Nicholas Jardine, *The Birth of History and Philosophy of Science: Kepler's* A Defence of Tycho against Ursus *with Essays on Its Provenance and Significance* (Cambridge, 1984), argues that Kepler's *Apologia* (unpublished until the nineteenth century and consequently unknown to Naudé) was forged within the disciplinary conflict and loose institutional status of astronomy in the late sixteenth century, and that it actively solicited the disciplinary validation of astronomy through historical argument. Joy, *Gassendi* (above, n.18), argues that one of the motives for Gassendi's life of Peiresc was the retrieval of a viable *modus philosophandi*, while he turned to a history of Epicureanism (with Naudé's advice and encouragement) to provide historical justification for non-Aristotelian natural philosophical positions. Most helpful has been Brian P. Copenhaver, "Natural Magic, Hermeticism and Occultism in Early Modern Science," in *Reappraisals of the Scientific Revolution*, ed. D.C. Lindberg and R.S. Westman (Cambridge, 1990), 261–301, who makes essential distinctions between doxographical, theoretical, and empirical motives in the early modern historiography of learning.

26. Naudé, *Instruction* (above, n. 6), 18. Cf. Lactantius, *Divinae institutiones* 1, 23, 8, "primus autem sapientiae gradus est falsa intelligere, secundus vera cognoscere." Pintard and those who follow him view Naudé's projected *De censura veri* and *Elenchus rerum hactenus falso creditarum* (which he describes in letters to Cassiano dal Pozzo and to the Dupuys) as typical libertine projects, yet another manifestation of Naudé's anticipation of Bayle. Cf. Pintard, *Libertinage* (above, n. 4), 457. However, this is more profitably seen as related to Bacon's proposals for a *Kalendarium dubitationum* and a collection of *Placita antiquorum philosophorum*, discussed below, and as a manifestation of Naudé's methodological interest in historical pyrrhonism.

27. Kristeller, "Naudé as an Editor" (above, n.4), 60.

28. *Advis*, 49–[50]—mispaginated as 34 = sig. Gi$^v$. Cf. *NO* I, 396.

29. *Advis*, 53–[54]—mispaginated as 38 = sig. Giii$^v$. Cf. *AS* I, 457, where Bacon's purpose is to isolate ancient "founders" from a rejected commentary tradition, while Naudé thinks both are important.

30. *Advis*, [55]–56—mispaginated as 39 = sig. Giv$^r$. Cf. AS I, 456.

31. *Advis*, 84. Naudé's list is as follows: Sidonius, Cassiodorus, and Boethius on the one hand, with Pico, Poliziano, Ermolao Barbaro, Theodore Gaza, Filelfo, Poggio, and George of Trebizond on the other. Cf. NO I, 186–87; AS I, 458–69, where Bacon derides extreme veneration of both antiquity and novelty, allowing for an evaluation of classical thought, but there is no medievalism.

32. *Advis*, 87. Cf. NO I, 184–85.

33. *Advis*, 87–88. This is one of Naudé's more comprehensive and interesting catalogues of the history of philosophy: Albertus Magnus, Agostino Nifo, William Aegidius (William Gilliszoon), John of Saxony, Pietro Pomponazzi, Alessandro Achillini, "Hervié" (?), Guillaume Durand, Marcantonio Zimara, and Ludovico Boccadiferro. Compare this with the catalogue of sixteenth- and seventeenth-century natural philosophers provided later in the *Advis*, 135: Antonio Telesio, Francesco Patrizi, Campanella, Bacon, William Gilbert, Giordano Bruno, Gassendi, Jacques Besson, Bernardinus Gomesius Miedis, Jacques Charpentier, and David Gorlaeus—"qui sont les principaux d'entre une milliace d'autres."

34. *Advis*, 89. Naudé here provides two lists, one for dominant medical thinkers, followed by a catalogue of medieval medical authors and natural philosophers whom he judges to have been unduly neglected. Those deemed to be dominant in the medical schools are: Amatus Lusitanus, H. Thriverus, Girolamo Capodivacca, Joannes Baptista Montanus, Valescus de Taranta [sic]. The neglected medieval authors are: Hugh of Sienna, Jacobus Forliviensis, Jacobus de Partibus, Valescus de Taranta, Bernard Gordon, Tommaso del Garbo, Dino del Garbo, and unspecified Avicennists. Cf. AS I, 456; NO I, 186.

35. AS I, 564. Cf. AS I, 456–57 and 461 on Gilbert and the alchemists; NO I, 157, 169, 182, 187–88, the strongest manifestation of Bacon's rejection of such empirical investigations, due to their lack of "art and theory" and "natural philosophy." See also *Historia sulphuris, mercurii, et salis*, in *Works* II, 82–83.

36. *Advis*, 52-53. The *sententia*, "I am not bound to swear allegiance to any master," is a favorite of Naudé's, from Horace, *Ep.* 1,1,14. Usually taken as a banner of Naudé's *libertinage*, as is clear from this context it signals more an adaptation of Baconian views.

37. *Advis*, 164–65. Cf. Blum, "*Bibliotheca Memmiana*" (above, n.4), 225ff.

38. The account of *historia litterarum* is much longer and more detailed in *De Augmentis* than it is in the *Advancement*. Cf. Bacon, *Advancement*, in *Works* III, 330; AS I, 503.

39. This study was made possible through the support of the Social Sciences and Humanities Research Council of Canada. I would like to thank Luc Deitz, Christopher Ligota, Harvey Mitchell, Orest Ranum, and Nancy Struever for discussion and criticism.

*Part II*
*Philosophy and History*

# THALES PHILOSOPHUS
## The Beginning of Philosophy as a Discipline[1]

### Constance Blackwell

By the mid-eighteenth century, a history of philosophy[2] was agreed upon which claimed that Thales was the first real philosopher. This had not always been thought to be true, indeed that philosophy began in the Middle East with the Chaldeans, Egyptians, and Hebrews was almost universally maintained until well into t-e seventeenth century and was Newton's belief.[3] A careful reading of early eighteenth-century histories of philosophy and in particular the work of Johann Jacob Brucker,[4] reveals that the shift of the origins of philosophy from the East to Greece was not merely due to new philological evidence which questions dating Hermetic texts to the time of Moses; the process of rethinking philosophy's past was a long and complex one, drawing on critiques taken from both historians of law and medicine as well as from the philologists. The result not only redefined philosophy's history, but identified what topics could be included within the discipline. This is a complex story which needs telling at this time when the question of the origins of philosophy and Western civilization are beginning to be debated again, in part because of the publication of *Black Athena*.[5] In 1992 part of an issue of *Isis*[6] devoted substantial space to the debate.

### The Philological Critique of the *Prisca Heologia* and Rewriting the History of Scholarship

The first version of the history of philosophy based on Diogenes Laertius' *Lives of the Philosophers* in the Renaissance appears in Giovanni Tortelli's *De orthographia*, (1448–1452) under the word *philosophia*, where the definition includes themes that remain with the definition of philosophy until Brucker's *Historia critica philosophiae* in 1742: that the first philosophers were called *magi* or in Egypt *sacerdotes* or *prophetae*, and that philosophy first arose among the Persians.[7] Tortelli reworked the text of Diogenes Laertius' preface freely. He could certainly read that Diogenes Laertius' preface did not say that philosophy originated in the East but merely that some had said so.[8] Tortelli organized the information chronologically and began philosophy with Zoroaster. Tortelli's simple chronological exercise rewrote the history of philosophy so that it said that both the priest and the philosopher could practise philosophy and that philosophy had been part of civilization from the beginning of time. A distinction between the methods for discovering truth by the priest and the philosopher was only definitively articulated when Jacob Brucker distinguished between religious knowledge gained by inspiration by the patriarchs and philosophical knowledge acquired through reason by the Greeks after Thales.[9]

The best known tradition that philosophy began in the East has been associated with the Florentine Neoplatonists and their tradition of *prisca theologia*. The term was used and popularized by Pletho,[10] and then incorporated into the writings of Marsilio Ficino[11] and Pico della Mirandola,[12] who also accepted the authenticity of the newly translated texts which had been attributed to Hermes Trismegistus. The tradition got another impetus with the work of Francesco Patrizzi.[13] But it was not just the Neoplatonists who held that philosophy began with either Zoroaster or Moses. A survey of all the historians of philosophy discussed in the first volume of Santinello's *The Models of the History of Philosophy*,(66–100) reveals that with a few exceptions, most of them[14] accepted the idea of some sort of *prisca theologia*. For example, the Jesuit Aristotelian, Pereira, took up the theme and quoted references from Plato and Aristotle which supported the view that philosophy began in the East.[15] His fellow Jesuit, Francisco de Toledo, hinted that there was a dispute about this question, and that Epicurus had said that the Greeks invented philosophy while others had claimed that it arose from Adam, to Noah and the Patriarchs, then to the Egyptians, from whom it came to Greece. At this point Toledo refuses to discuss the point further.[16]

Also like the Neoplatonists, these Aristotelians equated *philosophia* with *sapientia* drawing on classical precedents. Throughout Pereira's discussion *philosophia* and *sapientia* are synonymous. He describes Anaxagoras, Thales, and other early Greek philosophers as *sapientes* who sought divine knowledge, not knowledge which was useful to human beings.[17] Toledo specifically denies that philosophy is about one branch of knowledge or only about being, claiming it is universal, including a variety of sciences within it. "Philosophy is the likeness to God," Toledo said, "as much as that is possible for us. That is, philosophy is a double type of activity. For in God there are two types of actions, the knowledge of things, in as much as he reflects about everything and the other is the activity by which he governs."[18]

The view that the Hermetic texts were contemporaneous with Moses was challenged by Isaac Casaubon,[19] who asserted that they were third-century A. D. forgeries. This had a decisive effect on certain seventeenth- and eighteenth-century historians of philosophy: Jonsius (1624–1659), Morhof (1636–1691) and Brucker (1696–1770). They set out not only to criticize the notion of a *prisca theologia* but to write it out of the history of scholarship. The earliest survey of previous histories of philosophy was by Johannes Jonsius (1624–1659).[20] In the first chapter of the *De scriptoribus historiae philosophiae*[21] he clearly establishes that the words *philosophus* and *sapiens* have identical meanings and include within their purview the whole *encyclopedia* of knowledge: the contemplative, active, and effective sciences.[22] Although Jonsius identifies *philosophia* with *sapientia* and does not distinguish *philosophia* from *theologia*, he does not accept the evidence which ascribed ancient philosophy or theology to Hermes or Zoroaster. Indeed one can almost say that with Jonsius the tradition of eliminating these texts from the history of philosophy begins. Books 1 through 3 on ancient philosophy include no mention of either Hermes or Zoroaster. Pletho is mentioned only in connection with the Aristotle-Plato controversy, Ficino for his praise of Plato,[23] although they had introduced the *prisca theologia* to Italy, and while Jonsius commends Patrizzi's scholarly diligence, he writes that Patrizzi was audacious in support of the *prisca theologia* tradition.[24]

Toward the end of the 1680s Georg Daniel Morhof[25] surveyed the field of the history of philosophy in his massive bibliography cum history of the disciplines, the *Polyhistor*. Volume 2, book 1 of the *Philosophico-historicus* begins with *De historia philosophiae in genere ac in specie: barbaricae, populorum orientalium et septentrionalium*, where Morhof surveys and comments on histories of philosophy. Here again the tradition of *prisca theologia* is rejected or either directly or indirectly criticized. For example, Morhof criticizes Hornius's *Historia philosophica* for its lack of historical accuracy and Gale's *Philosophia generalis* for taking material from Steucho (whom Morhof does not even include among the historians of philosophy) and for attributing to the Hebrews what had originated in other nations.[26] Morhof's most devasting criticism comes in his discussion of Patrizzi. Morhof comments: "all is obscure and of uncertain evidence," as many "barbarian" texts had been mutilated, interpolated, or were plainly false, and he ends his section on Hermes with a pointed criticism of Patrizzi and with blanket caution about the validity of texts claiming to be on "barbarian philosophy."[27] He was equally doubtful about the claims concerning the antiquity of Zoroaster because the evidence came from the "recent Platonists"[28] who corrupted their sources. Morhof then adds a new theme when he criticizes the syncretism of certain Neoplatonists from the fifteenth, sixteenth and seventeenth centuries. He particularly objected to their belief that the doctrine of the Trinity had been known in a primitive form to certain pagans, a view that had been introduced by Pletho when he claimed to find the Trinity[29] in the sayings of Zoroaster.[30] The attempt to make Zoroaster's writings contain the Trinity in writings was reason enough for suspicion that it was a "pious fraud to make people think that the Gentiles had been saved."[31] While Morhof may have given space to the position that philosophy started in the East, he implies that those who held it either were not good scholars, or worse, corrupted the texts to prove their case.

In 1742 the most complete survey of historians of philosophy until recent studies in the twentieth century was published in the *Dissertatio praeliminaris* [32] to the the *Historia critica philosophiae* by the German Lutheran minister, Jacob Brucker.[33] Brucker was influenced by the philosophy of Locke in his youth[34] and by Christian Wolff later on,[35] and studied at Jena under two eclectic philosophers, J.F. Buddeus and J.J. Srybius.[36] The *Dissertatio praeliminaris* begins with praise for two general bibliographical works: Jonsius' *De scriptoribus historiae philosophicae* and Fabricius' *Bibliotheca Graeca*,[37] Brucker then names ancient histories of philosophy beginning with Diogenes Laertius whom he chastises for not getting into the minds of the philosophers but praises for preserving important information found nowhere else.[38] After mentioning Plutarch's *De placitis philosophorum* and Galen, he commends Sextus Empiricus for demonstrating the uncertainty of scientific proof, and the necessity of doubting and suspending judgment in the history of philosophy, as well as for defining the principal doctrines of the philosophers.[39] Among the Church Fathers, while he praises St. Epiphanius's epitomes against heretics and pseudo-Origen's *Philosophumena veterum*, he does not include either Clement of Alexandria[40] or Philo—both of whom also believed in versions of the *prisca theologia*.[41]

Brucker begins his list of later historians of philosophy with scornful comments about Walter Burley's ignorance[42] and praise for Lodovico Vives' attack on scholastic learning.[43] Given that Brucker had not included Clement of Alexandria or Philo among his ancient sources of the history of philosophy, it is perhaps not surprising that he, like Morhof, does not mention Steuco's *Philosophia perennis*.[44] Perhaps even more significantly, when Brucker came to publish his additional notes in 1767, he added Gassendi's name to the historians of philosophy, praising his vast *Syntagma philosophicum* not only for including important philosophical doctrines but also because it ascribed the invention of philosophy to the Greeks when all others attributed it to the barbarians.[45] Brucker includes some works with which he does not agree, saying about Hornius that he lacked judgment and used confused evidence and fallacious etymologies,[46] and of Gerard Vossius' *De Philosophia et Philosophorum Sectis* that it showed great learning although it did follow the eclectic method of Potamon and made "barbarian" philosophy equal to Greek.[47] He also thought Gale used too much *auctoritatis praeiudicium*—awe of ancient sources.[48] Brucker's praise for Thomas Stanley's *Historia philosophia* is for a work rather different from the original English version printed in 1655. The Latin translation by Olearius incorporated Jean Le Clerc's critical notes, which questioned the dating of texts attributed to Zoroaster.[49] Brucker ends with his fellow Germans—Christian Thomasius,[50] Hieronymus Gundling[51] and Christoph August Heumann;[52] while none of the men was fanatically anti-Catholic, each taught a decidedly Protestant version of the history of philosophy.

## The Critique Taken from the History of Medicine and Law

The rejection of *prisca theologia* was followed by the rejection of the related belief in *prisca sapientia*,[53] which held that natural philosophy was begun by the ancients and was well developed before the time of the Greeks. These rejections not only changed the interpretation of the history of philosophy, but how the history of Western and Middle Eastern civilization has been told. It must be borne in mind that a belief in *prisca theologia* went hand in hand with the assumption that was that there was one all-encompassing truth which did not change.[54] As Rossi has remarked of Kircher:

"History is not growth, development or progress. Progress lay in a return: truth and the way to truth required the retrieval of a truth that had long lain buried."[55] However the motivations for the concept of a *prisca sapientia* were more complex for it was created in part to bolster support for contemporary scientific innovation by maintaining that new concepts had ancient precedents. The formula was used in particular by the anti-Aristotelian philosophers who held that matter was made from atoms, and by non-Galenic Paracelian medical doctors who supported herbal medicine. Brucker knew and criticized Borrichius's *De Ortu et Progressu Chemiae Dissertatione*(1668)[56] where the latter claimed that chemistry began with Tubalcain, who is identified with Vulcan(Psalm 11); that Genesis provides evidence that it was known how to make gold; and that the story of the Golden Fleece was evidence of knowledge of chemistry.[57] This concept of *prisca sapientia* was maintained by some and questioned or discredited by others during the course of the seventeenth and eighteenth centuries.[58] Others like Cudworth and Mosheim,[59] (the later translated the *Intellectual System* into Latin, adding extensive critical notes to the text) believed that the Hermetic texts should be studied for evidence of Egyptian culture.[60] The big question then became how to assess this evidence? The most important initial text criticizing Egyptian science was *De Hermetica Medicina*, a Galenic-Aristotelian attack on Kircher and the Paracelsians by Hermann Conring(1606–1681).[61] Kircher[62] had claimed that the Egyptians invented chemistry, which Conring rejected, arguing that Egyptian philosophy and science [63] did not have advanced knowledge in mathematics and natural philosophy because: First, the Egyptians were superstitous and followed their priests, not their reason; Second, their script of symbolic or allegorical hieroglyphs was unsuited to thinking natural philosophy; Third, Egyptian doctors did not use demonstration to verify their scientific discoveries;[64] and Fourth, using the Aristotelian distinction between practical and theoretical knowledge, Conring rejects Egyptian mathematics and medicine as merely the practical arts. He explains it was not enough to be able to measure in geometry, but knowledge of its subtleties was necessary, and this only happened when Archimedes and Apollonius Pergaeus developed mathematics.[65] The argument was rephrased by Daniel le Clerc(1652–1728)[66] to conform with late seventeenth-century scientific practice advocated by his brother, Jean le Clerc as well. Le Clerc wrote a history of medicine which established that there had been three stages of development in medicine: the first level was natural medicine which used experience as its guide, the second was attained when observations were collected which made it possible to make rules identify illnesses by rules that had been developed; and the third level was reached when the rules and observations were tested by physics,[67] which happened only in Greece with Hippocrates. Brucker incorporated aspects of both critiques into his survey of "barbarian" philosophy.

Brucker made an equally important distinction between those he called *sapientes* and those he called *philosophi*, which he took in part from Christian

Thomasius.[68] Thomasius had an important influence on Brucker who not only lists him among the important historians of philosophy but places among his philosophical stars—the Eclectic philosophers in volume 5, along with Bacon, Descartes, Hobbes, and Leibniz.[69] Brucker read Thomasius carefully: indeed, we can see him at work underlining his own copy of Thomasius' dissertation on Pierre Poiret's *De eruditione solida, superficiaria et falsa.*,[70] where he marked the distinctions that Thomasius made between the method used to acquire knowledge through natural reason—the correct method for philosophy and law[71]— and the method used by religion which was by divine light.[72] Brucker applied these distinctions to philosophy and established that the method for reaching truth in theology was incompatible with the philosophical method. These three traditions, the philological critique of ancient texts, the critique from the history of medicine and the narrower definition of philosophy which grew out of the history of law were brought together by Brucker to establish a coherent definition of philosophy which identified it with the current approved inductive of reasoning used in natural philosophy.

## Gundling's View on the Origins of Philosophy

The first attempt known to me to bring these three traditions together was made by Nikolaus Hieronymus Gundling (1671–1729).[73] Gundling's *Historia Philosophiae Moralis* attacked the notion of *prisca sapientia* and the Biblical origin of philosophy. He argued from both religious and scientific-historical grounds developing some of the more general arguments of Conring, Le Clerc and Thomasius. Like Morhof, he claimed it idolatry to say that the first Egyptians maintained the idea of the Trinity, as Kircher had said, or that there had been books of divine contemplation in Egyptian writings as Iamblicus claimed.[74] Gundling goes on to condemn two editions of Hermes Trismegistus,[75] and Kriegsmann's association of Hermes Trismegistus with the early Germans[76] and demolishes the claim that the Chaldeans had wisdom by first quoting Pico's letter to Ficino that Pletho's Zoroastrian oracles had come from the Chaldeans. He questions whether such evidence ever existed, noting that on Pico's death, the text on which Pico had based his evidence Ficino found to be completely unreadable. Gundling exclaims, "Reader, look at the audacity of these men!"[77]

Gundling then redefines *philosophia* and *sapientia*–a redefinition which Brucker was to develop. He explains that the confusion between two words, *sapientia* and *doctrina*,[78] arose because those who identified philosophy with *sapientia* grouped wise men with philosophers, while those who understood what *doctrina* was knew the difference. *Doctrina* was *scientia* — that is, logic, chemistry and mathematics. At this point a notion of scientific progress comes into the discussion. Drawing on Daniel Le Clerc's *Historie de la medicine*, Gundling points out that while Le Clerc had written that Adam may have learned how to treat some illnesses, the treatment had been based on his own personal experience and not on scientific experiment.[79] This argument is then applied to the medical knowledge available to Moses in Egypt. While Gundling acknowledges that Moses may have been instructed by the Pharaoh's daughter, the Egyptians, like Adam, based their medicine on custom, not experiment. While the Egyptians included the triangle in their hieroglyphs, used numbers, and knew certain methods for the improvement of health, Gundling maintains that what they knew could not be defined as chemistry or theoretical knowledge.[80] Not surprisingly, Gundling takes a dim view of the Cabala and those who supported it as a source of ancient wisdom, criticizing Pico della Mirandola, Reuchlin, Henry More, and Knorr van Rosenroth.[81] Gundling ends the book with a curious fudge, as Ulrich Schneider has pointed out,[82] it is one which was used off and on until the nineteenth century. On the one hand, he criticizes Diogenes Laertius for saying that the Greeks invented philosophy, and on the other, he admits that early man inquired into the nature of things, but states that they did this without intellectual subtlety, chemistry, and scientific instruments. Brucker took up this theme and developed it into a clear historical and philosophical distinction between knowledge gained by relgious inspiration and that gained through the senses and through imperfect reason and applied it to the history of philosophy. The result was a clear rejection that the barbarians had true philosophy and proof that the Greeks with Thales developed theoretical knowledge. While Brucker did not deny the "barbarians" a hearing, and, as Schneider has pointed out, "barbarian civilization" was discussed into the nineteenth century, after Brucker it was judged that their thought could not be considered to be truly philosophical.

## Brucker's Classical Sources for the New History of Philosophy

After Brucker listed his sources for the history of philosophy in which he eliminated texts supporting the *prisca theologia*, in his *Dissertatio Praeliminaris*, he wrote an introduction to the main body of his work in which he dramatically reworked material from Clement of Alexandria, who supported the claim that philosophy arose in the East at the beginning of time, and turned him into a supporter for the position that philosophy began with the Greeks. This intellectual sleight of hand was done by selecting a quotation which noted that when Clement divided philosophy between the barbarians and the Greeks, he said that one tradition must be selected which most freed the soul from ignorance. This, Clement admitted, was the Greek tradition, which used an *artificiosa philosophandi methodus*, while in contrast the "barbarians" used a way of speaking unsuited to philosophy.[83] In the additional notes, Brucker also adds references to Gassendi, Lactantius and Diogenes Laertius, all of whom also held that philosophy began with the Greeks.[84] He concludes with a telling remark that while Aristotle was not the first philosopher, he was one of the first who set out a "systematic method of philosophizing."[85] Brucker then proceeded to write a history of philosophy which identified which philosophers should be included in the canon and all those selected were like Aristotle in the sense they set out complete philosophical systems. These were philosophers and not wise men.

As Brucker said, the beginning of philosophy must be separated from the increase of learning that came from habit[86] and an accurate method for determining causes and principles had to be segregated from the *sapientia* (including the wisdom of the *Bible*) handed down through the authority of parents and masters. However, while there is nothing in the Scripture that can shed light on how to philosophize, it can identify important issues in ecclesiastical history and the history of heresy. These topics Brucker did not hesitate to include in his work, interweaving it with discussions of Atheism, Gnosticism and Spinozism.

Brucker then tells his version of the cultural evolution of knowledge. With Adam only "the first rudiments of philosophy are understood only by using a method of meditation and from experience or as empirical philosophy as a learned man[Heumann] said."[87] The vocabulary that Brucker uses here is important. Experience, as Le Clerc had said, was not enough for science. Real science only came after the knowledge about the natural world was collected and tested through the methods of natural philosophy. Brucker maintained that Adam after the fall had *sapientia*, a notion of God, his own happiness and justice which Adam received through divine light; this was religious wisdom but not philosophical knowledge.[88] Brucker denies that the names that Adam gave to the animals were scientific, as the Hebrew names of the animals do not give any indication of their essence, and Brucker also comments that the nature of the original primeval language is uncertain.[89] He also denies Adam had books since Moses did not mention any, which he certainly would have done, had it been the case.[90] Adam's knowledge of God and his own virtue derived from revelation; he was a prophet and virtuous man, not a philosopher. Brucker goes on to deny that Tubalcain was identical to Vulcan as Borrichius had maintained[91] or that Noah was a philosopher because he built the ark.

Brucker was able to write a longer section about Moses' learning, as there was more information about Moses' education from both the Jewish and Christian traditions.[92] He first lists those ancients who supported the assertion that Moses was a philosopher: both Clement of Alexandria and Philo claimed Moses learned arithmetic, geometry, music and, above all medicine, from the Egyptians; there was also an old Jewish tradition preserved in Eusebius which held that Moses knew logic,[93] and Borrichius, Sennert and Dickenson[94] believed he knew chemistry. Following Conring and Le Clerc, Brucker denied that Moses knew chemistry because he smashed the golden calf and mixed its gold with water for the Israelites to drink. Moses was a *vates*, or prophet, and who followed the divine light; he had been intstructed by God's revelation to form a theocracy, but although he did this he was not truly a moral philosopher and prophets who followed him also were not, properly speaking, philosophers.[95] Brucker ends the section by quoting from Thomas Burnet to back up his view.[96]

When he comes to discuss Egyptian philosophy, Brucker shifts his emphasis, focusing his critique on the extent and method of Egyptian scientific learning.

Taking his lead from Cudworth that there might have been something of value in the Hermetic writings, but following the line of thinking further, he paraphrases Mosheim's comment[97] with approval, and maintains that if one is to take the Hermetic books as valid, it has to be proved that the Egyptians also treated them as such. However, Brucker notes that the only person who did take these texts seriously was Iamblicus, whose critical judgment Brucker questions. "At that time [of Iamblicus] the whole of Egyptian philosophy had degenerated from its early form and constitution and had become greatly corrupted by the philosophers who, as was their wont, used foul syncreticism and mingled together Oriental, Egyptian and Platonic doctrine. It is very likely that [sc. syncretism] was from the activity of these men, most delinquent and ingenious in foisting books at great men of a former age, gave birth to such a false offspring, especially where the Christians allowed themselves to be deceived by them and rather incautiously fell into their nets."[98]

Despite this criticism, he does read critically what evidence there was about ancient Egypt in the Hermetic texts. The poor Egyptians do not get off lightly. Fundamental to Brucker's argument is his assessment of the level of learning attained by the three men called Hermes. The first used symbolic language rather like Jesus, a type of symbolic language Brucker approved of, but after the first Hermes' death floods and decay ensued, and the pillars[99] on which he wrote his knowledge were scattered; although the second Hermes ordered the pillars to be collected and housed in temples, the knowledge they contained became corrupted. After that there were floods and wars in Egypt and almost all knowledge was lost. Then the third Hermes appeared who looked for the key to the hieroglyphs, developed knowledge and increased the colleges of priests: he was called Hermes Trismegistus. Brucker took these stories and constructed a history of a corrupt, or barbaric, civilization. For example, the religion was corrupt because stories about the early kings were retold as stories about gods by the priest who wanted to keep the truth from the common people.[100] Brucker, well aware that the Greeks recognized Egyptian learning, attributes this to Orpheus' influence, since Orpheus brought aspects of Egyptian religion to Greece by creating Greek mythology. But mythology was not the stuff of philosophy for Brucker.[101]

Brucker, like Conring and Le Clerc before him goes on to question the development of advanced mathematics among the Egyptians. He notes that it was said that they could measure fields, calculate the movement of stars and practise herbal medicine. Taking his argument from Conring and Le Clerc and imposing the Aristotelian classification of practical and theoretical science on the Egyptians, Brucker devalues what the Egyptians were said to have known by classifying it as practical or empirical knowledge and therefore not the matter of philosophy. Brucker held that geometry was a theoretical discipline and used in philosophical reasoning, while arithmetic was not. Thus Brucker noted that while Egyptians may have measured the stars, Thales predicted the eclipse, which indicated that Thales understood the geometric theory behind astronomy.

Egyptians measured the fields using only arithemetic, while Thales taught them how to measure a pyramid. This meant Thales knew the theory behind geometry. This association of Thales with theoretical geometry was set out clearly by Thomas Stanley in his discussion of Thales when he refers to Proclus' commentary on Euclid as evidence of the Greeks' awareness of the theoretical basis of geometry and unhistorically attributes this to Thales.[102] It should also be noted that the same hierarchy in mathematics was maintained by the contemporary philosopher whom Brucker so admired, Christians Wolff.[103] Brucker then used the same arguments as Conring and Le Clerc when he criticized Egyptian medicine for being based on herbal medicine learned through custom and chance, and not by demonstration and scientific method.[104]

It then is with Thales then, not the earlier Greeks, Orpheus or Homer that philosophy begins. Brucker rejected Orpheus as a philosopher, because the knowledge which he brought from Egypt was secret and told in allegorical or symbolic fables which were methodologically like hieroglyphs and thus unsuited for open investigation of nature.[105] While he admits that Orpheus' fables may have helped to civilize the rude Greeks and supply them with the basic ingredients of their religion, these fables were not the matter of philosophy. Brucker opens Book 2 with the bold statement that it was with the *secta Ionica* that the human *ingenium* began to develop philosophy by using the correct approach toward the truth of things divine and human by employing meditation and reflective reasoning.[106] Taking his lead from Diogenes Laertius and from later works by Scipio Aquiliano[107] and Thomas Burnet, Brucker says that while infomation is incomplete about early Greek philosophy,[108] it is with Thales and especially with Thales' student, Anaximander, that the scientific method was introduced to Greece and philosophy was initiated.[109]

Thales may have been the first real philosopher but as such was not able to philosophize as completely as later philosophers. A very real concept of scientific progress developed not only among such medical historians as Daniel Le Clerc but also among historians of philosophy like Heumann who in 1723 made a measured and historically acute assessment of Thales's knowledge of mathematical science.[110] Brucker wrote his own version of Thales, describing him as a philosopher who had not made a complete break with the mythological works on the theogonies because Thales held that the beginning was from water, while the doctrine on the immortality of the soul may have come from the Egyptians.[111] Brucker was also rather unfairly critical of Thales' belief that the floods of the Nile were cause by the wind: Brucker notes that they were really caused by floods in Ethiopia.[112] The fact is that it was still unclear in the early eighteenth century what caused the floods on the Nile. He also objected to the unhistorical claim by Plutarch that Thales said bodies could not be infintiely divided, but would divide into atoms. Brucker believed that atoms had not been discovered by Thales' time.[113]

Brucker's interest in classifying levels of intellectual sophistication led him to overestimate the superiority of Thales' mathematical and astronomical knowledge. In this Brucker followed those who gave a particularly high place to mathematical reasoning, like Descartes and Wolf. He argued that when Thales went to Memphis in Egypt to study mathematics, he came with a better philosophical training than the Egyptians, which is further proved by the fact that Thales also knew how to bisect a circle and understood the principles of an equilateral triangle. In astronomy he divided the sky into five zones, divided the year into 365 days and four seasons and forecast the eclipse of the sun. Although Thales left no system of moral philosophy, but merely a few sayings, Greek philosophy developed from that time forward because he initiated philosophy as an open field of enquiry with facts which could be openly arrived at and tested. Thales' student, Anaximander, opened a philosophical school and taught others publicly what he had learned privately from Thales and wrote the first book on natural philosophy.[114] Finally Brucker rejected his teacher Buddeus' claim that Thales was an Atheist, and maintained that Thales took his idea that God was the soul of the world from the Egyptians and Greek theogonies.[115]

Brucker's definition of philosophy and with it Greek philosophy as a rational pursuit is not dissimilar to recent thinking by such historians of Greek philosophy as Geoffrey Lloyd.[116] Should we group Brucker then with the current historians of Greek science and philosophy whom David Pingree claims are gripped by Hellenophilia? When Brucker was writing, history of philosophy in Germany served as introductory chapters to a course in natural philosophy, ethics and logic. For example, Jean Le Clerc's *Physica* opened with a quick survey of the history of physics from Thales to Copernicus,[117] while Buddeus' *Elementa Philosophiae* was written as an introduction to his philosophy text which included sections on logic, ethics and natural philosophy and only appeared as a discrete volume after his death.[118] These works were designed not to be objective surveys of all philosophical culture but as background to current philosophical problems in natural philosophy, ethics, metaphysics and logic.

David Pingree[119] rightly points out that while for his own research on early Babylonian, Iranian and Indian science, he includes not only various forms of astronomy and mathematics, but astral omens, astrology, law, medicine and magic, a modern scientist doing research would be correct to exclude most of these fields from their investigation. Given the assessment of the trustworthiness of non-Greek texts and the model of scientific thought he was heir to, Brucker could not help but engage in a kind of Hellenophilia. His notion of scientific method which was an amalgam of earlier Aristotelian demonstration, contemporary classification about levels of mathematical complexity and the inductive methods taken from Locke and Boyle came from both his sources and his school tests. Brucker was not original, but instead a genius at turning disperate traditions into a coherent story.

Nevertheless, the result of his methodological bias should be pointed out. Pingree's own vision has Babylonian astronomy and mathematics as the taproot of science and mathematics, linking Egypt, Greece, Syria, Iran, and India—

it is a tradition in which he sees great and fertile variation.[120] Brucker's view severely criticizes the philosophies of the ancient Middle East, the Cabala and the Neoplatonists and all those Renaissance and seventeenth-century philosophers who revived these traditions. For Brucker there is one correct method of thought, the model of inductive reasoning as he understood it; those philosophical systems which did not use it were excluded from his canon. It has not been until the 1960s that the Hermetic tradition and with it the relationship between magic and philosophy has been studied in Renaissance and seventeenth-century philosophies and much still needs to be done. The question of whether knowledge started in the East is alive as it was in the seventeenth-century, but for different reasons. Today many scholars are interested in how thinking in natural philosophy developed, and seek to investigate this by examining how the various philosophers thought rather than by imposing what is considered the correct logical method on the texts. This interest includes Cardano and Campanella who were interested in magic, the late Aristotelian tradition and its interest in scientific method as well as Brucker's eclectic philosophers.. Clearly how philosophy and correct reasoning in philosophy is defined have determined which philosophies have been included in the canon of philosophy.

## Notes

1. Versions of this paper have been given in the following seminars: Donald Kelley's "History and the Disciplines" seminar at the Folger Library, Christopher Ligota's "History of Scholarship" seminar at the Warburg Institute, Richard Popkin's "History of Philosophy" seminar at Emory University, Nancy Streuver's seminar at the Johns Hopkins Centre in Florence, Jean Moss's Humanities Seminar at Catholic University, and to the History of Philosophy Department in Padua directed by Giovanni Santinello and Gregorio Piaia. I have greatly benefited from discussions with Antonio Clericuzo, Michael Fend, Sarah Hutton, Martin Mulsow, Richard Popkin, Ulrich Schneider and Charles Webster. I am particularly grateful to Michael Hunter, Christopher Ligota and Clive Strickland for their close reading and thoughtful comments on the text, and to Donald Kelley for his interest in Brucker and the formation of the canon of philosophy.

2. On the history of philosophy as a genre see the preface to Émile Bréhier, *Histoire de la philosophie*, (Paris, 1926 nine editions by 1967); Lucien Braun, *Histoire de l'histoire de la philosophie* (Paris, 1973); M. Gueroult, *Dianoématique: Histoire de l'histoire de la philosophie*, 3 vols. (Paris 1984–86); for the most complete treatment see: *Storia delle storie generali della filosofia*, ed. Giovanni Santinello (vol. 1: *Delle origini rinascimentali alla "historia philosophica,"* [Brescia, 1981]; vol. 2: *Dall' età cartesiana a Brucker*, [Brescia, 1979]; vol. 3: *Il secondo illuminismo e l'età Kantiana*, 2 parts [Padua, 1988]; vol. 4: *L'età hegeliana*, part 1 [Padua, 1995] and part 2 [forthcoming]; and vol. 5: *Il secondo ottocento*, 5 vols. in 8 parts [forthcoming]. Volume 1 has been translated into English and appears as G. Santinello, *From Its Origins in the Renaissance to the "Historia Philosophica, Models of the History of Philosophy,"*, intro. and ed. C.W.T. Blackwell with P. Weller, vol. 1 (Dordrecht, 1993); Vol. 2: *From the Cartesian age to Brucker* (forthcoming 1998). Italian volumes will be referred to by *Storia*, the English version as *Models*.

3. J.E. McGuire and P.M. Rattansi, "Newton and the 'Pipes of Pan,'" *Notes and Records of the Royal Society of London* 21 (1966) 108–43.

4. His history first appeared in German as *Kurze Fragen aus der philosophischen Historie von Anfan der Welt, bis auf de Geburt Christi mit ausführlichen Anmerckungen erlautert*, 7 vols, (Ulm,

1731–36). Latin version is: *Historia critica philosophiae*, 5 vols. (Leipzig, 1742–44); in the 1767 edition a volume of additional notes was added to the reprint. It will be referred to as *Historia* from now on.

5. M. Bernal, *Black Athena, the Afroasiatic Roots of Classical Civilization*, vol. 1, *The Fabrication of Ancient Greece*, 1785–1985 (London, 1987).

6. *Isis* 83, no. 4 (1992), "Special Section: The Culture of Ancient Science": F. Rochberg, "Introduction", 547–53; D. Pingree, "Hellenophilia versus the History of Science", 554–63; G.C.R. Lloyd, "Methods and Problems in the History of Ancient Science: The Greek Case", 564–77; H. Von Staden, "Affinities and Elisions, Helen and Hellenocentrism," 578–95; M. Bernal, "Animadversions on the Origins of Western Science", 596–607. A recent article by Josine H. Bock, "Proof and Persuasion in Black Athena: The Case of K.O. Müller", *Journal of the History of Ideas*, 57, no. 4, (1996), 705–24, convincingly attacks the evidence that Bernal claims to have found in German nineteenth-century histories. For how later German historians of philosophy discussed "barbarian" knowledge see Ulrich Johannes Schnieder, *Die Vergangenheit des Geistes, Eine Archäologie der Philosophiegeschichte* (Frankfurt am Main, 1990), 227–64.

7. Giovanni Tortelli, *De orthographia* (Venice, 1471), n.p. "Nam philos amor, & sophia sapientia dicitur. Et apud persas primum claruisse: ac eorum sapientes magos fuisse dixerunt. Apud babylonios vero: & assyrios deinde fluxisse chaldaeos. Apud indos aut gymnosophistas: quos pro sapientibus habuerunt. Nam ut Clearchus Solensis affirmat: a magis gymnosophistae perfluxerunt. Ex his ut aiunt ad Aegyptios philosophia manavit. Cuius rei antistites: sacerdotes ac prophetas appellant. Nam teste Aristotele li. primo de philosophia aegyptiis antiquiores sunt magi. Inde apud celtas: seu gallos dryidae sapientes fuere: de quibus in dicione dryidae vidimus. Sed magorum principem Zoroastrem Persen fuisse: memoriae perditum est. Hunc Dion ex interpretatione sui nominis: astrorum asserit fuisse cultorem." The work was quite popular, published at least seven times during the fifteenth century, and served as a type of dictionary cum reference work. It was used by Bernard André for his education of Arthur, son of Henry VII, in the 1490s, by Vives in the 1520s and by Thomas Elyot in his *Dictionary* (1536). See Constance W. T. Blackwell, "Creating Defintions for Words: the *Ortus vocabulorum* (1500) versus Vives (1523) and Elyot (1538)" in *Italia ed Europa nella Linguistica del Rinascimento*, ed. Mirko Tavoni, vol. 2 (Ferrara, 1996), 235–56.

8. Diogenes Laertius, *Lives of the Philosophers*, trans. R.D. Hickes (London, 1972). The prologue to book 1 first lists those who say that philosophy began with the "barbarians" but states flatly–prologue, paragraph 3, that "These authors forget that the achievements which they attribute to the barbarians belong to the Greeks, with whom not merely philosophy but the human race itself began". And states in paragraph 4 , " thus it was from "the Greeks that philosophy took its rise: its very name refuses to be translated into foreign speech."

9. The separation of religion from philosophy began with Pomponazzi's assertion that for Aristotle the discussion of the soul belongs to physics and that it is impossible to prove its immortality. See C. Lohr, "The Sixteenth-Century Transformation of Aristotelian Natural Philosophy," in *Aristotelismus und Renaissance, In Memoriam Charles B. Schmitt*, ed. E. Kessler, C. Lohr, and W. Sparn, (Weisbaden, 1988), 89–100, and "Metaphysics and Natural Philosophy as Sciences: The Catholic and The Protestant view in the Sixteenth and Seventeenth Centuries," in *Sixteenth- and Seventeenth-Century Philosophy and its Conversation with Aristotle* ed. Constance Blackwell and Sachiko Kusukawa (forthcoming).

10. G. Pletho, *Traité des lois*, ed. C. Alexandre, trans. A. Pellissier ( Paris, 1858, reprint 1966). For Pletho, see C.M. Woodhouse, *George Gemistos Plethon, The Last of the Hellenes* (Oxford, 1986).

11. Marsilio Ficino, *Opera omnia*, reprinted (Turin, 1959), 50–51 and *Marsilio Ficino: The Philebus Commentary*, ed. and trans. M.J.B. Allen (Berkeley, CA, 1975). On the intellectual history of the Hermetica, see D.P. Walker, *The Ancient Theology: Studies in Christian Platonism from the Fifteenth to the Eighteenth Century* (London, 1972); C.B. Schmitt, "Perennial Philosophy from Agostino Steuco to Leibniz" (I) and "Prisca theologia e philosophia perennis: due temi del Rinascimento italiano e la loro fortuna" (II) in *Studies in Renaissance Philosophy and Science* (London, 1981). For a useful discussion of Ficino and the *prisca theologia* as well as additional bibliography, see also C.B. Schmitt and B. Copenhaver, *Renaissance Philosophy* (Oxford, 1992), 136–52.

12. For Giovanni Pico della Mirandola and the *prisca theologia*, see Walker, "*The Ancient Theology*," Schmitt, "*Perennial Philosophy*," and Schmitt and Copenhaver, *Renaissance Philosophy*, 163–76.

13. F. Patrizzi, *Summi philosophi Zoroaster & eius 320 Oracula Chaldaica Asclepii dialogus & Philosophia magna* (Hamburg, 1595), 16–17. "Joannes quoque Picus ait, nomen hoc Magus, idem esse apud Persas quod apud Latinos sapiens & apud Graecos philosophus. Et Pico anteriores Philo & Hermias Magos in hunc describunt modum Prior ille; veram quidem illam Magniam philosophiam, hoc est contemplativam scientiam per quam clarius naturae opera cernuntur, ut honestam atque expetendam, non vulgus solum sectantur; sed maximi etiam regum reges, precipue Persae tam sunt harum artium studiosi ut regnare nemo possit nisi sit inter Magos iudicunt. Hoc vero ita magi sunt, qui de singulis philosophantur." See also Schmitt and Copenhaver, *Renaissance Philosophy* 184–95.

14. A clear exception is Chytreus (Santinello, *Models*, 87–88). David Chytreus, *Tabula philosophica: series philosophorum, et sectae eorum praecipuae: A Thalete et Socrate imprimis usque ad Ciceronem deducta* (Berlin, 1581).

15. Benito Pereira, *De communibus omnium rerum naturalium principiis et affectionibus liber quindecim. Qui plurimum conferunt, ad eos octo libros Aristotelis, qui de Physico auditu inscribuntur, intelligendos* (Rome, 1576), 125–26.

16. Francisco Toledo, *Commentaria una cum quaestionibus, in octo libros Aristotelis de physica auscultatione* (Venice, 1593) 1–2.

17. Pereira, De communibus omnium rerum, 1.

18. Toledo, 1–1v. "Nec oportet existimare, philosophiam aliquam esse peculiarem scientiam, quae aliquod ens particulare contempletur, non enim sic modo consideramus, sed ut universale genus est, varias sub se scientias complectens . . . ut philosophus non aliud sit, quam sciens omnia, ut sciri possunt, ut dicitur 1 Met. cap. 2, hoc autem non unico habitu sit, sed multis . . . Philosophia est Dei similitudo, quo ad fieri a nobis potest, id est, philosophia est per quam Deo similes sumus, in Deo enim duplex operationum genus constituitur circa res has: altera operatio est rerum cognito, qua omnia speculatur: altera actio, qua omnia gubernat."

19. A. Grafton, "Protestant versus Prophet; Isaac Casaubon on Hermes Trismegistus," *Journal of the Warburg and Courtauld Institutes* 46 (1983), 78–93 and *Forgers and Critics, Creativity and Duplicity in Western Scholarship* (Princeton, 1990), 75–103. See also I. Casaubon, *De rebus sacris et ecclesiasticis exercitationes XVI ad Cardinalis Baronii Prolegomena in Annales* (London, 1613), 70–87. Grafton points out that Casaubon was influenced by Porphyry's attack on the Hermetic writings, an attack Porphyry made in an attempt to keep the Platonic tradition pure.

20. Santinello, *Models*, vol. 1, 388–98.

21. Joannes Jonsius, *De scriptoribus historiae philosophicae libri 4*, ed. J. Christophorus Dornius, pref. J. Gotthelfius Struvius (Jena, 1716), 5." Deinde philosophia idem notat quod sapientia, quae principiorum intellectum & scientiam continet. Quoniam vero substantia reliquis Categoriis prior est, earumque quasi principium; hunc tot saltem erunt philosophiae & sapientiae, quot substantiae. At substantiae tres sunt una immobilis & incorruptibilis ut Deus: altera mobilis & incorruptibilis ut coelum: tertia mobilis & corruptibilis, ut sublunia, vide Aristotelem Metaph. lib xii c. 1 & 6 prima ad Theologiam pertinet; duae posteriores ad Physicam, quae gemina propterea est sapientia Astrologia & Physica sublunaris . . . Deinde quia sapientia principia prima nosse debet, hinc eae tantum scientiae, quae principia prima considerant, sapientiae & philosophiae erunt."

22. Ibid., 4. "Sunt philosophiae sive scientiae contemplativae, Theologia, quam posteriores Metaphysicam nomine antiquis ignoto appellant, Physica & Mathematica. Activae sunt Oeconomica & Politica, cuius prior pars est, quam Ethicam nuncupamus, quae principium Politicae, optimam scilicet vitam. Effectivae vero scientiae & philosophiae sunt Analytica, Dialectica, Rhetorica, quarum illa veris propositionibus scientiam, ista probabilibus opinionem, haec verisimilibus fidem in animo auditoris efficere & producere intendit. Effectiva quoque scientia est Medicina juxta Autorem Eudemiorum il. 1, c.5. Poetica item, Grammatica, Statuaria, Pictoria."

23. Ibid., 309. "Eodem tempore GEORGIUS GEMISTUS PLETHO opusculum edidit, qua in re dissideant inter se Aristoteles & Plato . . . Extremo hoc seculo MARSILIUS FICINUS Platonis, quem admirabatur, vitam scripsit, quae initio operum ejusdem plerumque legitur."

24. Ibid., 311. "FRANCISCUS PATRICIUS audacis homo ingenii ut nomen sibi compararet, Aristotelis philosophiae refutandam sibi sumpsit Tomis quatuor discussionum . . . Verum in superbo hoc opere diligentiam ejus laudamus, judicium desideramus, audaciam improbamus."

25. Georg Daniel Morhof, *Polyhistor*, ed. J. Moller, 2nd ed. (Lübeck, 1714). The last two books, the *Polyhistor philosophicus* and the *Polyhistor practicus* were published from notes left after Morhof's death by Joannes Moller, the first complete edition being printed in 1708. A very useful discussion of Morhof's development of the genre of *historia literaria* can be found in the Johns Hopkins dissertation by Paul Nelles, "The Public Library and Late Humanist Scholarship in Early Modern Europe: Antiquarianism and Encyclopaedism" (Baltimore, 1994), 365–71. A collection of articles will be edited by Constance Blackwell and François Waquet, *Morhof and the Polyhistor*, forthcoming.

26. *Polyhistor philosophicus*, 4–5. "GEORGIUS HORNIUS, Scripsit & ille Historiam Philosophicam libris 7., in qua veterum philosophorum doctrinam proponit, & cum novis quoque confert, sed in veteribus indiligens & inaccuratus, nonnumquam in digressiones supervacuas dilabitur . . . THEOPHILUS GALEUS . . . Galeus quidem laudandus ob illum conatum, qui hoc in argumento multos tamen antecessores habet, quibus non pauca debet, e.g. Augustinum Steuchium in *libris de perenni philosophia* . . . Sed & ille tamen nimium interdum partibus suis favere videtur, & ad Scholam Judaicam trahere, quae aliunde originem habent."

27. Ibid., 7, "FRANCISCI PATRICHII HERMES & ZOROASTER. Aegyptiorum & Chaldaeorum prima apud omnes fama est, quibus adjungendi quoque Thraces. A Barbaris illis prima sapientiae initia, sive ab Hebraeis ad illas pervenerint, sive e traditionibus Majorum, quorum cum Patribus primaevis conversatio fuit, sed obscure omnia & incerta fide referuntur. Id quidem verissimum est, extitisse illos Philosophos, & fuisse illorum scripta diligenter a posteris servata, sed postmodum ob iis, in quorum manus inciderant, interpolata, vel plane falsa illis supposita. Fuerunt tamen, qui fragmenta aliqua ejus Philosophiae collegerunt. Hermes quidem quem antiquissimum alias Aegyptiorum Philosophorum faciunt fuit ejus nominis Magnus apud illos Philosophos; sed, an genuina sint, quae sub ejus nomine circumferuntur, affirmari non potest."

28. By this Morhof means Pletho, Ficino, Pico della Mirandola, Steuco and Patrizzi. The distinction between the Neoplatonists and Plato was made in the seventeenth century. For its history see E. N. Tigerstadt, "The Decline and Fall of the Neoplatonic Interpretation of Plato," *Commentationis humanarum litterarum* 52 (1974), 5–108. Brucker gave a great deal of attention to the issue in his youthful *Historia philosophica doctrinae de ideis* (Augsburg, 1723). See Constance W. T. Blackwell, "Epicurus and Boyle, Le Clerc and Locke: 'Ideas' and their redefinition in Jacob Brucker's Historia Philosophica Doctrinae de Ideis, 1723" in *Il vocabolario della République des Lettres. Terminologia filosofica e storia della filosofia. Problemi di metodo*. Atti Convegno internazionale in memoriam di Paul Dibon (Napoli, 17–18 maggio 1996) ed. Marta Fattori (Florence, 1997) 77–92.

29. For the earlier tradition that the pagans had a concept of the Trinity, see: M.J.B. Allen, "Marcilio Ficino on Plato, the Neoplatonists and the Christian Doctrine of the Trinity," *Renaissance Quarterly* 37 (1984), 555–84. Morhof knew of Pletho through a French edition: G. Pletho, *Magica Zoroastri Oracula, Plethonis commentariis enarrata*, interpret, ed. J. Martano (Paris, 1539).

30. S. Hutton, "The Neoplatonic roots of Arianism: Ralph Cudworth and Theophilus Gale," in *Socinianism and Its Role in the Culture of the Sixteenth to the Eighteenth Centuries*, ed. L. Szczucki (Warsaw/Lodz, 1983), 139–45.

31. Morhof, *Polyhistor philosophicus*, 8. "Persae ipsum [sic Zoroastrem] pro Patre Magorum habent, ac praeceptorem Pythagorae facit Apulejus. Mentio ejus multa, & magna apud antiquos Autores fama. Sic recensentur passim ap. Platonem, & Platonicos, praecipue recentiores (quod suspicionem alicujus corruptionis arguat) ejus sententiae. Hae sententiae, undecunque collectae, uno volumine editae fuerunt superiore seculo in Gallia. Sexaginta enim Oracula, a Juliano quodam Chaldaeo Graece facta, Parisiis excusa sunt anno circiter 1530, in quibus ipsa pene Christiana fides, mysteriumque Trinitatis proponitur; quam ob causam suspicati fuerunt multi, pia quadam fraude his actum fuisse, & supposita quaedam a Christianis ad convincendo Gentiles. Clario enim illic Trinitatis notitia, quam in ipso V Testamento."

32. Brucker, *Historia*, vol. 1, 31–38. This list does not appear in the original German version.

33. Santinello, *Storia*, vol. 2, 529–635. A contemporary life of Brucker appears in Johann Heinrich Zedler, *Grossen Volstaendiges Universal-Lexicon aller Wissenschaften una Kuenste*, 4th supp. (Leipzig, 1754), 744ff. From the tone of the article it seems as if Brucker wrote a great part of it himself.

34. On the influence of Robert Boyle and John Locke on Brucker's *Historia philosophica doctrinae de ideis* (Augsburg, 1723), see Constance Blackwell, "Epicurus and Boyle, Le Clerc and Locke 'Ideas'.

35. Brucker did not include Wolff among the eclectic philosophers in volume five of the *Historia critica philosophiae*, as he was still alive when this volume was written, but he did write a long entry about Wolff in the additional notes, *Historia* (Leipzig, 1767) vol. 6, 878–902.

36. Brucker praises both men in *Historia*, vol. 5; for Buddeus, see 528–32, for Syrbius see, 541–2.

37. J.A. Fabricius, *Bibliotheca Graeca, notitia scriptorum veterum Graecorum*, 3rd ed. (Hamburg, 1717–27). Fabricius collected information on sources for the concept of *prisca sapientia*, not only mentioning the editions in which their works were published but summarizing their doctrines and listing the inventions associated with their names in the *Bibliotheca*, vol. 1: Zoroaster, 242–52; Hermes, 46–8, Orpheus, 110–36.

38. Brucker, *Historia*, vol. 1, 32. "Licet enim supinam in eo neglegentiam et aliquam quoque in attendendo ad mentem veterum philosophorum incuriam, iudicii tum in tanto argumento proditam paupertatem, nimiamque credulitatem culpaverint viri docti; conservavit tamen nobis tot monumenta veterum philosophorum . . . " Brucker used Giles Menage's edition of Diogenes Laertius, *De vitis dogmatis et apophthemgatis philosophorum libri x* (London, 1664), which included notes by H. Estienne, I. Casaubon, and Menage himself.

39. Ibid., 33. " . . . .qui [sc. Sextus Empiricus] etsi eo potissimum respexit, ut omnem dogmaticam philosophiam oppugnando, et scientiarum incertitudinem demonstrando, necessitatem dubitandi et suspendendi iudicium, scepticaeque philosophiae pretium evinceret, tot tamen modis omnem dogmaticorum philosophiam exposuit, tot nobis de veterum dogmatibus relationes conservavit, ut recte adhibentibus et sobrie legentibus plurima inde in historia omni philosophica lux oriatur." Brucker used *Sexti Empirici Opera Graece et Latine*, notes added by J.A. Fabricius (Leipzig, 1718).

40. In his additional notes published in 1767, Brucker adds fulsome praise to S. Augustine and his criticism of the Neoplatonists. Brucker *Historia*, vol 6, 13. " . . . .sed has putidas scaturigines produxisse, immanemque et imprimis a D. Augustino (De. C. D. L. IX, c. 16; L X, fere toto) reprobatum errorem Platonicorum Alexandrinorum, de variis bonorum geniorum et daemonum classibus, quibus velut mediatoribus ad Deum utendum sit, progenuisse: quod imprimis a Pseudo-Dionysio ad D. Thomam et ab hoc ad reliquos scholae doctores traductum esse eleganter demonstrat." Brucker knew the writings of Clement of Alexandria well; see *Historia* vol 6, 518–28.

41. Brucker may have omitted Philo in the *Praefatio*, but he discusses him extensively in the *Historia*, vol. 2, 767–802.

42. Walter Burley's *De vita et moribus philosophorum* was not published after 1515. Brucker took his reference from C.A. Heumann, *Acta philosophorum* (Halle, 1723), vol. 13, 282–98. Heumann wrote this work between 1713 and 1727 in 18 volumes. It will be referred to as Heumann, *Acta* from now on.

43. Brucker, *Historia*, vol. 1, 34. "Inter quos omnium calculo LUDOVICO VIVI primas merito deferimus; certe is inter primos fuit, qui abstersurus horrendum, quo disciplinae hactenus pressae fuerant, squalorem philosophica historia ad resitituendum scientiis nitorem in doctissimis de corruptis artibus et tradendis disciplinis libris egregie usus est. . . . "

44. A. Steucho, *Opera quae extant omnia* 3 vols. (Paris, 1577–78); 2nd ed. 3 vols. (Venice, 1590–91); reprint, *De perenni philosophia*, intro. C.B. Schmitt (New York, 1972). On the tradition of Steuco and perennial philosophy see C.B. Schmitt, ibid., introduction v-xvii. This concludes with a useful bibiography. See also C. B. Schmitt, "Perennial Philosophy: from Agostino Steuco to Leibniz," *Studies in Renaissance Philosophy and Science*, (London, 1981), I, reprint from the *Journal of the History of Ideas* 27 (1966), 505–23.

45. Brucker, *Historia*, vol. 6, 25. "c. 5 exponit, qui praecipue philosophati praeter Graecos olim fuerint, ubi omnis barbaricae philosophiae ambitus exponitur . . . ." Cf. Petrus Gassendi, Liber Prooemialis, Syntagma, *Opera omnia* vol. 1 (Lyon, 1658), 6–7.

*Thales Philosophus* 77

46. For Georg Hornius see Santinello, *Models*, vol. 1, 236–57.
47. For Vossius see Santinello, *Models*, vol. 1, 22–35. Brucker, *Historia*, vol. 1, 35: " . . . sed et philosophiae barbaricae pariter atque graecae ad sectam potamonicam sive eclecticam usque historiam methodo ipsi peculiari adspersis ubique observationibus ex multa eius lectione in scriptis veterum petitis delineavit. Quod si fata superstiti permisissent, argumentum hoc eatenus perpolire, ut opus non fuisset, ex aliena culpa exemplum petere, dubium non est, quin pro ut magna eruditione in literis erat versatus, perfectum quid nobis reliquisset." Brucker rejected the eclecticism of the Church Fathers, which indiscriminately mixed Egyptian and Neoplatonic philosophy with Christianity, redefined as a philosophy which selected individual ideas from ancient philosophers, and added these selectively with new systems of natural philosophy: his model eclectics were Bacon, Descartes, Hobbes, Locke, Leibniz and Thomasius. Also cf. Heumann, *Acta Philosophorum* (Leipzig, 1729), vol. 3, 711–45.
48. T. Gale, *The Court of the Gentiles*, 4 vols, vols 3 and 4 (London, 1667), vols. 1 and 2 (Oxford, 1669, 1671), *Philosophia generalis* (London, 1676). Cf. *Models*, 292–330.
49. On Stanley see Santinello, *Models*, vol. 1 163–203. It is important to note that *The History of Philosophy* was first introduced to Europe by Jean Le Clerc when he translated the part on Oriental thought: *Historia philosophiae orientalis recensuit, ex anglica lingua in latinum transtulit, notis in Oracula chaldaica et indice philologico aunxit Johannes* Clercus (Amsterdam, 1690). Le Clerc, an early supporter of Locke's philosophy, was critical of those who held the view of the *prisca philosophia*. He omitted Psellus' and Pletho's comments on the Oracles which had been in the original English version. For further comments on the differences between Olearius' Latin version and Stanley's original see *Models*, 178. On Brucker's interpretation of the Neoplatonic elements in Zoroasterizn oracle in the *Historia philosophia doctrinae de ideis*, see Constance Blackwell, "Epicurus and Boyle, Le Clerc and Locke.
50. Christian Thomasius was one of the seventeenth-century philosophers whom Brucker called an eclectic. Brucker, *Historia critica*, vol 5, 447–520. For Thomasius see the collection of articles published in *Wissenschaftliche Zeitschrift der Martin-Luther Universität Halle-Wittenburg* (1954–55), 493–569.
51. Brucker groups Gundling, a student of Christian Thomasius at Halle, along with the other minor eclectic philosophers: Brucker *Historia*, vol. 5, 447–520. See also Martin Mulsow's article on Gundling in this volume.
52. Brucker, *Historia*, vol. 1, 38. For Heumann, see Santinello, *Storia*, vol. 2, 437–76. Heumann's history of philosophy essays appeared in the *Acta philosophorum*, Each essay was a self-contained unit and the essays were not arranged chronologically.
53. Heumann, vol. 8, (Halle, 1717), 173–204 which includes a chapter, "Catalogus scriptorum de philosophia veterum in Oriente barbarorum" that attests a great interest in non-Western philosophy during the seventeenth century, although Heumann holds that the barbarians did not have philosophy. For a discussion of the concept of *prisca sapientia* and chemistry see A. Clericuzio, "Alchemia Vetus et Vera, Les Théories sur l'origine de l'alchimie en Angleterre au XVII siècle", eds. D. Kahn and L. Matton, *Alchimie: Histoire et Mythes* (Paris, 1995), 613–24.
54. see C.B. Schmitt, intro. to Augustinus Steuco, *De Perenni Philosophia*, (New York, 1973). See also Malusa "The Renaissance Idea of Concordism in Philosophy", *Models*, 26–38.
55. P. Rossi, *The Dark Abyss of Time*, trans. L.G. Cochrane (Chicago, 1984), 123. For Athanasius Kircher, see T. Leinkauf, *Mundus Combinatus. Studien zur Struktur der barocken Universalwissenschaft am Beispiel Athanasius Kirchers S.J.* (1602–80) (Berlin, 1993) with a complete bibliography.
56. Olaus Borrichius, *De ortu et progressu chemiae dissertatione* (Copenhagen, 1668) surveys ancient science 1–42, history of Middle Ages including Arab science 121–43.
57. ibid., 84.
58. For the complex story of the debate in England see S. Hutton, "Edward Stillingfleet, Henry More and the Decline of Moses Atticus: A Note on Seventeenth-Century Anglican Apologetics" in R. Kroll, R. Ashcroft, P. Zagorin, *Philosophy, Science and Religion in England 1640–1700* (Cambridge, 1993), 68–83.
59. For Mosheim see *Johann Lorenz Mosheim* (1693–1755). *Theologie im Spannungsfeld von*

*Philosophie, Philologie und Historie.* eds. Ralph Häfner, Martin Mulsow, Florian Neumann, Helmut Zedelmaier, forthcoming Weisbaden, 1997(Wolfenbütteler Forschungen).

60. I am using the edition Brucker used of Mosheim's Latin Translation and commentary, R. Cudworth, *Systema Intellectuale huius Universi seu de Veris Naturae Rerum Originibus, Commentarii quibus omnis eorum Philosophia,* ed. and trans. J. L. Mosheim (Jena, 1736) 374–5, n. 52. "Recte & praeclare animadvertit vir magnus, non esse dubitandum, quin libri Hermetis, qui vocantur, aliqua ex vetere disciplina Aegyptica contineant." Others like Bentley and Stillingfleet did not agree with Cudworth's opinion, see A. Grafton, *Defenders of the Text* (Cambridge, Mass. 1991), 15–21.

61. H. Conring, *De Hermetica Medicina Libri Duo* (Helmstedt, 1667).

62. Athenasius Kircher, *Oedipus Ægyptiacus* (Rome, 1652).

63. Not only was Conring's scholarship on Egyptian medicine widely known to Brucker and Gundling but it received extensive diffusion in Heumann's *Acta*, vol. 12, (Halle, 1720), 660–697.

64. Conring, 169 ,172 and 173 " . . . Earum prima est ac praecipua ingens populi superstitio . . . Alteram causam non minoris ponderis arbitror, scribendi nempe illam hieroglyphicam rationem . . . Intelligo autem hieroglyphicarum literarum omnes, omnino symbolicas dictas . . . (172) Causam tertiam ruditatis Aegyptiacae merito dixeris illud quod Aegyptii doctores ea quae ex sensu perceptis erant demonstranda & demonstrari poterant, citra demonstrationem docuerint. Id certe eos fecisse, patet cum ex libris Hermetis & Aesculapii quae feruntur, tum ex universa Hieroglyphica doctrina, quae demonstratione omni vacat. (173) Sunt autem haec ab omni ratione remotissima. Unicum videlicet verae scientiae humanae parandae artificium est demonstratio: & citra demonstrationes ad scientiam nemo pervenerit."

65. Ibid., 145. "Longe tamen illos fuisse infra Archimedis & Apollonii Pergaei, imo et aliorum Graecorum subtilitatem argumento est, quod nusquam legere sit, eximiam quandam hujus scientiae doctrinam Graecos ab Aegyptiis, vel ipsis Ptolomaeorum regum temporibus, acceptisse, ex adverso omnis laus Geometricae peritiae penes solos fuerit Graecos a Pythagorae vel certe ab Eudoxi usque Platonis aetate."

66. P. Von Röthlisberger, "Daniel Le Clerc(1652–1728) und sein *Histoire de la médecine Gesnerus*, 21, 1964, 126–41. I am grateful to Heikki Micheli for this reference. Daniel Le Clerc, *Histoire de la medicine ou l'on voit l'origine & le progrès de cet art, de siècle en siècle, depuis le commencement du monde* (Geneva, 1696), 2nd ed. (Amsterdam, 1702). A much lengthened treatment was printed as *Nouvelle édition augementée . . . d'un plan pour servir à la continuation de cette histoire depuis la fin du siècle ii jusque au milieu de xvii* (Amsterdam, 1723).

67. D. Le Clerc, 3–5. "L'expérience seule a presque suffi à ceux qui ont inventé la Médecine au prier de ces trois sens, & il ne leur a pas fallu de raisonnement plus recherché que celui que fournit le sens commun. Les seconds ont été obligez de pousser le raissonement un peu plus loin, appuyez d'ailleurs sur la meme expérience. Les troisièmes ont dû non seulement raisonner, mais joindre encore l'étude de la Physique à celle de la Médecine." For his brother Jean Le Clerc's view of the progress in physics see his history at the beginning of his *Physica sive de Rebus Corporeis, Libri Quinque* (Amsterdam, 1696), 6–7.

68. Historia, vol. 5, 447–520. A true eclectic, Brucker took what he liked from Thomasius. In his chapter on the philosopher, Brucker criticizes him for his belief that the ancients had wisdom. 477.

69. For Brucker and Eclecticism see my article: Constance W.T. Blackwell, "The historiography of Renaissance philosophy and the Creation of the Myth of the Renaissance Eccentric Genius—Naudé through Brucker to Hegel", *Girolamo Cardano, Philosoph Naturforscher Arst,* ed. Echard Kessler, Wolfenbütteler Abhandlungen zur Renaissanceforschung, Bd. 15 (Wiesbaden, 1944); "The case of Honoré Fabri and the historiography of Sixteenth- and Seventeenth-Century Jesuit Aristotelianism in Protestant History of Philosophy : Sturm, Morhof and Brucker", *Nouvelles de la République des Lettres* (1995–1), 49–77.

70. I am indebted to Martin Mulsow for directing me to the few books owned by Brucker which remain in the Augsburg Stadtbibliothek.

71. One of the clearest expositions of the distinction between natural and supernatural light can be found in Thomasius' *Historia Juris Naturalis* (Halle, 1719), 3. "Vll In hoc latiori significatu

differentia luminis naturalis & supernaturalis sequenti modo perspicue congnoscitur: Lumen naturale quidem ope sanae rationis etiam sine revelatione divina agnoscit miseriam humanam intuitu hujus vitae; imprimis vero idem manuducit humanum genus ut quilibet praecipue de propria miseria sit sollicitus."

72. Christian Thomasius, *Dissertatio ad Petri Poiret* in Petrus Poiret, *De Eruditione Solida, Superficiaria et Falso, Hac Nova Editione Auctores et Correctiores* (Frankfurt, Leipzig, 1708), 12–13. "Unde duplex lumen eruditionis oritur *unum naturale, seu recta ratio, alterum supernaturale*, quod diverso respectu revelationis divinae scripturae sacrae, item fidei nomine insigniri solet. *Illud unice ad felicitatem hujus vitae rendit, hoc vero ad consequendam felicitatem futuri seculi opus est. Atque his opus eruditionis et prudentiae ut ista lumina inter se non confundantur, sed distinctum proponantur illae scientiae & prudentiae quae ex lumine naturali & quae ex lumine supernaturali seu revelationis divinae deducuntur.*" (Words in italics are underlined by Brucker.) Brucker also quotes extensively from the *Dissertatio* in his chapter on Thomasius, *Historia*, vol. 5, 472–74.

73. For a more complete description of Gundling see Martin Mulsow's article in this volume.

74. N.H. Gundling, *Historiae Philosophiae Moralis, pars prima* (Halle, 1706), 10–11, "VI Fidem autem omnem superant, quae de Trinitate in una Deitate* mysterio priscis Ægyptiis cognito fabulatur Kircherus. Ineptiis ille Mercurii Trismegisti innititur, quae Abrahami . . . venditant, librosque complures de divinarum rerum contemplatione literis Hieroglyphicis descriptos ei adfingunt (k) [Brucker's footnote reference]. Iamblichus ineptiens 36529. ei libros adscribit, quorum ne unus hodie supersit."

75. François de Foix, Duc de Candale, *Hermetis Mercurii Trismegisti Pimandras utraque lingua restitutus* (Burigale, 1574), which is a Latin and Greek edition. François de Foix also wrote a French translation entitled *Le Pimandre de Mercure Trismegiste, de la philosophie Chrestienne* (Bordeaux, 1579). Gundling also lists a gigantic commentary by Frater Hannibal Rosseli, *Pymander Mercurii Trismegisti, cum Commento Fratris Hannibalis Rosseli* (Cracow,1585), which Gundling probably knew through the Cologne edition of 1630.

76. Wilhelm Christoph Kriegsmannus, *Conjectaneorum de Germanicae gentis origine ac conditore Hermete Trismegisto . . . in Taciti de Moribus Germanorum, liber unus* (Tübingen, 1684).

77. Gundling, *Historia Philosophiae Moralis*, 20. "Quod doctis admodum familiare est, fuitque nisi fallor Joanni Pico Mirandulae Comiti, qui Marsilio Ficino in epistola, quae stat in eius operibus p. 249 persuadere temere voluit, se omnia Zoroastris oracula Chaldaice scripta habuisse, *in quibus* inquit, *& illa quoque quae apud Graecos mendosa, & mutila circumferuntur, leguntur integra, & absoluta; tam est in illa Chaldeorum sapientia brevis quidem & salebrosa, sed plena mysteriis interpretatio. Deus bone*, pergit, *quam Pythagoricae, quam plena priscorum dogmatum & secretioris disciplinae inuosit statim animum efficax votum, me posse haec per me absque interprete evoluere & perscrutari, atque hoc nunc ago, hoc indefessus assiduusque saxum voluo.* Putasses eum non invenisse, quod pueri in fabula. Sed quis vidit illum thesaurum: Affirmat Ficinus post mortem inter eius manuscripta & cimelia fuisse repertum. Sed tamen adeo fatetur detritas, luctuque difficiles fuisse chartas, ut verisimile sit, ipsum Picum legere illas non potuisse. Quis ergo adserere certo potest, Zoroastris fuisse oracula? Vides, lector, quanta sit horum hominum audacia!"

78. Ibid., 60, b) "Est hoc nostrae aetatis hominibus vitium fere commune, ut eodem pede antiquos metiantur, quo nostros. Crevit eruditio per temporum spiramenta; nec veluti Noachi columba uno actu in robem est emissa. Unde tot fabulas videas propagatas, tot defensa deliria, ut risum vix teneant amici tui. Erroris initium fuit confusio spaientiae & doctrinae. Quos enim sapientes crediderunt, eos etiam Syllogisticam, Chemicam, Mathesin, omnemque scientiam cognovisse crediderunt."

79. Ibid., 7–8.
80. Ibid., 64.
81. Ibid., 91–3.
82. See Ulrich Johannes Schnieder, *Die Vergangenheit des Geistes, Eine Archäologie der Philosophiegeschichte* (Frankfurt am Main, 1990), 227–64.
83. Brucker, *Historia*, vol. 1, 48. "Unde Clemens Alexandrinus omnem eruditionem in barbaram atque Graecam dividit, et ex his unam eligendam esse monet. Et hoc ad ipsam quoque philosophiam,

quae animum ab igorantia et barbarie maxime liberat, traduitum est; Graeci enim artificiosa philosophandi methodo usi, cum apud barbaras gentes simplici traditione sapientiae praecepta propagari cernerent, hanc philosophiam non esse rati." The edition Clement Brucker used was *Stromata l, Opera omnia*, ed. John Potter, Bishop of Oxford (Oxford, 1715), 350, paragraph 299. "Omnia autem sunt Graeca & barbara: alterutra autem non sunt utique omnia. Recta autem sunt iis, qui sensum volunt accipere. Eligite disciplinam & non argentum . . . " This interesting edition is heavily annotated by Potter, who often comments critically on Clement's scholarship and rejects Clement's belief that there had been a *prisca theologia*.

84. Brucker, *Historia*, vol. 6, 35–6.

85. Ibid., 36. "Quod magna postea incrementa cepit, ubi philosophi systematica ratione coeperunt philosophari, et peculiaria philosophiae aedificia extruere: inter quos si non primus, primorum certe unus fuit Aristoteles." For Brucker and the history of the reputation of sixteenth- and seventeenth-century Aristotelian method, see "The reputation of the Jesuits in Protestant history of philosophy, the case of Honoré Fabri," in *Nouvelles de la République des Lettres*, 1–1995, 49–77, and "The Aristotelian method versus the New Philosophy, Johannes Sturm, Georg Morhof and Jacob Brucker" in *Order and Method in Renaissance Aristotelian Commentaries*, ed. Eckhard Kessler, (Variorum, Aldershot, 1997) forthcoming.

86. Brucker, *Historia* vol. 1, 29–31. Here Brucker is clearly developing a classification of knowledge similar to the one suggested by Daniel Le Clerc for the history of medicine.

87. Ibid., 52. "Quae, si prima nascentis foetus stamina et formam aliquam rudissimam consideres, cum intellectu humano eiusque usu nata esse censeri suo modo potest, ne tamen philosophiam talem, qualem hodierna nobis ingenii cultura exhibet cogitemus, sed philosophiae prima rudimenta, et aliqualem rerum divinarum et humanarum ex rationis meditatione et experientia cognitionem, quam philosophiam empiricam vir quidam doctus non inepte dixit, intelligamus." Brucker then refers the reader to Heumann, *Act. philos.* vol. 1, 760.

88. Ibid., 54. "Consistebat enim sapientia protoplastorum in cognitione Dei et sui ipsius et in efficaci atque practica notitia earum rerum quae ad suam felicitatem in concreata iustitia sitam necessaria habebant: quae hominem quidem sapientem, et ad consequendum verae felicitatis finem aptis mediis intellectus et voluntatis instructum, non vero philosophum, eo sensu, quo in historia philosophica sumitur efficiunt . . . "

89. Ibid., 55. " Quae enim pro physica eius scientia affertur onomatothesia (the giving of names), non dum rem conficit, cum non tantum ex quorundam nominum significatione, saepe contorta longiusque in gratiam hypotheseos quaesita, quae cum natura bestiarum convenire dicitur, physica cognitio non queat inferri, sed et omnis illa, de primaeva lingua disputatio incertissma sit, frustraque hodie inde argumentum sumatur . . . "

90. Ibid., 56.

91. Ibid., 59–60.

92. The English Neoplatonists debated whether Moses was a philosopher. See S. Hutton, "Edward Stillingfleet, Henry More, and the decline of Moses Atticus: a note on seventeenth-century Anglican apologetics," in *Philosophy, Science and Religion in England 1640–1700*, eds. Richard Kroll, Richard Ascraft, and Perez Zagorin, (Cambridge, 1992), 68–83.

93. I. Struve, *Rudimenta logicae Ebraeorum* (Jena, 1697).

94. Borrichius, *De Ortu et Progressu Chemiae*, 47 ff; D. Sennert, *De Chymicorum cum Aristotele et Galeno Consensu* (Wittenburg, 1619), 46; Edmund Dickenson, *Physica Vetus et Vera* (London, 1702), 475.

95. Brucker, *Historia*, vol. 1, 84–5. "Tandem legis quoque Mosaicae perfectionem non negamus, civitatis formam atque regiminis ab eo introduci praestantiam ultro fatemur, verum Mosi eam vati divinitus edocto, omnemque legem et constitutionem a summo legislatore accipienti adscribimus, qui nec dux proprie populi, nec legislator proprie dictus dici, sed minister potius Dei summi regis, qui peculiari regiminis forma, theocratiam recte appellant viri docti, populum divinum rexit, et salutaribus legibus a Deo ipsi revelatis instruxit. Haec decebant vatem, qui typus esse debebat maximi prophetae . . . non vero inter philosophos ei locum assignant." Brucker examines the method by which Moses had attained his knowledge, concluding that Moses had been instructed

by God's revelation to form a theocracy, thus he was not a moral philosopher. On the concept of theocracy Brucker refers to John Spencer's *De Legibus Hebraeorum Ritualibus Eorumque Rationibus* (Tübingen, 1732), 226–230 and to S. Pufendorf *De Habitu Religionis Christianae ad Vita Civilem* (Bremen, 1687), 34.

96. Thomas Burnet, *Archaelologiae philosophiae* (London, 1692), 363. Cf. Brucker, *Historia*, vol. 1, 86. "Notum est in disciplinis mathematicis aut philosophicis numquam praecelluisse hanc gentem, neque in caeterarum artium studiis, aut id genus n ullo humani ingenii eximio foetu . . . Quae autem apud ipsos erant scholae atque academiae pristinae, non tam ad encyclopaediae studia, ut solent hodie, formatae et compositae erant, quam ad religionis instituta et dona prophetica imbibenda."

97. Brucker, *Historia*, vol. 1, 246–7. He refers to Mosheim's note: Cudworth, p. 375, n. 52, "Recte & praeclare animadvertit vir magnus, non esse dubitandum, qui libri HERMETIS, qui vocatur, aliqua ex veteri disciplina Aegyptiaca contineant. Nam, qui eos composuit, homo callidus & veterator facile prospicere potuit, nullam prorsus haec volumina fidem esse inventura, nisi aliquid ex illis scitis & dogmatibus admistum haberent, quae inter Aegypitos pro Hermeticis habebantur. Quod vero addit, fieri nullo modo potuisse, ut placerent Aegyptiis & magnam inter eos auctoritatem obtinerent, nisi res ita esset, id multo plus haberet momenti, si testibus constaret idoneis, Aegyptios Deorum cultui deditos aliquod libris istis pretium posuisse. Nusquam vero, quod sciam, perscriptum exstat, volumina haec, quum prodiissent, vel reiecta & repudiata ab Aegyptiis, vel etiam recepta publice approbataque fuisse. Degeneres quidam Christiani & recentiores Platonici certarunt olim veluti utri antiquis sapientiae professoribus callidius ingenii sui commenta possent adscribere. Nam utrique commodis suis consentaneum in primis decebant fore, si vulgo crederentur dogmata sua omnibus omnium aetatum philosophis placuisse. Quocirca utroque in confingendis libris Hermeticis curam posuisse aliquam, existimem."

98. Brucker, *Historia*, vol. 1, 263. "Recte enim notavit Moshemius vir summus, hoc quidem a Platonicis et haereticis, non parum praesidii rebus suis in eiusmodi libris quaerentibus, factum esse, ab Aegyptiis vero illis tantum auctoritatis tributum esse, id Cudwortho et sequacibus esse probandum. Addimus nos his: fieri potuisse ut hi libri, vel qui huius commatis olim fuerunt obtrusi, in Aegypto eo tempore, quo Iamblicus scribebat, aliquam habuerint fidem; verum eo tempore Aegyptiaca philosophia tota a pristino habitu et constitutione sua defecit, et a philosophis, orientalia, Aegyptiaca et Platonica dogmata foedo syncretismo more suo miscentibus vehementer fuit corrupta. [ Brucker's comment (Conferendus omnino Conringius qui hactenus dicta mire illustrat, quemque frustra in hac caussa Borrichius oppugnavit)] Ex quorum hominum in supponendis magnis aevi prisci viris libris studiosissimorum et ingeniosissimorum officinis eiusmodi foetus adulterinos prorepsisse, longe est verisimillimum: praecipue ubi Christiani incautius in haec retia incidentes ab iis decipi se passi sunt."

99. Festugière remarks on the popularity of this story. See: A.J. Festugière, *La Révélation d'Hermès Tresmégiste* (Paris, 1950–54), vol. 1, 74–5.

100. Ibid., 252–60.

101. Ibid., 374–82.

102. Thomas Stanley, *The History of Philosophy*, 2nd ed. (London, 1687), 8–10. Stanley states that the Egyptians knew geometry, but then quoting Proclus' commentary on Euclid he says somewhat ahistorically that there was much in Euclid that had been held by Thales.

103. Christian Wolff, *Elementa Matheseos Universae, commentationem de Praecipuis Scriptis Mathematicis, commentatione de Studio Mathematice Recte Instituendo*, 2nd. ed. (Halle, 1741), vol. 5, 169–71. Wolff makes detailed distinctions between the different levels of complexity needed for arithmetic and geometry, while in his *Psycologia Rationalis Methodo Scientifica Pertractata* (Verona, 1734), Wolff says that rational psychology is based on sense perception and reason.

104. This attack on things Egyptian seems to have affected assessments of the writings of Hermes Trimegistus in the twentieth century, see comments by Festugière in *La Révélation d'Hermès Trismégiste* 2nd ed. (Paris, 1950). For a more favourable evaluation see Peter Kingsley, "Poimandres, The Etymology of the Name and the Origins of the Hermetica" *Journal of the Warburg and Courtald Institutes*, 56, 1993, 1–24.

105. On Orpheus, see Brucker, *Historia*, vol. 1, 379–80. On Orpheus' method: "Usus vero est, ut ex dictis colligere est, Orpheus modo docendi arcano, fabulisque et allegoriarum vel obscurato. Cuius rei ratio ex eius circumstantiis hactenus enarratis facile peti potest. Erat enim ex Thracia oriundus; Hyperboreorum vero sacra allegoricis traditionibus constitisse, et arcana methodo fuisse propagata, suo loco audivimus. In Aegytium postea delatus mythologoumena quod Diodorus testatur didicit, arcanaeque sapientiae atque imprimis dogmatum sacrorum factus particeps, modum occultandi dogmata fabulis, et hieroglyphicis literis, et, obscurando mysteriorum recondito sensu populum ab interior epoptia arcendi assecutus est; et disciplinam sacerdotum arcanam inspexit."

106. Ibid., 457 "Vagientem hactenus in cunis, vel puerascentem etiam Graecorum consideravimus philosophiam, ad illud tandem tempus delapsi, quo ingenium humanum iusto habitu philosophari et meditationibus atque ratiocinationibus de veritate rerum divinarum et humanarum sollictum esse coepit, quem honorem Graecorum industriae deberi, in limine historiae nostrae, ubi de originibus philosophiae disquisivimus, demonstratum est."

107. Scipio Aquilianus, *De placitis physicis veterum philosophorum ante Aristotelem* (Venice, 1623). This was edited and republished by Jacob Brucker and his son Carol Friederic Brucker, Leipzig, 1756. The book appears under the editorship of his son in the title, but it is quite clear from the personal comments in the footnotes that Jacob Brucker was actively involved in the edition.

108. Brucker, vol. 1, 464. "Non deferunt quidem inter recentiores, qui ut physicorum, fere omnium qui ante Aristotelem vixerunt, ita et Ionicorum praecepta de natura rerum explicaverunt, inter quos imprimis nominandi sunt Scipio Aquilianus and Thomas Burnetius, aliique."

109. Ibid., 458. " Debetur itaque introductae methodo scientifica et exornatae justa facie habituque philosophiae inter Graecos, Thaleti Milesio, Ionicae sectae conditori, . . . "

110. Heumann, (Halle, 1723), vol. 14, 160–80.

111. Brucker, vol. 1, 273.

112. Ibid., 473.

113. Ibid., 472.

114. Ibid., 478. " Initia philosophiae naturalis inter Graecos felici auspicio facta continuauit et auxit Anaximander; qui privatam Thaletis doctrinam disciplina publica exposuit, sicque sectam in cathedra quasi formavit et constitutuit."

115. Ibid., 470, "Nec ignorare potest, qui ea legit quae supra de theologia Aegyptiorum atque theogonia scriptorum Graecorum fuse diximus, in ea sententia omnes hos fuisse esse Deum animum muni huncque peculiari ratione ea Deo emanasse, sicque ex Deo producta materia illo animante motumque largiente cuncta generata esse."

116. Lloyd does not ignore the relationship between myth and science see: G.E.R. Lloyd, *The revolutions of Wisdom, Studies in the claims and practice of Ancient Greek Science*, (Berkeley, 1987).

117. Jean Le Clerc, *Physica sive de Rebus Corporeis, Libri Quinque* (Amsterdam, 1696), 6–7.

118. J.J. Buddeus, *Elementa Philosophiae Instrumentalis (philosophiae theoreticae) seu Institutionem Philosophiae Eclecticae* (Halle, 1706). The introduction on the history of philosophy was published in a greatly expanded version, *Compendium Historiae Philosophicae Observationibus Illustratum*, ed. J.G. Walch (Halle, 1731).

119. D. Pingree, "Hellenophilia versus the History of Science" *Isis* 83, no. 4 (1992), 554.

120. Ibid., 563.

# ECLECTICISM AND THE HISTORY OF PHILOSOPHY

## Ulrich Johannes Schneider

### Ante Portas

Eclecticism has only recently been identified as a philosophy preferred by many thinkers, many of them German, in the last two decades of the seventeenth and the first three decades of the eighteenth century. Among historians of philosophy, philosophical eclecticism has been recognized as an important phenomenon of the early enlightenment (*Frühaufklärung*), and interpreted in the context of the critique of prejudice.[1] Eclecticism has been found behind the injunction "to think for oneself"[2] and has been also thought of as related to conceptions of certainty and "judgment" (*iudicium*).[3] A larger context has been opened up by interpreting eclecticism as an early modern *philosophia christiana*.[4] Where twenty years ago little was known about eclectic philosophy[5], today we know almost too much, as witnessed by the recently published comprehensive inventory of texts which use the term.[6] It is quite stunning how a whole new world of thought has emerged and been interpreted by historians of philosophy: *terra incognita* discovered!

It is possible, however, that the new historical knowledge about eclecticism has adapted too rapidly to the traditional conception of the history of philosophy. There is a common assumption that every "-ism" indicates some position and eclecticism is consequently understood as a—rather weak—philosophical position which existed only for a very limited period, possibly a period of crisis.[7] In this perspective eclecticism was held to disappear when contradicted and superseded by systematic philosophy. Even after its "discovery" today, historians of philosophy will insist that eclecticism lacked any raison d'être and was therefore never again taken up as a part of a philosophical program, with the single exception of that of Victor Cousin.[8] Even in the eighteenth century Christian Wolff and Johann Gottlieb Fichte accused eclecticism of being a philosophy without foundation.[9] Not much later Georg Wilhelm Friedrich Hegel tried to discern the "principles" implied in every philosophical system and called eclecticism "altogether meaningless and inconsequent," which was as much as to say: no philosophy at all.[10]

This philosophical verdict on eclecticism persists in today's histories of philosophy. Eclecticism is presented as something to feel sorry about, and eclectic thinkers have difficulty in winning the respect of historians of philosophy.[11] Most of them still find it all too easy to sympathize with the way the early claims of eclecticism were ignored by systematic thinkers such as Christian

Wolff or Immanuel Kant, and they also feel that Victor Cousin's self-proclaimed "eclecticism" was obviously weaker than Hegelianism. This understanding rests on distinctions between important figures in the history of philosophy and minor ones, between thoughts which last and ones which do not, between the peaks and the valleys in the landscape of thought. In order to achieve historical understanding, however, one has to be careful not to apply these distinctions to explicit opinions only. To think of the history of philosophy in terms of conflicting positions or in terms of a hierarchy of ideas means thinking of it as a kind of intellectual battleground—which of course it was, and is. But still there is a danger of missing the relevant point at issue and in the case of eclecticism, this includes the idea of not having a philosophical position at all.[12]

## In Medias Res

In 1745, when Hieronymus Georg Gloeckner, professor of *Weltweisheit* at Leipzig, started to question the existence of the philosopher Potamon of Alexandria, he practically concluded a debate which had been going on for nearly sixty years. Many German academics, including some important ones, had participated in a debate about the eclectic philosophy, whose founding father Potamon was thought to be. Gloeckner had one of his students defend a dissertation in which he tried to reconstruct the so-called "younger Platonic philosophy," using material mainly drawn from Johann Jakob Brucker's monumental history of philosophy, then just published. The aim of Gloeckner's dissertation was to establish whether Potamon was indeed a Neoplatonic thinker. Gloeckner tried to solve the problem by examining the available sources. There were no surviving writings by Potamon himself, and Gloeckner quickly determined that Diogenes Laertius had provided the most reliable account.[13] This account, however, turns out to be very brief indeed. Diogenes Laertius called Potamon the founder of a sect which professed no particular doctrine, but chose whatever it pleased from other sects.[14] Diogenes Laertius considered the emphasis on choice to be the distinctive feature in Potamon's philosophy and consequently called him "a chooser"—in Greek: *eklektikos*, in Latin: *electivus* or *eclecticus*. Diogenes, writing at the end of the third century A.D., added that the eclectic sect was a recent phenomenon and therefore did not go into detail.

The source thus provided not so much information as a blank space. It was this space—a bare indication of the existence of a new and special sect—that was transmitted by Diogenes for more than a thousand years along with other miscellaneous information about ancient philosophers. As we can see from humanist editions of Diogenes Laertius and commentaries on him, no one paid much attention to it till the end of the seventeenth century.[15] In the middle of the eighteenth century, scholars like Gloeckner inaugurated a new period of indifference, pushing Potamon back into oblivion again. In between these dates however, eclecticism flourished as a fervently professed and debated philosophy. All was started off by the little note in Diogenes Laertius.

## Secta Eclectica

Early in the seventeenth century, Justus Lipsius, a famous scholar from Leyden University, added a little commentary to the reference in Diogenes Laertius, saying that he liked the eclectic sect.[16] Half a century later, in 1657, another great Dutch scholar, Gerhard Johannes Vossius, published a new version of Diogenes Laertius, which devoted the whole final chapter to the eclectic sect, whereas Diogenes had mentioned it only in his preface. After all the sects comes, for Vossius, a sect which is not itself another sect, but which regards them all as its sources. Vossius, like Lipsius, praises this philosophy and seems to include himself among its followers, by concluding that one should indeed choose the best from all existing sects.[17]

It is of course true that if Vossius had had time to continue his history of philosophy, the eclectic sect would have become one episode amongst others, so losing its privileged place at the end.[18] Another history of philosophy, published almost at the same time by Vossius's colleague Georg Horn, illustrates this point: when modern philosophy was included the eclectic sect lost its prominence.[19] Vossius probably had a different opinion of modernity in that he still clung to the "prejudice of antiquity," not granting much originality to later thought. But it is precisely in this perspective that his appraisal of eclecticism becomes important, as can be seen in reference to yet another contemporary history of philosophy, by the Englishman Thomas Stanley. Stanley confined his history to ancient philosophy, but never even mentioned the eclectic sect.[20] In stressing the active and methodical character of eclecticism, Vossius was clearly attempting to justify his own work as an historian who no longer believes in the possibility of importing the knowledge of things past into the present, but who takes responsibility for representing and interpreting the past using his own judgment. In the eclectic sect Vossius recognized a philosophy that might justify the work of an historian.[21]

It was with a similar intention that Gottfried Olearius, philologist and professor for theology at the University of Leipzig, translated Stanley's history of philosophy into Latin in 1711, adding a new concluding chapter on the eclectic philosophy. Olearius too wanted to show that "the eclectic manner of philosophizing" represents a decisive step away from the sectarianism of ancient philosophy. Olearius's way of recommending the eclectic method to his contemporaries may also be understood as a reflection on the task of an historian. He highlights specific eclectic virtues such as "modesty, justice, caution and courage,"[22] and seeks to avoid scepticism by means of the simple eclectic maxim "to investigate carefully the propositions and opinions of the philosophers, and to discern the principles and guidelines in which they agree."[23] So in describing the eclectic sect the historian was indirectly explaining his general relationship to his object: past thought.

When in 1690 a second and enlarged edition of Vossius's little history of philosophical sects was printed, more than thirty years after the first, the book became part of an already vivid debate about eclecticism.[24] The fascination was not confined to scholars. In 1686 Johann Christoph Sturm, professor of physics at the University of Altdorf, published a series of dissertations defending eclecticism in an overtly programmatic way.[25] In 1688, there came the *Philosophy of the Court* (*Philosophia Aulica*), of Christian Thomasius, a professor of law first in Leipzig, and then, from 1690, in Halle. Thomasius too unmistakably advocated eclecticism. The heated debate triggered off by Vossius, Sturm and Thomasius was to last well into the eighteenth century, and important new contributors included Arnold Wesenfeld, professor of ethics, logic and metaphysics at Frankfurt (Oder), and Johann Franz Budde, professor first at Halle and then, from 1705, professor of theology at Jena.[26]

The debate over eclecticism was connected with some wider issues at the time: freedom of teaching and research; independence from authority, both political and theological; and the conditions for forming responsible judgments and reasonable forms of discussion. To the eclectics, membership of a sect meant repeating opinions without examination or understanding and they welcomed eclecticism as an additional weapon in the fight against prejudice. Even without knowing much about eclecticism, the general idea of an unprejudiced selection of the best opinions suited the purpose of enlightened thinkers. So the claim to be eclectic was associated unproblematically with the plea for a more rational, moral and juridical style of philosophy, as put forward for instance by Christian Thomasius.[27] In many ways the urge for sect-free philosophizing was linked to the critique of prejudices, discussed at great length around the same time by the same people.[28]

At first glance, then, it seems that Vossius did no more than hand over the term "eclectic" to the philosophers. And indeed the theoretical significance of the idea of eclectic thinking as superseding sectarianism was not lost on enlightened intellectuals among the German academics, who were happy to march under the banner of eclecticism. However, the historical meaning of eclecticism never quite disappeared, and in the period 1680 to 1740 it was present in all contemporary writings on eclectic philosophy. The idea of the eclectic sect as nourishing itself from others was always at work even in the philosophical debate, and was indeed at the very heart of the modern fascination with eclecticism.

Eclecticism is not equivalent to enlightenment, and although many enlightened thinkers cherished eclecticism, they would use the term to epitomize their claims. The critique of pedantry or Aristotelianism, of prejudices in general, could dispense with all reference to eclecticism. Wherever this reference is made, however, an historical problem becomes evident in terms of the relation of present and past (modernity and antiquity) as well as in terms of the present reflection on past doctrines. Viewed from an eclectic perspective, any critique of sectarianism or authoritarianism is a position which must come at the end of a history of opinions and arguments. At the same time, one has to realize from this privileged final position that present opinions are not fundamentally different from opinions of the past, and that consequently it is no use simply to reject the past. Rather, present thinking must relate positively to the past if it is to avoid repeating it. If the present is a consequence of history, and if historical understanding is a mediation for the present, then the philosopher and the historian are not very different one from another anymore.[29]

So behind the philosophical debate about eclecticism lies something like a problematization of the burden of history, and it is probably through this debate that, in the period around 1700, this problematization became a concern for philosophers. From the eclectic position, the historical existence of philosophy is a positive phenomenon, even though this may not mean much more than having a past, being contingent, and taking many different forms. The field opened up by eclecticism was different of course from any nineteenth-century dialectics of spirit and time; rather it comprised all the handed-down facts about philosophy and philosophers, the mass of miscellaneous information about their life and opinions. In itself, the idea of representing tradition was not of interest for any seventeenth-century philosopher; but every radically modern philosopher finds himself indirectly concerned with it, since tradition proves all former philosophers to have failed. This is precisely the meaning of the term "sect"—an unsuccessful philosophy of interest only to a few followers. Eclecticism is a permanent discussion of sects and sectarianism, and thus allows for enlightened thinkers to rethink their proper aim; by way of respecting the historical multitude of philosophical intentions, philosophers may find themselves related to historical reality, not opposed to it. In this view, seventeenth-century eclecticism represents within the history of modern European philosophy a first and fundamental recognition of the relevance of historical knowledge to philosophy, without yet developing a proper historical interest. The historiography of philosophy remained, up to Hegel, a scholarly enterprise. Even Leibniz, a philosopher and a scholar, only projected the writing of a history of philosophy.[30]

*Philosophia Eclectica*

There appears to be an important analogy in the debate about eclecticism in Germany between the 1680s and the 1740s concerning eclecticism as a sect which once existed, and eclecticism as involved in the formulation of every great philosophy.[31] The transition from *secta eclectica* to *philosophia eclectica* marks the point of actual philosophical interest at the time. Furthermore, many concluded not only from the existence of an eclectic sect to the possibility of an eclectic philosophy, but also to the identity of eclecticism and philosophy itself.

The transition from the (historical) *secta eclectica* to the (theoretical) assumption of a *philosophia eclectica* can be found even in histories of philosophy, as for instance in the mid-eighteenth-century history of philosophy by Johann Jakob Brucker, scholar and pastor in Augsburg and member of the Berlin Academy of Sciences.[32] Brucker's history (published in Latin from 1742 onwards) bears witness to the generalization of the phenomenon. In his view, eclecticism is at work everywhere in history—not of course labelled as such, but in terms of what it stands for. Long before Brucker, Sturm had been the first to argue that Aristotle was an eclectic, taking Aristotle's constant critique of predecessors as proof.[33] This retrospective ubiquity of eclecticism can also be found in many dissertations on eclecticism, as for instance in Johann Heinrich Zopf, master at the University of Jena, who included Thales, Pythagoras, Socrates and Plato among eclectic thinkers, as well as Zeno and Epicurus, "concluding the series of the old eclectics."[34] Other authors had yet other lists of names in order to prove that basically every great thinker of the past had been an eclectic.[35] These interpretations all turn on the idea that these thinkers had predecessors whom they could not ignore. On this view, philosophers who made reference, critical or affirmative, to other philosophers formed the history of philosophy into a series of different inventions, each inaugurated by a new thought. It is this view which is fully integrated in the work of Brucker.[36] While dealing with schools and sects Brucker was always concerned with underlining the achievements of individual authors. The formation of schools and sects thus represents forms of decadence or deviation from original intentions.[37] At any rate, the authentic moment of philosophical thinking is identified exclusively with the distance taken towards the ideas and thoughts of predecessors.

The generalization of eclecticism is of course not so much an historical insight as a philosophical claim. Enlightened thinkers were not historians of philosophy. They interpreted the history of philosophy as the history of "honest heretics," as Heumann said.[38] Only exceptionally did they believe that eclectic philosophy might comprise nothing but the history of philosophy, an opinion held by the Jena moral philosopher Ephraim Gerhard.[39] When Heumann designates Aristotle, Luther and Descartes as great eclectics, he does not state historical facts, but deliberately points to these philosophers as examples.[40] There are trivial forms of this attitude, as when Protestantism is identified with the revolutionary struggle against authority. The schoolmaster Johann Heinrich Stuss of Gotha declared Luther an "arch-eclectic," since he overthrew the greatest authorities of his time, Aristotle and the pope.[41] The title "eclectic" serves here as an index for revolutionary intentions behind major historical changes. This "intentionality" of implicit eclecticism points to the supposed radicality of eclecticism—a radicality attributed to it as a method.

For Arnold Wesenfeld the fascination of the Alexandrian philosopher Potamon stemmed from the fact that he had recommended just a method and, consequently, that he professed no doctrines of his own.[42] (The lack of information about Potamon was thus turned into a positive argument.) This is a quite radical point of view: the eclectic character of true philosophy consists in disrupting the dogmatic tradition and attempting to organize it anew. The moment of disruption und reorganization is described by some as *meditatio* (Budde), by others as examination (Teller, Thomasius) or as fruitful doubting (Feuerlin). It is the moment the philosopher appears as somebody who "uses his own eyes instead of those of others," as Thomasius put it.[43] It remains however a paradox that intellectual independence, whether claimed in the present or attributed to the past, cannot itself have a tradition except the one from which it tries to disrupt or emancipate itself: the tradition of philosophical doctrines.

In this way, radical implicit eclecticism was thought to be identical to philosophical method. Budde called meditatio "the exact and diligent accomodation of thoughts to laws in order to find the truth."[44] Meditatio should be involved in every operation of the mind. Budde's pupil, the Jena theology professor Johann Georg Walch, defines meditatio as "the examination of truth."[45] Teller calls it (following Cicero) "examination by free judgment":[46] an examination which subjects all existing philosophies to test in the light of the eternal search for truth.[47] Sturm extends his definition of eclecticism as a method into a whole scientific research program, to which all scholars and scientists should subscribe if they want to "grasp the truth freely and be purified in their judgments."[48] Wesenfeld takes up this scientific formulation and writes that the eclectic must recognize truth in tradition (through reading) as well as in nature itself (through observation).[49]

It is clear from these statements that the definition of eclecticism as a method cannot establish a concrete working program; as put forward by German writers of around 1700, eclecticism is primarily an intellectual activity. Only the really active mind is protected against sectarianism, Heumann says,[50] and accordingly the real strength of a philosophy called eclectic depends entirely on the way of disrupting and re-organizing knowledge (constituted by tradition and experience).

On the one hand, eclecticism is self-determined thinking, but on the other, this self-determination is always endangered by parroting followers. Many contributors to the debate about eclecticism tried to encompass both sides of the phenomenon, both affirming the ideal and recognizing the inevitability of losing it time and again. The ideal was formulated for instance by Christian Thomasius who held that philosophy cannot be conceived as being dogmatically fixed.[51] To explain why this ideal was constantly betrayed, and philosophy perverted into dogma, he turns back to the history of philosophical doctrines. It seems to be no coincidence that these analyses, undertaken by many writers, are often rather lengthy reflections on what today we might call the dialectical moment inherent in every formulation of philosophy. Arnold Wesenfeld dealt with this problem most diligently in the longest essay ever written on eclecticism (his four dissertations fill 170 pages in small quarto), an essay which for some reason did not find any echo in his time. Wesenfeld's main point is that there can only be relative freedom from sectarianism, never absolute. In other words: eclecticism cannot dissipate sectarianism, and has to admit the historical possibility of it at any time, just as it presupposes its own possibility at any time in history. As a consequence of admitting that eclecticism and sectarianism are both ubiquitous, the question arises how to relate one to the other. For the modern radical eclectic the problem was how to interpret sects and sectarians once their existence has been accepted not only as unavoidable, but as constituting the history of philosophy itself. This problem is primarily a hermeneutical one.

*Interpretation of Philosophy*

As soon as eclecticism is defined as a method of thinking, of philosophy, the difference between scholarship and philosophy begins to melt away. Philosophizing in an eclectic way means maintaining constant discussion with already formulated philosophical systems and thoughts, and thinking against their very formulation. Mistrusting the dogmatic character of philosophical propositions is simply the reverse of the conviction that all great philosophers have been eclectics. In order to transform this conviction into a practical attitude, a way of dealing with traditional philosophy has to be found, if scepticism is to be avoided. Budde says at one point that the philosopher is forced to play the role of the historian:[52] he therefore has to have rules. Of course the eclectic attitude towards other philosophies does not lead to a fully historical conception of philosophy: it does not turn present philosophy into the historiography of philosophy (even in the case of Brucker, eclecticism is not the main motive for historical work). However, the eclectic attitude can lead to a limited acceptance of past philosophies without ending in despair about philosophy's fate and without being paralyzed by the multitude of forms it is expressed in. A typically eclectic interpretation would be not reproductive, but productive.

In all late seventeenth- and early eighteenth-century dissertations dealing with eclecticism, rules of interpretation are put forward. These are only basic rules, even in a dissertation as long as Wesenfeld's, and fall well short of the hermeneutical considerations included in logical treatises of the time.[53] In most cases, the often quoted motto from Horace, "not to swear on any master's words" serves as an explanation of what "selecting" exactly means, providing some sort of positive version of the quotation in form of general precautions. Freedom from prejudices and freedom of judgment are said to be one and the same: Wesenfeld sees the main evil of sectarian philosophizing in the hasty formation of opinion, influenced by contempt or adoration. On the other hand, perfection in philosophy would be "to know how to select with reason."[54] The expression "to know" in this definition shows that the eclectic selection of the best represents a problem of knowledge and cognition. Consequently, Wesenfeld narrows down his definition in saying that the eclectic method consists in "understanding the thoughts and the works of others not as the final result of cognition, but as an instrument for further perfection."[55] What follows from this definition may very well be called a hermeneutical program. Once the general rule is accepted that no result of cognition, no end of thinking (*terminum cognitionis*) can be stated just because some proposition has been formulated and believed, there is no limit to further interpretation. So eclecticism takes a vital interest in interpretation, since no selection is possible without it, as Olearius affirms.[56]

Eclectics do not stop at single terms or propositions when they realize their "selecting" relation between present and re-presented past thinking. This is obvious from the widespread criticism of compilation: the mere putting together of different doctrines and *dogmata* is not sufficient to ground an independent judgment. There is general acceptance of Vossius's thesis that eclectics cannot be *miscelliones*.[57] Wesenfeld attenuates this thesis only insignificantly in saying that although compilations do not constitute understanding, they may serve as its means.[58] So even if the eclectics recognize the pedagogical and even scholarly merit of compilations, they refuse to identify it as their method of selecting. Selection is no exterior procedure, but requires interpretation, since there is always a difference between the original, authentic philosophy and its dogmatic form. Interpretation alone transcends the outward form, and the superiority of the "selector" resides not in his capacity to subtract or add parts of this or that doctrine, but in his reflection and close analysis of doctrines. Only through interpretation does the work of internal differentiation, selection and appropriation become possible.

The critics of eclecticism always had an easy job dismissing it, as long as they neglected its constitutional interest in interpretation. It would indeed be very simple to think of eclecticism as a pure selection among given things, choosing without reading. However, it is very clear that philosophical eclecticism is from the outset characterized and distinguished by its hermeneutical insight that interpretation is unavoidable, and can only be perfected. Above all, it was their concern with interpretation which led the eclectics to criticize sectarianism. A judgment (*iudicium*) could become a prejudice (*prae-iudicium*), only once it had lost its original force as an authentic reflection and interpretive discussion of alternative opinions, and seems therefore to depend solely on unregulated inclinations and external influences (these are the *praeiudicia praecipitantiae et autoritatis*). From a sectarian point of view, this observation is impossible to make: it is precisely his incapacity to interpret the judgments of others that prevents the sectarian from regarding judgments as interpretations.

The eclectic stance is clear for instance from Wesenfeld's response to the Dutch scholar Johannes de Raey, a Cartesian. De Raey had attacked eclectic philosophers, saying that they confused the freedom of thinking with the simple rejection of sectarianism. Wesenfeld replied by saying that philosophical freedom cannot be achieved simply by rejecting sectarian authorities.[59] He says: "Eclectic thinkers do not by any means hold that free philosophizing is nothing but the renunciation of sects, rather they combine freedom of thinking with respect for the writings and doctrines of the others."[60] Respect is taken here in the literal sense of looking back: nothing within the tradition should be overlooked that may help to reveal the truth. In this explanation of eclecticism the philosophical program mutates backwards into the scholarly enterprise which generated the whole debate. And Vossius was well aware that it would require great industry (*industria*), intelligence (*ingenium*) and judgment (*iudicium*) to attend to a limited number of sects, let alone to all of them.[61] Although it seems utopian, Wesenfeld's program of respecting all sects was probably meant to be applied only to the main rival sects of his time, Aristotelian school philosophy and Cartesianism. But then, how could one respect both of them equally?

*Conciliatio*

Eclecticism does not definitively privilege one doctrine or another: this attitude is often represented as the position of "not only, but also."[62] In the German writings around 1700 this position is called one of "reconciliation" (conciliatio)—with the exception of Budde and Gottlieb Stolle, professor of philosophy at Jena, both of whom denied eclectic thinkers the title *conciliatores*.[63] All other authors explicitly included the task of taking "seemingly contradictory propositions" together within the hermeneutical program of eclecticism.[64] No simple compilation was intended, and no catalogue of what could be reconciled was established. The principle of method was summed up by Olearius when he said that everybody should be heard (*omnes audiantur*).[65] Thomasius put it differently, declaring that eclectic philosophy "is not a partisan philosophy, but extends the same love to everybody." Sturm pleaded that war among philosophers has to end, and no proposition should be rejected out of hand.[66]

The reconciliation or mediation involved in conciliatio does not ignore differences—on the contrary. Mediation presupposes a recognition of diversity and devotes itself to searching for a "third term" to enable the opposing nature of the differences to be relativized.[67] This third term, in which a common measure was to be found, was truth. With respect to truth, the eclectic program of interpretation amounted to a radically simple will to dismember tradition, and put together whatever could be put together. An eclectic course can therefore be designed as an encyclopedia of disciplines, as proposed by Jakob Wilhelm Feuerlin, professor of logic at Altdorf University.[68] It is however important to realize that there is no proper eclectic canon, but rather a continual examination, as Wesenfeld warns. When eclectics call the truth neither Christian nor pagan, they do not want to define how to approach it, but to indicate what ways are leading towards it, especially what ways out of the past.[69]

It was of great importance for the eclectics to find ways to truth in general, and to accept and try those ways: ways others have found, in other times, at other places. Eclecticism may appear to traditional minds as revolutionary, since it does not respect the letter of any doctrine; but at the same time to moderns it will appear as traditional, since it does not share the belief in methodological exclusivity. A fine example is Wesenfeld's treatment of Francis Bacon. Quite often he quotes from Bacon with approval. But sometimes his quotations break off abruptly when Bacon starts to get angry about ancient philosophers and begins to attack the tradition. For instance Wesenfeld quotes a long passage by Bacon criticizing scholars, but omits the last sentence where Bacon refers to the "previous poverty of spirit." Instead Wesenfeld quotes another passage in Bacon where reference is made only to the "poverty of things which have previously occupied the mind of men."[70] The accusation of tradition is transformed into an accusation of a traditional attitude. So when Wesenfeld quotes Bacon mocking the scientific achievements of the ancients, he is leaving out the famous last sentence where Bacon calls truth "the daughter of time" (*veritas filia temporis*). Instead Wesenfeld alludes to the humanist thinker Ludovico Vives and proposes the truly eclectic thesis that time cannot change truth.[71] Of course such attenuation of criticism of ancient and medieval philosophy did not prevent the eclectic thinkers from identifying with modernity and agreeing with Bacon and Descartes in discarding traditional metaphysics. Eclecticism was however very much keen to play down the opposition between ancient and modern science. Wesenfeld did not quote Bacon's opinion that one cannot learn anything from Greek philosophy, but replaced it with his own opinion that one can profit even from the examination of the most abstract scholastic terminology.[72]

With respect to its principle that truth has to be freed from hiding in tradition and experience, eclecticism can be understood as part of the "querelle des anciens et des modernes."[73] Eclecticism may even count as evidence that this "querelle" took place in philosophy as well. However, the eclectic obligation towards truth leads to a certain appeasement, a kind of coexistence of old wisdom with modern scientific rigorism, which eclectics also see as *studio sapientiae*.[74] It is for this reason that Wesenfeld often refers to Jakob Thomasius.[75] The impartiality of eclecticism is yet again based upon the double feature of philosophy which eclectic thinkers are articulating all along: for them, philosophy as a whole has to be interpreted on the one hand as something prone to dogmatic fixation (sectarianism) and on the other hand as something originally undetermined (*libertas sentiendi*). In this sense the recognition of an inner conflict in the "secta eclectica" as stated by Vossius can be said to have led to a deep reflection of the character of (modern) philosophy and a problematization of its claims in the light of what might today be called the historicity of reason.

## Extra Muros

There are several other aspects of the role and function of eclecticism within the history of philosophy which have to be explored in order to explain the rapid career of the term. There are perhaps three main areas of interest for an historian of philosophy.

Without much doubt eclecticism has to be considered firstly as a part of the history of philosophical hermeneutics. There is as yet no adequate history of hermeneutics before Schleiermacher.[76] If we think of Schleiermacher's general hermeneutics as making explicit the inarticulate rules presupposed in every process of thinking, then eclecticism could be recognized as an early form of it. Its insight into the fact that one cannot go beyond the structure of prejudices, and that prejudices have practical meanings, makes eclecticism a kind of hermeneutical philosophy quite distinct from other hermeneutical theories of the seventeenth and eighteenth centuries (logical ones as in Johannes Clauberg, historical ones as in Martin Chladenius, and biblical ones as in Richard Simon and Baruch Spinoza).[77] The eclecticism of the seventeenth century should probably be understood as "an effort of thinking in a late culture"[78] and therefore a sort of reaction against Aristotelianism and Cartesianism.

Secondly, eclecticism belongs to philosophy's "querelle des anciens et des modernes."[79] From the study of eclecticism one can conclude that the birth of modernity takes place in the acknowledgment of being related to tradition, or—more precisely—to antiquity. The very emergence of eclecticism as a philosophical concept, as well as its repeated though sporadic appearance throughout European intellectual history, is one important indication that modernity has a problem with antiquity. The eclectic belief that eclecticism has always been practiced proves the gap between antiquity and modernity to be almost unbridgeable for the moderns themselves. The fact that eclecticism grants "older" philosophers the right to be heard, and that it bases knowledge on interpretation, indicates the gap which has to be bridged by special efforts. The traces within modern thought of this reflection upon older figures became fully legible only in the nineteenth century, when the history of philosophy functions as an integral part of philosophy itself, for example in Hegel. But already in the eighteenth century, when the history of philosophy first emerged as a literary genre, emancipating itself from the erudite restraints of "historia litteraria," the problem of the relation between ancient and modern thought represented a major difficulty. The narrative construction of philosophy as one (single and identical subject) of "its own" history was then the most successful conciliatio between philosophy and its many historical articulations.[80]

Thirdly, eclecticism is opposed to systematic philosophy. Next to scepticism and theology, eclecticism was probably the most important enemy of the philosophical style in thinking and writing inaugurated by Descartes. To a far greater degree than all others, eclectics developed an affinity with the historian's attitude and even with an empirical and historical conception of thought itself. In its empirical conception, eclecticism aims not only at criticizing the weaknesses of the human mind, but also at investigating them in the hope of formulating a theory of the actual process of cognition. In its historical conception, eclecticism tries to transform the critique of sects and doctrines into historical knowledge about tradition, in order eventually to construct a history of philosophy which would be a sort of philosophy of history at the same time. Already very early in the eighteenth century, such empirical and historical conceptions were clearly contradicted by systematic philosophy, as in Christian Wolff, or, later, Immanuel Kant.[81] However, even if overruled as a model for philosophy, eclecticism belongs in its very antisystematic intention to the same period the canon of great philosophers labels as the modern times.

There are two further ways of exploring eclecticism, both concerned with extra-philosophical features which might help to explain why its popularity in Germany was restricted mainly to the period around 1700.

Fourthly, to start an investigation of eclecticism in terms of intellectual history rather than the history of philosophy we have to care about the religious and/or theological subtext of the claim for free and independent—secular—scholarship. To what extent this subtext influenced the interest in eclectic concerns may be learned from the philosophy of Leibniz—who did not use the term eclecticism or related ones—and his efforts to reunite all Christian churches.

In his early writings, Leibniz insisted on several occasions on the validity of forgotten philosophies: "when we penetrate to the foundations of things, we observe more reason in most of the philosophical sects than commonly believed."[82] In his *New Essays Concerning Human Understanding*, he characterizes his own system as a synthesis of "Plato with Democritus, Aristotle with Descartes, the Scholastics with the Moderns, theology and morals with reason," and in his essay on the *Theodicée* he claims to have found "that in the case of disputes among men of high merit, there is reason on both sides, albeit with respect to different aspects."[83] Clearly these claims belong to the eclectic vocabulary of the time, and many may find no difficulty in aligning Leibniz with eclecticism.[84] More interesting still is the fact that Leibniz was not only theoretically close to eclecticism but also exercised conciliatio in practice, as witness his project of uniting the Protestant and Catholic churches, which he based on his insight into the insignificance of most controversies between Protestants and Catholics and for which he argued in his letters to Bossuet.[85] After the failure of this project, he concentrated his efforts on the reunification of the Lutheran and the Reformist churches, again basing himself on the assumption that theological differences were insignificant. The whole philosophy behind Leibniz's "politics for European peace"[86] emphasizes that the sense of philosophical or theological divergence is not simply a matter of having to deal with the history of philosophy and its many sects. It is rather the other way round, and the attention paid to the history of philosophy and its many sects is motivated—in Leibniz as well as in eclectic writers—by the idea of truth being widely distributed, so that one has to look for it everywhere. The assumption of the ubiquity of truth goes always with the idea of being able to reassemble it and thus to reveal stable ground shared by different dogmatisms, religious as well as philosophical.

The example of Leibniz is not intended as a model of how to extend the concept of eclecticism beyond its explicit formulation, but rather to orientate the historian's attention to the political, theological and religious dimensions of what eclecticism meant around 1700. It is no surprise that Leibniz, never being affiliated to any university after he left Leipzig and Altdorf as a student, could do what no thinker within the university could do, that is, openly attempt to connect theory with practice, philosophy with politics, theology with conciliation of belief. We have to look much closer into the histories of universities to determine whether there were similar tendencies at work and exactly what part eclectic thinking may have played in them.

University teachers might have had another reason for favoring eclecticism, one which stemmed from their very practice of teaching. So the fifth context would be the situation of university teachers at German universities in the late seventeenth and the early eighteenth centuries. The philosophical faculty was then considered the lower faculty, and having a university career usually meant moving upwards and getting a professorship at one of the "higher" faculties, of jurisprudence, of medicine, or of theology.[87] The exact measures every university member had to take in order to be able to make that upward move depended on the churches and on the state where the university was situated. But they also depended on the students: inasmuch as the philosophical faculty was designed to train students in knowledge of a kind which today would be described as philological or historical, teachers could hardly prevent the students from getting involved in debates. It was (and still is) impossible to teach how different philosophical "sects" argued about matters of the highest intellectual interest without kindling curiosity and a desire to investigate the differences further or even to settle the disputes. We can indeed observe that eclecticism was especially popular among young academics, that is to say among "authors who had not yet reached the end of their career. Most of them later became theologians and never again produced anything eclectic, as for instance Budde who became theology professor in Jena in 1705."[88] It would be interesting to investigate this question further, and find out more about academic life in Jena, Halle, and Leipzig, where most of the dissertations dealing with eclecticism were produced. It is highly probable that the success of early eighteenth-century eclecticism can at least partly be explained by "the structure and the needs of school-philosophy as taught in universities."[89]

The degree to which historical knowledge was used in the teaching of philosophy at the end of the seventeenth century and the beginning of the eighteenth is difficult to determine. At least to the eclectic historians of the early eighteenth century, eclecticism made it possible to have an attitude of detachment to the contents of every philosophy, yet it made it also necessary to establish what exactly that content was.[90] Thus eclecticism could not fail to affect the method of teaching philosophy, which had to include indications of the origins of current opinions, and reflections upon the differences between them.[91] It is at any rate very important not to disregard the historical situation in which eclecticism was popular and openly affirmed, since this situation could tell us more not only about the motives behind the promulgation of eclecticism, but also about the reasons for its subsequent decline.

## Notes

1. Werner Schneiders, "Vernünftiger Zweifel und wahre Eklektik. Zur Entstehung des modernen Kritikbegriffs," *Studia Leibnitiana* 17 (1985): 142–61; see also W. Schneiders, *Aufklärung und Vorurteilskritik* (Stuttgart, 1983).

2. Norbert Hinske, ed., *Eklektik, Selbstdenken, Mündigkeit* (Hamburg, 1986); see also Helmut Holzhey, "Philosophie als Eklektik," *Studia Leibnitiana* 15 (1983): 19–29.
3. Wilhelm Schmidt-Biggemann, *Topica Universalis. Eine Modellgeschichte humanistischer und barocker Wissenschaft* (Hamburg, 1983), 249–92.
4. Horst Dreitzel, "Zur Entwicklung und Eigenart der 'eklektischen' Philosophie," *Zeitschrift für historische Forschung* 18 (1991): 281–343.
5. The entry "Eklektizismus" in *Historisches Wörterbuch der Philosophie*, ed. J. Ritter, vol. 2 (Basel, 1972), cols. 432–33, gives no indication of the historical importance of eclecticism and discusses it mainly in connection with J.J. Brucker and Victor Cousin.
6. Michael Albrecht, *Eklektik. Eine Begriffsgeschichte mit Hinweis auf die Philosophie- und Wissenschaftsgeschichte* (Stuttgart, 1995).
7. When Paul Hazard spoke of the "Crisis of the European Mind" (*La Crise de la conscience européenne 1680–1715*, Paris, 1961), eclecticism was unknown to him.
8. When Victor Cousin tried to use the term "eclecticism" in opposition to Hegelianism, his success was limited in France, and he was heavily criticized in Germany; Amadeus Wendt called the term "most inconvenient": see his review of Cousin's *Fragments philosophiques* in *Göttingische Gelehrte Anzeigen* 25.Sept. (1834): 1539. On Cousin see No. 18/19 of the journal *Corpus*, (Paris, 1992).
9. On C. Wolff see Albrecht, *Eklektik*, 536, esp. 546; on Fichte, W. Schmidt-Biggemann, *Theodizee und Tatsachen. Das philosophische Profil der deutschen Aufklärung* (Frankfurt am Main, 1988), 215–17.
10. See G.W.F. Hegel on the "Alexandrian philosophy" in *Lectures on the History of Philosophy*, vol. 2, trans. E.S. Haldane and Frances H. Simson (London, 1894), 401. An exception to the general verdict is of course Denis Diderot; see his articles on "éclectisme" in the *Encyclopédie*, vol. 5 (1755), 270–92, and on Christian Thomasius, vol. 16 (1765), 284–94.
11. W. Schmidt-Biggemann views eclecticism from previous encyclopedic systems such as polyhistoria. See his *Topica Universalis*, 255 and 288–92; M. Albrecht regrets that "no philosopher has ever published a truly eclectic work" and that "no great philosopher was ever won over by the idea of eclecticism" (*Eklektik*, 475, 666); Dreitzel speaks of an "eclectic syndrome" ("Zur Entwicklung und Eigenart," 300 and passim).
12. Some of the material presented in the following essay was used in an article which appeared in French and in German; see U.J. Schneider, "L'Éclectisme avant Cousin: la tradition allemande," *Corpus* 18/19 (Paris, 1991): 15–27, and "Über den philosophischen Eklektizismus," *Nach der Postmoderne*, ed. A. Steffens (Düsseldorf, 1992), 201–24.
13. Cf. H.G. Gloeckner, *De Potamonis Alexandrini Philosophia Eclectica Recentiorum Platonicorum Disciplinae Admodum Dissimili Disputatio* (Leipzig, 1745), §§5–12, 12–20.
14. Cf. Diogenes Laertius, *Life and Opinions of Famous Philosophers*, preface. There are only two other places where the Greek word *eklektikos* is used in a similar way: the *Stromateis* of Clement of Alexandria and the so-called "Suida" of the tenth century. Cf. Schmidt-Biggemann, *Theodizee und Tatsachen*, 217.
15. Cf. the commentated editions of the history of philosophy by Diogenes Laertius by H. Stephanus (Paris, 1570), Th. Aldobrandini (Rome, 1594), Is. Casaubonius (Leyden, 1595), Aegidius Menagius (London, 1663), and H. Wetstenius (Amsterdam, 1692).
16. Justus Lipsius, *Manoductionis ad Stoicam Philosophiam Libri Tres*, (1st ed. 1604, Lugduni Batavorum, 1644), Book I, Dissertatio V, 22f.; Johann Konrad Dannhauer shares the praise a little later in his *Epitome Dialectica* (Strasbourg, 1634), Praefatio; cf. also H.-E. Hasso Jaeger, "Studien zur Frühgeschichte der Hermeneutik," *Archiv für Begriffsgeschichte* 18 (1974): 61.
17. G.J. Vossius, *De Philosophorum Sectis Liber* (The Hague, 1657) ch. 21, §16.
18. Vossius died in 1649. His book on the philosophical sects was published by his son Isaac.
19. Cf. G. Horn, *Historiae philosophicae Libri VII, Quibus de origine, sectis et vita philosophorum ab orbe condito ad nostrum aetatum agitur* (Lugduni Batavorum, 1655); on Horn see the first volume of *Storia delle storie generali della filosofia* (*Dalle origini rinascimentali alla "historia philosophica,"*) ed. Giovanni Santinello (Brescia, 1981), 252ff.; esp. 261, 269f.
20. Thomas Stanley, *The History of Philosophy* (London, 1655; 2d ed. 1687; 3d ed. 1702).

98    ULRICH JOHANNES SCHNEIDER

21. See W. Schmidt-Biggemann, *Topica Universalis*, 255ff., where the whole conception of Vossius's thought is characterized as eclectic.

22. See Gottfried Olearius, *De philosophia eclectica*, in *Historia philosophiae vitas, opiniones, resque gestas, et dicta philosophorum sectae cuiusvis complexa, autore Thoma Stanleio, ex anglico sermone in latinum translata, emendata, variis dissertationibus atque observationibus passim aucta* (Leipzig, 1711), 1219; 2d ed. (Venice, 1731), vol. III, 358. Olearius taught Greek and Latin until 1708, when he took over the chair of theology from Seligmann.

23. Ibid., 1207f.; (1731 ed.), vol. III, 346).

24. Johann Jakob Ryssel, *Continuatio in Gerardi Johannis Vossii Librum de Philosophorum Sectis* (Leipzig, 1690, 2d ed. 1705).

25. In the previous year three smaller dissertations had been published: Georg Paul Rötenbeck, *Principii Aristotelici: Impossibile est idem simul esse et non esse, et Cartesiani: Cogito ergo sum, amica methodoque eclecticae conformis collatio principii simpliciter et absolute primi dignitatem ab utroque removens* (Altdorf, 1685); Johann Friedrich Titius, *Exercitatio Academica de modo philosophandi electivo* (Helmstedt, 1685); Gottlob Friedrich Seligmann, *De philosopho conciliator* (Rostock, 1685).

26. On Budde and his eclecticism cf. Johann Gottlieb Buhle, *Geschichte der neueren Philosophie* vol. 4 (Göttingen, 1803), 661, and also Horst Dreitzel, "Zur Entwicklung und Eigenart," 292f.

27. Cf. Ernst Bloch, *Christian Thomasius, ein deutscher Gelehrter ohne Misere* (Berlin, 1953), Ernest Bloch, *Gesamtausgabe* vol. 6 (Frankfurt am Main, 1961), and Werner Schneiders, ed., *Christian Thomasius 1655–1728. Interpretationen zu Werk und Wirkung* (Hamburg, 1989).

28. Cf. Manfred Beetz, "Transparent gemachte Vorurteile. Zur Analyse der 'praejudicia auctoritatis et praecipitantiae' in der Frühaufklärung," *Rhetorik* 3 (1983) and W. Schneiders, *Aufklärung und Vorurteilskritik* (Stuttgart, 1983), where eclecticism is treated as a part of the problem of prejudices.

29. Cf. Helmut Zedelmaier, *Bibliotheca universalis und Bibliotheca selecta. Das Problem der Ordnung des gelehrten Wissens in der frühen Neuzeit* (Vienna and Cologne, 1992), esp. ch. 4.

30. Cf. Lucien Braun, *Histoire de l'histoire de la philosophie* (Strasbourg, 1973), 91f.; on Gottfried Wilhelm Leibniz's idea of research see Wolfgang Hübener, "Leibniz—Ein Geschichtsphilosoph?," *Studia Leibnitiana*, Sonderheft no. 10 (Wiesbaden, 1982), reprinted in W. Hübener., *Zum Geist der Prämoderne* (Würzburg, 1985).

31. Cf. Ryssel, *Continuatio in Gerardi Johannis Vossii Librum De Philosophorum Sectis*, §1: "Differt Eclecticum esse, et Sectam Eclecticam profiteri"; the same sentence is found in Christian Thomasius, *Introductio ad philosophiam aulicam* (Halle, and Magdeburg, 1688); 2d ed. 1702, in German published as *Einleitung in die Hof-Philosophie* (Frankfurt and Leipzig, 1719), ch. 1, §36.

32. Cf. J.J. Brucker, *Kurze Fragen aus der Philosophischen Historie* vol. 7 (Ulm, 1736), 3; *Historia Critica Philosophiae* vol. 2 (Leipzig, 1742), 189f.; vol. 3 (Leipzig, 1743), 421 and passim.

33. Cf. Johann Christoph Sturm, "De philosophiae electivae priscis modernisque cultoribus," ch. 4 in *Philosophia Eclectica* (Altdorf, 1686), 43ff.

34. Cf. Johann Heinrich Zopf, *Exercitatio Historico-Philosophica de Origine Philosophiae Eclecticae* (Jena, 1715), 33: "Agmen Eclecticorum veterum claudet Epicurus, instaurator scholae Epicurae"; Zopf became director of a Lutheran school in Essen in 1719. See also Thomasius, *Introductio ad Philosophiam aulicam*, ch. 1, esp. §93, and Ryssel, *Continuatio*, §4.

35. Budde mentions for instance Socrates, Plato, Pythagoras, Aristotle, Zeno, Descartes, and Grotius; see J.F. Budde, *Elementa Philosophica Instrumentalis seu Institutionem Philosophiae Eclecticae*, tomus primus (1st ed. 1703) (Halle, 1722), 96, (ch. 6, §42).

36. On Brucker cf. Braun, *Histoire de l'histoire de la philosophie*, 119–39; Constance Blackwell, "The Historiography of Renaissance Philosophy and the Creation of the Myth of the Renaissance Eccentric Genius," *MS* (1991).

37. Cf. Brucker, *Historia Critica* vol. 2 (Per. 2, Pars 1, Lib. 1, Ch. 2, Sect. 4), 190, on the negative effects of ambition.

38. Cf. Christoph August Heumann, "Von dem Ingenio Philosophico," in *Acta philosophorum, das ist: Gründliche Nachrichten aus der Historia philosophica, nebst beigefügten Urteilen, von denen dahingehörigen alten und neuen Büchern* (Magdeburg, 1715), P. 4 (1716), 595f.

Eclecticism and the History of Philosophy    99

39. See E. Gerhard, Introductio praeliminaris in historiam philosophicam (Jena, 1711), 32; cited in Storia delle storie generali 2:434; see also Mario Longo, Historia Philosophiae Philosophica. Teorie e metodi della storia della filosofia tra Seicento e Settecento (Milan, 1986), 61ff.
40. Cf. Heumann, Acta philosophorum, P. 4 (1716), 605ff.
41. Cf. J.H. Stuss, Exercitatio Historico-Philosophica de Luthero Philosopho Eclectico (Gotha, 1730), 11 (§17). Stuss remained rector of the school in Gotha until his death in 1768.
42. Arnold Wesenfeld, Dissertationes Philosophicae Quatuor Materiae Selectioris de Philosophia Sectaria et Electiva (Frankfurt, 1694), Diss. 3, Ch. 1, §2, 5.
43. Thomasius, Introductio ad Philosophiam aulicam, Ch. 1, §90; see also the echo of this formulation in Heumann, Acta philosophorum, P. 4 (1716), 582.
44. Budde, Elementa Philosophica, P. 1, Ch. 4, §5, 159f.; see also his De Cultura Ingenii (Halle, 1699, 2d ed. Jena, 1723), Ch. 3, §10 (p. 92–94).
45. Io. Francisci Buddei Observationes in Elementa Philosophiae instrumentalis editae cura et studio Io. Georgii Walchii, ed. Johann Georg Walch (Halle, 1732), 201.
46. Cf. Romanus Teller, Disputatio de Philosophia Eclectica consensu Amplissimae Facultatis Philosophicae in celeberrima Lipsiensium Academia, publicae placidaeque Eruditorum disquisitioni (Leipzig, 1674), Ch. 3, §7; cf. also Ch. 1, §4. In Ch. 2, §8, Teller quotes Cicero, Tusculanae disputationes, Liber IV, 7. Teller was school rector in Wurzen near Leipzig.
47. Teller, Disputatio de Philosophia Eclectica, Ch. 1, §2; cf. also §9.
48. Sturm, "De philosophiae electivae necessitate, utilitate et praestantia," ch. 2 in Philosophia Eclectica, 14.
49. Wesenfeld, Dissertationes de Philosophia Sectaria et Electiva, Diss. 3, Ch. 2, §6 (p. 12). Cf. also §10 (p. 18).
50. Heumann, Acta Philosophorum, P. 4 (1716), §8 (p. 579) and §9 (p. 581).
51. Cf. Thomasius, Introductio ad Philosophiam aulicam, §98.
52. Cf. J.F. Budde, Isagoge historico-theologica ad theologiam universam (2d ed. Leipzig, 1730), Praefatio, quoted in M. Longo, Historia Philosophiae Philosophica, 58; cf. Storia delle storie generali, ed. Santinelli, 2:383.
53. Wilhelm Risse informs that in Johann Clauberg's Logica vetus et nova quadripartita (Amsterdam, 1654) hermeneutics was introduced as a department of logic and that this tradition lasted down to Christian Wolff. Cf. W. Risse, Die Logik der Neuzeit, 2:1640–1780 (Stuttgart, 1970), 62, 600; other traces of the same problem can be found in the treatises of Protestant teachers such as J. K. Dannhauer, Idea boni interpretis (Strasbourg, 1630), Ludwig Meyer, Philosophia Scripturae interpres (Amsterdam, 1666), and Johann Heinrich Ernesti, Compendium hermeneuticae profanae (Leipzig, 1699). On the origins of hermeneutics see the debate between H.-E. Hasso Jaeger (Frühgeschichte der Hermeneutik, a. a. O.) and Hans-Georg Gadamer, "Logik oder Rhetorik? Nochmals zur Frühgeschichte der Hermeneutik," in Arch. f. Begriffsgesch. 20 (1976), and esp. Manfred Beetz, "Nachgeholte Hermeneutik. Zum Verhältnis von Interpretations- und Logiklehren in Barock und Aufklärung," Deutsche Vierteljahrsschrift für Literaturwissenschaft und Geistesgeschichte 55 (1981).
54. Cf. Wesenfeld, Dissertationes de Philosophia Sectaria et Electiva, Diss. 1, §18 (p. 31) and Diss. 4, §1 (p. 3).
55. Ibid., Diss. 4, §14 (p. 24).
56. Cf. Olearius, De philosophia eclectica, 1218 (1731 ed. 3:357).
57. Cf. Vossius, De Philosophorum Sectis Liber, §1; Sturm, Philosophia Eclectica, ch. 1, 6ff.; Wesenfeld, Dissertationes de Philosophia Sectaria et Electiva, Diss. 3, §1 (p. 4); Olearius, De philosophia eclectica, 1207 (1731 ed. 3:345).
58. Wesenfeld, Dissertationes de Philosophia Sectaria et Electiva, Diss. 3, §11 (p. 22).
59. Wesenfeld quotes Raey's Dissertatio de Libertate et Servitute, from the appendix of his De Interpretatione, §25 ("Errat, quisque solos Eclecticos putat esse liberos Philosophos.") and divides his answer to six points, see Dissertationes de Philosophia Sectaria et Electiva, Diss. 4, §17 (p. 30).
60. Wesenfeld, Dissertationes de Philosophia Sectaria et Electiva, Diss. 4, §17 (p. 30): "Nec Eclecticorum assertio est, "liberè philosophari (absolutè) nihil aliud esse quàm nulli se sectae etc. tradere"; sed tantùm de libertate sentiendi in respectu ad aliorum scripta et dogmata loquuntur."

61. Vossius, *De Philosophorum Sectis Liber*, Ch. 21, 11; see also Johann Baptist Röschel, *Exercitatio Academica ex Historia Philosophica De philosophia conciliatrice*, Wittenberg 1692, §18.

62. Cf. Alexandre Kojève, *Essai d'une histoire raisonné de la philosophie païenne*, vol. 3: *La Philosophie hellénistique, les Néo-Platoniciens* (Paris, 1973), 213ff.

63. Cf. Budde, *Elementa Philosophiae Instrumentalis*, Ch. 6, §44 (p. 96f.); G. Stolle, *Anleitung zur Historie der Gelahrtheit* (1st ed. Halle, 1718, 2d ed. 1724, 3rd ed. Jena, 1727), P. 2, §108, (p. 421).

64. Cf. explicitly Johann Jakob Höfler, *Conciliatorum et eclecticorum diversa ratio philosophandi* (Altdorf, 1742), and G. F. Seligmann, *De philosopho conciliator*.

65. Cf. Olearius, *De philosophia eclectica*, 1218 (1731 ed. 3:357).

66. Thomasius, *Introdutio ad philosophiam aulicam*, §92; Sturm, *Philosophia Eclectica*, Ch. II, 26f.

67. Cf. Röschel, *De philosophia conciliatrice*, §16.

68. Cf. J.W. Feuerlin, *Cursus philosophiae eclecticae* (Altdorf, 1727), Table 2; Feuerlin had previously published *Medicina Intellectus sive Logica* (Nuremberg, 1715). Later he also taught metaphysics and Oriental languages.

69. Cf. Wesenfeld, *Dissertationes de Philosophia Sectaria et Electiva*, Diss. 3, Ch. 1, §6 (p. 12); Ch. 2, §5 (p. 33).

70. Ibid., Diss. 3, Ch. 2, §10 (p. 19f.); Bacon is quoted with 85 from the first book of *Novum Organum*.

71. Cf. Wesenfeld, *Dissertationes de Philosophia Sectaria et Electiva*, Diss. 1, 15 (p. 30); with respect to 84 in Bacon (first book of *Novum Organum*) he says: "Temporis vero per se nullam efficaciam esse." The quotation of Juan Luis Vives is from *De causis curruptarum artium* (1st ed. 1531), which Wesenfeld quotes frequently.

72. Wesenfeld, *Dissertationes de Philosophia Sectaria et Electiva*, Diss. 3, Ch. 1, §9 (p. 18): the critique of *Logodaedalia*, although first taken from Bacon (§71 of the first book of *Novum Organum*) is later attenuated: "quae merè subtilia primò apparent, cum cautione tamen tractanda, nec statim rejicenda sint."

73. See Joseph Levine, "Ancients, Moderns, and History," *Humanism and History* (Ithaca, 1987), 155–77.

74. Cf. Wesenfeld, *Dissertationes de Philosophia Sectaria et Electiva*, Diss. 4, §1 (p. 3); Olearius, *De philosophia eclectica*, 1207 (1731 ed., 3:345).

75. Cf. J. Thomasius, *Programma XXXIX adversus philosophos libertinos* (1665) and *Programma XLI adversus philosophos novantiquos* (1665), both in *Dissertationes LXIII varii argumenti*, ed. Chr. Thomasius (Halle and Magdeburg, 1693); cf. also: J. Thomasius, *Oratio XV de Syncretismo Peripatetico* (1664), in *Orationes* (Leipzig, 1683).

76. Wilhelm Dilthey's investigations of the question were not published by himself: see W. Dilthey, *Leben Schleiermachers*, vol. 2: *Die Hermeneutik vor Schleiermacher*, ed. Martin Redeker (Berlin, 1966). A major study of seventeenth- and eighteenth-century hermeneutics is to be found in a book by the Russian philosopher and disciple of Husserl, Gustav Shpet. Originally written in 1918, his *Hermeneutics and Its Problems* is now available in Russian, published in the journal *Kontekst* (Moscow 1989–92) and in German: *Die Hermeneutik und ihre Probleme* (Munich, 1993). Shpet does not mention eclecticism.

77. See the introduction by Hans-Georg Gadamer, *Seminar philosophische Hermeneutik*, ed. H.-G. Gadamer and G. Boehm (Frankfurt am Main, 1976).

78. "Eine philosophische Denkanstrengung in kulturellen Spätzeiten": Günter Abel, *Stoizismus und frühe Neuzeit. Zur Entstehungsgeschichte modernen Denkens im Felde von Ethik und Politik* (Berlin, 1978), 71.

79. Cf. Beetz, "Nachgeholte Hermeneutik," and Peter K. Kapitza, *Ein bürgerlicher Krieg in der gelehrten Welt. Zur Geschichte der querelle des anciens et des modernes in Deutschland*, (Munich, 1981).

80. Cf. U. J. Schneider, *Die Vergangenheit des Geistes. Eine Archäologie der Philosophiegeschichte* (Frankfurt am Main, 1990).

81. Wolff stresses the superiority of systematic over eclectic thought at one point, but otherwise does not set them in opposition to each other. See *Vernünfftige Gedanken von Gott . . .* , Part 2 (Frankfurt am Main, 1724), 377 (*Werke* [1983]. Ser. 1, vol. 3, 411); and Albrecht, *Eklektik*, 541.

*Eclecticism and the History of Philosophy* 101

On the concept of system and its idealistic interpretation see Martin Heidegger, "Was heißt System und wie kommt es zur Systembildung in der Philosophie?," *Schellings Abhandlung über das Wesen der menschlichen Freiheit* (Lecture 1936, Tübingen, 1971), 27ff.

82. Cf. Leibniz, *Clarification of the Difficulties which Mr. Bayle Has Found in the New System of the Union of Soul and Body*, in *Leibniz. Philosophical Papers and Letters*, ed. Leroy E. Loemker (Dordrecht, ²1969), 496. Loemker actually translates the original French ("lorsqu'on entre dans le fond des choses, on remarque plus de la raison qu'on ne croyoit dans la plupart des sectes des philosophes") differently, making it hard to understand: ". . . more reason than most of philosophical sects believed in." See also *Discourse on Metaphysics*, §11, ibid., 309 (rehabilitation of Ancient philosophy) and §22, ibid. 318f. (combination of different perspectives).

83. Leibniz, *Nouveaux essais sur l'entendement humain*, Book 1, Ch. 1, and *Essais de Théodicée sur la bonté de Dieu, la liberté de l'homme et l'origine du mal*, Preface.

84. Albrecht cites several such opinions, with which he disagrees, cf. *Eklektik*, 295.

85. Cf. Eric J. Aiton, *Leibniz. A Biography* (Bristol, 1985), 180ff.

86. Cf. André Robinet, *G. W. Leibniz. Le Meilleur des mondes par la balance de l'Europe* (Paris, 1994).

87. Cf. Friedrich Paulsen, *Geschichte des gelehrten Unterrichts auf den deutschen Schulen und Universitäten vom Ausgang des Mittelalters bis zur Gegenwart*, 2 vols. (Leipzig, 3d ed. 1919).

88. Albrecht, *Eklektik*, 472; cf. also 383–84; Albrecht is right in his general assessment, although Budde himself affirmed eclecticism even after becoming a theology professor, cf. his preface to Martin Musig, *Licht der Wahrheit . . . , nebst dessen* [i. e. Budde's] *Approbation und Vorrede von der Welt- und Schul-Gelahrtheit*, Part 1, (Frankfurt and Leipzig, 1st ed., 1709, 2d ed. 1716), §49.

89. Dreitzel, *Zur Entwicklung und Eigenart*, 306–7; cf. also Max Wundt, *Die deutsche Schulmetaphysik des 17. Jahrhunderts* (Tübingen, 1939).

90. Cf. *Storia delle storie generali della filosofia*, ed. G. Santinelli, vol. 2: *Dall'età cartesiana a Brucker* (Brescia, 1979), 504–10, 527–635. Cf. also Donald R. Kelley, "History and the Encyclopedia," *The Shapes of Knowledge from the Renaissance to the Enlightenment*, ed. D.R. Kelley and R.H. Popkin (The Hague, 1991), 7–22, esp. 12.

91. There are also examples of those contrary to this assumption, such as Heumann, who, while advocating eclecticism as the only way of philosophizing, wanted to restrict teaching to dogmatic methods and contents. Cf. Heumann, *Poecile sive Epistolae miscellaneae ad literatissimos aevi nostri viros Accedit appendix exhibens dissertationes argumenti rarioris*, Tomi I, Liber I (Halle, 1722), 136.

# GUNDLING VS. BUDDEUS
## Competing Models of the History of Philosophy

Martin Mulsow

Translated from the German by Charlotte Methuen

## A Dialogue in the Kingdom of the Dead

In the years after 1729, a short anonymous work circulated at the universities of Halle and Jena. Its title was *Examen rigorosum zwischen Budde und Gundling im Todtenreich*.[1] The *Examen* features two professors, both of whom had died in 1729 and both of whom were among the pioneers of the early German enlightenment. The former had left philosophy to pursue a career in theology; the latter had turned from philosophy to law. The unknown author of this dialogue of the dead was employing a genre of text which had regained popularity after the publication of Fontenelle's *Nouveaux Dialogues des Morts*. David Faßmann had been instrumental in encouraging its use in Germany from 1718,[2] but his *Gespräche in dem Reiche derer Todten*, which appeared prior to 1739, seems to represent only the tip of an iceberg the greatest part of which is formed by a mass of short, rapidly printed works. The fictitious *Examen* between Gundling and Buddeus is one of these.

Until 1705, Nicolaus Hieronymous Gundling[3] and Johann Franz Buddeus[4] were colleagues at the University of Halle; as fellow students and colleagues they worked closely with Christian Thomasius; together they edited the journal *Observationes selectae ad rem literariam spectantium*. It was, therefore, perfectly reasonable to bring them together in a dialogue of the dead. More problematic, however, was the fact that the differences in the development and tendencies of the two men's thought meant that they represented two wings of the Thomasian school, wings which gradually drifted apart. Gundling had come to Halle in 1699; under Thomasius's influence he had given up his study of theology, turned to law and soon became Thomasius's star pupil, embarking on a career as professor of both eloquence and natural law. This combination of reflection through *historia literaria* with moral philosophical legal theory is found in the thought of virtually no other. Gundling represents the liberal wing of the Thomasians, which stood for the separation of church from state, of philosophy from theology. Buddeus's development was quite different. He had been in Halle since 1693 and was deeply influenced not only by Thomasius but also by the pietist August Hermann Francke. Although he taught moral philosophy in Halle until 1705, theology always remained his primary interest. He used the new impulses of Thomasius's doctrines of reason and of morals to enrich his theology, drawing also upon the contemporary search for historical origins and on pietist themes. In 1705 Buddeus moved to Jena where he became professor for theology and developed a varied and influential course of teaching. Buddeus thus represented the theological wing of early Enlightenment thought: a moderate, modern combination of theology and philosophy.

Despite their differences, Gundling and Buddeus seem to have remained life-long friends and to have retained a high degree of respect for each other's views. It is, therefore, tempting to ignore or at least to underestimate the tensions in the early enlightenment which are thrown into relief by the differences between their theories, for these hardly ever erupted into open conflict. One reason for this may be that the "eclectic" ideology shared by many thinkers of the early Enlightenment, a program which aimed to achieve a historically aware philosophy free from dogma, had an inherent tendency to disguise dogmatic differences rather than to emphasize them.[5] Nevertheless, when the arguments of the texts are analysed in their own contexts and their implicit tactics sought out, it becomes possible to discern the tensions concealed beneath the intentions both of this programme and of the unitive spirit of Halle's early Enlightenment. These latent tensions and differences will form the main focus of the reconstruction that follows. In a way it will be a new version of the dialogue between Buddeus and Gundling in the kingdom of the dead. This reconstruction is intended to make clear the consequences of these tensions and differences for the early phases in the development of just such a critical history of philosophy as that eventually realised by Jakob Brucker.

The formative phase of the discipline "history of philosophy" and with it the first attempts at writing a history of philosophy as the history of a discipline should be seen as taking place during the first decades of the eighteenth century in Germany.[6] This phase can only be understood in the context of three conditions, each of which had a fundamental role in shaping what was only later to grow into a discipline. The first was the influence of Christian apologetic in the historical research of the seventeenth century, which had the consequence that the discussion of earlier philosophical ideas initially took place only as part of a *historia atheismi*. Interest centered on identifying which philosophical trends had been the forerunners of dangerous contemporary forms of atheism and differentiating these from the philosophy which could safely be taught in the universities as compatible with Christianity. However, around 1700 this approach began to change and another trend appeared under the influence of thinkers as various as Pierre Bayle and Gottfried Arnold. Although this new approach was still theological, it began to rehabilitate thinkers who had previously been outlawed: instead of condemning them as "heretics" it undertook the defense of thinkers who had been falsely branded as atheists, the *atheismi falso suspecti*.[7] The second important condition was the *historia literaria*. This discipline was very influential in Germany, but the pattern of its influence, like that of Christian apologetic, was beginning to change. Just as Bayle in his *Dictionnaire* had used his refurbishing of the history of human knowledge to identify and criticize the mistakes and prejudices of the past, so too in Germany this history was put through the purifying process of historical scepticism. Earlier attempts to amass encyclopaedic knowledge were giving way to the selection of historically tested facts. However, philological and historical criteria were not the only factors which influenced this selection in the early Enlightenment, for the context was shaped too by a third condition, a new concern with natural law and an associated interest in the theory of the passions. Through this concern, psychological and moral considerations also gained in importance. The philosophical past could be read in terms of its moral corruptions, whether in terms of philosophy's contribution to the corruption of Christianity in "Papism" or as a typology of possible philosophical mistakes arising from psychological one-sidedness. The resurrection of the theory of temperaments by Christian Thomasius equipped this historical scepticism about philosophical prejudice with a terminological tool.

These were the ingredients from which history of philosophy was brewed. The thinkers of the early German enlightenment, as represented by the constellation of Gundling and Buddeus, differed as to the priorities they chose to give the different conditions which shaped the context in which they found themselves. The effect of these differences upon their individual formulations of the history of philosophy needs to be investigated. The discipline was still malleable enough to be bent in different directions according to these differences. In other words, in this early phase it was still worth fighting over the different ways of doing the history of philosophy, for these were both an expression of and a defense for the "symbolic capital" implicit in the different directions.[8] It was not yet clear what sort of "enlightenment" would result from this discipline. The task of our investigations is, therefore, to offer a cross-section of the latent tensions within the group of thinkers who made up the early Enlightenment in Germany. This was in no way a homogenous group, and so we need to inquire which "enlightenments" were under discussion in this phase in which the definition of the discipline history of philosophy was still open.

## Adam the Pedant and Adam the Farmer

In Seneca's treatise *De beneficiis* there is a passage which polemizes strongly against philosophers who are so concerned to compare human beings with the gods that they end up hating nature (*naturam oderint*) and unable to appreciate the range of faculties (*virtutes*) and skills (*artes*) available to humankind.[9] Gundling found it necessary to draw upon the reminder of the ancient Stoic to counter a tendency which was beginning to be widespread in Germany in the early eighteenth century: a revelling in the weakness of the human spirit and in the depths of fallen human nature. Gundling's essay *Reflexion über den Locum Senecae de Benefic. Lib. II. Cap. XXIX.* offers a corrective to this tendency, which was often found in circles which were either sympathetic to or which openly stood for Pietism. Moreover, Gundling's work is in effect a disguised review of Buddeus's work *Elementa philosophiae instrumentalis*, which had appeared in 1703. Contemporary readers would have recognized it as such, for although he never mentions his friend's name, the *O quantum est, quod nescimus!* cited three times by Gundling[10] is at the same time a parody on and an overobvious echo of the beginning of Buddeus's chapter on understanding in the *Elementa*, in which he discusses human ignorance and its consequences in just these terms.[11] Gundling ridicules the *cranks* who indulge themselves with complaints about the weakness of the human body and the human intellect. He is clearly referring to Buddeus and to those who shared this opinion, and probably in particular the pietist Joachim Lange, who in his *Medicina mentis*, published 1704, had repeated Buddeus's argument in a more developed form.

But not all those who held this view were pietists. Gundling puts a lament for the Fall into the mouths of those against whom he is writing: *Adam the first man was truly happy: and had he remained in this state of innocence*—so Gundling paraphrases the position of his opponents—*he would have brought up only educated children, who would have had needed no Praeceptor, no Orbis pictus, no Tyrocinius, no Portulae Seidelius*.[12] Thus Gundling enters the field upon which not only theologians but also proponents of natural law fought out the *condicio humana*: the question of Adam's prelapsarian state and the results of the Fall. The first problem is that of the opposite to the unknowing human being, and thus of the state of Adam in Paradise. Human ignorance is to be measured against the original perfect human state. Gundling suspects that a fantasy of perfection lies behind these speculations about Adam's original knowledge—*Don't get so excited. How would it help you / if you knew the essence of all things?* He formulates this, in the terminology of the passions, as a suspicion of the vice of avarice, analogous to the speculators who seek the stone of the wise only in order to make more gold. What he is referring to, however, is the wish for an omnipotence which would make it possible to see the innermost secrets of the world. The learned have always been easily seduced by this idea. Just as, according to Gundling, they have a tendency to impose upon their image of God exactly the human characteristics that are of greatest importance to them, so too they see Adam through the lens of the theses of their own narrow world, describing his knowledge in the terms of scholastic disputations and academic disciplines. Gundling's criticism ridicules the pedantry of scholars who see their own problems as the navel of the world. Gundling's comment that in reality *there was no faculty in which Adam could have become Doctor or Baccalaureas* shows this more than adequately. Ulrich Huber, professor in Franecker, whose *Oratio de pedantismo* had been printed by Christian Thomasius in an appendix to his *Hof-Philosophie* as an example of a manifesto for undogmatic thinking, had already ridiculed projections of this kind.[13] For Gundling the idea of Adam as a pedant or a scholastic portrayed an Adam who was purely simply the projection of pedants and was consequently quite absurd.

Although he does not say so explicitly, Gundling's discussions of the things which would not have been necessary to Adam's children had he remained in the state of innocence envisaged by Gundling's opponents are a gloss upon the study *De scholis antediluvianis*, published anonymously by Christian Thomasius in the *Observationes selectae* in 1700. Thomasius's intention had been to distance himself from the ideas of Adamist or antediluvian schools.[14] Gundling's parody of seventeenth-century school books, in which every picture of an object is allotted the appropriate designation, plays on Adam's assumed competence to name every individual thing. It pours scorn upon the view, then common in Lutheran Orthodoxy, that the first schools and academies had been founded directly after the Fall, complete with books and catechisms for the teaching of divine doctrine.[15] Thomasius criticised this view as a pious but irresponsible projection by theologians upon the first human beings; the theologians wished simply to find their own catechism at the beginning of time. Gundling's attack can thus only be understood as innuendo against the Lutheran theologians as a whole, and Buddeus's circle, whom he is really addressing here, is simply drawn into the feud.

The background to this invective is probably to be found in the tense situation in which proponents of natural law such as Gundling and Thomasius found themselves ranged against the theologians on the question of the Fall. Pufendorf, the founder of the school of natural law upon which the professors at Halle built their arguments, had indeed developed a theory about the natural human condition, according to which human beings were originally characterized by a weakness and simplicity which drove them to socialize. This theory served as a hypothesis for the establishment of legal norms. But he had couched this theory in the theological language of prelapsarian humanity and the effects of the Fall. Lutheran orthodoxy saw this theory as competition: here was a doctrine of human *natura corrupta* which not only made no reference to the text of revelation, but which associated it with social relationships and subjective rights in such a way as to undermine the Christian tradition. The resulting disputes were, therefore, thoroughly predictable.[16] This model of the Fall continued to be of great importance in the thought of Thomasius and his followers; its importance was, however, based in the fact that it presented a useful foundation for their understanding of natural law together with an anthropology which found in its thesis of the weakness of the human will a means of psychological moral criticism.[17]

In this theory the Fall is seen as affecting only the human will, not human intelligence.[18] For Gundling and Thomasius, therefore, the model of the Fall retains its full force in the sphere of the will. In the Fall, the will, originally correctly oriented, is transformed into its own opposite so that it acts against what the intellect has already recognized to be correct.[19] In this respect the model of the Fall appeared indispensable, since it made it possible to place the ethical norm and the correct state of mind in this world, even if only as projections back into an ideal state. Ethical perfection was not, of course, included in the accusations of pedantry, and so it seemed possible to restrict the criticism of fantasies of perfection and projections to the cognitive sphere.

For Gundling, however, such criticism was absolutely essential in that sphere. He mustered a whole range of arguments against those of Buddeus and his likeminded colleagues, who saw the effects of the Fall not only as having an adverse affect upon the will but also as a darkening of the intellect. Gundling's arguments include Seneca's reminder, but also the strategy of subsuming the arguments of Buddeus's group into Lutheran orthodoxy's fantasies about antediluvian schools. The arguments and counterideas employed by Gundling are thus of an extremely disparate nature. The convincing strength of his position is, however, to be found in precisely this combination of different arguments. First, the wisdom which must be attributed to the biblical patriarchs is not identical to erudition. As Gundling notes, this had been recognized even by the Socinians.[20] He agrees with Christian Thomasius that Adam was equipped with a *simplicitas sapientis prudentiae*.[21] However, it is interesting that the difference between the prelapsarian and postlapsarian Adam, elsewhere of great importance, has here been smoothed out. Gundling draws a picture of an Adam who possessed a simple wisdom, who knew as much about things as was necessary to him, and who was happy because he attempted to achieve nothing that was out of his reach. The cognitive aspect of this human being could belong as well to the time after the Fall as to the time before. This observation is congruent with the simplification of the image of the Fall implicit in the new understanding of natural law as taught by Pufendorf.

Adam in Paradise must, therefore, be seen as a simple man. After the Fall his intellect is affected only secondarily, in as far as it is injured by the impairment of the will.[22] This is the crucial point of Gundling's criticism of Buddeus. The *habitus* of the intellect, which does not change, is that it learns only with difficulty and on the basis of experience. Gundling indicates an interpretation of the relationship between simplicity and wisdom, between knowledge and the struggle to attain it, which makes possible two ways forward: the road to empirical philosophy and the road to an ethic of self-sufficiency. To begin with the latter: in the text against Buddeus it is striking how often Gundling raises the questions of what things are of use and what things bring happiness and of the relation between these. Adam's wisdom[23] can be understood above all as the ability to attain what is necessary to him.[24] The fulfilment of these needs is what makes him happy. Here it is clear that the pragmatic moral concerns of the early Enlightenment's proponents of natural law with their question *Cui bono?* have undoubtedly produced a counterpart and opponent to the baroque Adam, the polyhistorian who now appears to be only a pedant. "Would you be happier?" is the question posed by Gundling again and again as the criterion for true perfection. The Adam of the early Enlightenment thinkers in Gundling's camp is, therefore, Adam the simple farmer, who is sufficient unto himself.

The deciding argument upon which his position hangs can only be seen from the other road that leads from Gundling's understanding of the first human being. This originates in the thought of John Locke and was taken over and developed by Locke's friend Jean LeClerc. In his *Essay Concerning Human*

*Understanding*, Locke visualizes a (postlapsarian) Adam who, *with a good understanding, but in a strange country*, is capable of naming things.[25] Adam attributes these designations on the basis of his knowledge of the characteristics of different things. However, because he can never know all these characteristics, this process of designation can been seen to be a process which does not reflect the true essence of the things themselves. Adam differs from later human beings only in that he was the first to name things, but theoretically this naming is possible in every era and for every human being. The possibility that Adam had some kind of direct mystical understanding of real essence is thus excluded. This argument was then developed further by Jean LeClerc first in his *Logica*[26] and later, and more importantly, in his commentary on Genesis. Gundling argues similarly that one "would certainly be misled" if one were to believe that "Adam knew the essences of all things." For, "M. le Clerc did not without reason conclude in his Logick that one cannot say that one single Creature can know the substantias rerum / and it is also not necessary to conclude / that the angels must know it."[27] This latter conclusion draws upon Locke's argument that it is impossible to say anything about the angels' capability for knowledge, since, as human beings, we can only ever imagine rational creatures which have been created along the same lines as ourselves.[28] Therefore, Gundling concludes, it is also impossible to speculate about the possibility of perfect knowledge in angels. Just as in the case of his parodies of antediluvian school books, the impulse for this assimilation of passages from Locke and LeClerc comes once again from Gundling's opposition to the speculations of Lutheran orthodoxy, such as Adam's supposedly natural knowledge of the Trinity, the resurrection, salvation and other articles of faith. Although Buddeus had no part in it, this was one way in which religious and political legitimation was sought by Lutheran orthodox theologians. Gundling opposes this trend with a scepticism which can—and indeed must—distance itself from all philosophical speculation about theology. LeClerc's commentary on Genesis offered a model for the way in which the history of the beginnings of humankind could be told while remaining true to Locke's new epistemology. Here was a story of learning, of searching and of experiencing.[29]

LeClerc's interpretation also had the advantage that it could take into account new research into the problem of chronology. It allowed not only for the consequences of the work of John Locke, but also for those of the theories of John Spencer.[30] Spencer was one of the first to make possible the "Copernican revolution" from the model of history as the corruption of a perfect beginning to the model of history as the gradual perfection of a continual accumulation of knowledge. He integrated arguments showing Egyptian civilization to be older than Jewish with the understanding of the Jews' special status as the people of God by understanding the Jews to have been at the outset a simple and ignorant people to whom in the course of history God gave a particular education through their contact with the knowledge of the Egyptians. Underlying this understanding of history is a new doctrine of providence. It may well be referred to as accumulative history, for it assumes, as Gundling put it, "that trees do not suddenly become big; they achieve their height only gradually."[31] Thus the way was open for a view of human beginnings which dispensed with the necessity of perfection.

This approach did, however, have the effect of conclusively excluding from history all the exegetical arguments based upon biblical statement that Adam gave the animals their names, which seemed to support the hypothesis of a perfect knowledge and the beginnings of learning. The historical philological interpretation developed by the highly idiosyncratic theologian Hermann von der Hardt[32] offered itself as a replacement for these arguments, and Gundling gladly seized upon it as offering a solution which could at least be deemed possible. According to von der Hardt's interpretation, the scenes described in Gen. 2.18–20 are not a process of naming but of gathering the animals by calling their names. The passage can thus be ready as a reminder that the ancient Israelites should not seek to communicate with the spirits of the souls of the dead, as was often the custom, but should prefer instead the company of living animals. What was depicted here, then, was a process of gathering animals in order to avoid practicing magic with the dead; thus it was in no sense a cognitive exercise. Gundling amalgamates this argument with the others and presents a new picture. Its strength is in the many different levels upon which it works, drawing as it does upon exegetical, epistemological, pragmatic and critical considerations. This view had considerable influence and is to be found in later discussions. In 1715 and the succeeding years, Jakob Wilhelm Feuerlein developed the critical position of Gundling together with that of Reimmann further, summarizing them in a number of disputations.[33]

After this detour via the understanding of the first man and his skills we are now in a position to recognize the force of Gundling's implicit criticism of Buddeus's *Elementa philosophiae instrumentalis*. In the closing paragraph of the *Reflexion*, Gundling turns directly to those against whom he is writing. Although he does not mention the names of either Buddeus or Lange, he uses the second person singular: "So let it be at last / you are wretched / and you know that you are wretched / how will it help you? Listen to what M. Pascal says," and he cites the words of Pascal in which the greatness of humankind is derived from its recognition of its misery.[34] Gundling continues, "Or if you prefer something else / we still want to remain good friends: For myself I read with plaisir because of its beautiful Latin the book / which Mackenzee has written de Inbecillitate mentis humanae. May it go well with you."

The words have a drastic clarity, brushed with irony: the apparent encouragement, the hand stretched out with the suggestion of reading Mackenzee's book, a text which Lange had printed in an appendix to his *Medicina mentis*, are in reality a ruse, for in reading the book it is not the content that is important for Gundling here, but the style. It is amazing that his ridicule was not taken by Buddeus as exceeding the bounds of friendship. If one can believe the "we still want to remain good friends," then it must be said that the relationship between the colleagues, at least as far as Gundling's tone here is concerned, leaves nothing to be wished in terms of critical openness. It seems to have been possible for Gundling to include his friend among the "cranks" and to ridicule him in this way without ever, so far as we know, causing any serious break in their friendship. Moreover, Gundling continued to cite Buddeus's works with great respect right up until his last lecture. Life in the circle of contributors to the *Observationes selectae* seems to have been seasoned with a good dose of satire. Nevertheless, behind the satire there is certainly a serious point and a clear difference of opinion between the two friends. But if one takes all of Gundling's arguments together, one can see that he is moving in the direction of what Charles Taylor, speaking of the modern period, has called "the affirmation of ordinary life": a recognition of the normal state of humankind as opposed to a supposed state of perfection.[35] Buddeus, on the other hand, and despite the fact that he was open to many of the new perspectives introduced by Thomasius, is a theological thinker who still stands in a tradition which believes itself to be dependent upon drawing a sharp contrast between the ideal and the fallen condition.

## Natural Law, the Fall, and the Philosophy of the Hebrews

The new context within which questions about the Fall and about the natural state of humankind were raised after the time of Pufendorf made it possible to integrate empirical philosophy and the accumulative theory of history into a simplified model of the first human. Thomasius's study of the antediluvian schools had originated in the context of a movement which aimed to replace the theories of the original human state which had been produced by Lutheran orthodoxy and which had been accepted on the basis of *piae fabulae* by a doctrine of paradise and the Fall which had been accommodated to natural law. His aim was to free Protestantism from such "papist relics." Nevertheless the disagreements about the role of the intellect, which, as has already been mentioned, divided early enlightenment thought, were carried over into the discussions which took place within this framework and were real and deep. For the moment Buddeus held fast to his understanding that the Fall brought a darkening of the intellect. At this time his lead was also followed by a not insignificant number of his colleagues, by no means all of whom were Pietists. It was scholars such as these, the proponents of historical scepticism and of the so-called historical pyrrhonism, who saw their *Epoché* in questions of ultimate knowledge as an expression of a *docta ignorantia* and this in turn as Christian humility and a recognition of human weakness.[36]

Historically speaking, any attempt to preserve the Lutheran view of history as the history of corruption required the retention of a view of the development of philosophy which was not very advantageous for philosophy. In the Lutheran view only decline and increasing darkness were possible after the perfect beginning. Convinced Pietists such as Joachim Lange saw philosophy, therefore, as *philomoria*, love of madness. Johann Wilhelm Zierold had earlier placed an even more extreme emphasis upon the deficits of philosophy over and against religion.[37] Gundling wished to dissuade his colleague Buddeus from keeping company with these and other "cranks." Of course the pietists' anti-Aristotelian, or, more specifically, anti-scholastic, point of view suited the early Enlightenment's criticism of the philosophy of the orthodox academies well. The interpretation of syllogistic logic as the work of the devil[38] fitted neatly with the propaganda for a practical and simple piety as opposed to the sophism of the time of the confessional wars. But Buddeus's position is more interesting than the extreme views held by Lange and Zierold. Buddeus combined Thomasian influence, theological tradition and historical reinterpretation into a highly original creation of his own which cannot simply be subsumed into this tendency. It was important to him to use the resources of natural theology against the "unnatural" or, in his view, atheist, tendencies of his era.[39] For this reason philosophy plays an important role in his thought. Indeed, as Gundling once said, even his early theological works were in reality long philosophical digressions.[40] Following Vossius, but also, and perhaps more significantly, More and Cudworth, Buddeus developed an individual portfolio of interests which searched the writings of the ancient Hebrews for cabbalist traces of a natural theology while at the same time seeing European mysticism and the Christian cabbala as continuations of these traditions. The first pole is intended to refer to an original theology, while the second is meant to be based upon the moral theological ideal of an original ethic, namely the Christian *simplicitas*. In his programme Buddeus sets the concept of nature used by natural theology and that used by a natural ethic alongside that of the new natural law.

In his early years, Buddeus's interests in the history of philosophy were shaped by this programme. He consciously sought to consider traditions as a whole; in terms of ethical history, he issued warnings about the morals of the Stoa, for he recognised in Stoic thought the predecessor to Spinozism.[41] He reinstated cabbalist mystics and in 1702 wrote an introduction to Hebrew philosophy, *Introductio ad historiam philosophiae Ebraeorum*, the first example of a reconstruction of Jewish philosophy as a philosophical "sect" as a tradition upon which eclectic philosophy could usefully draw. His *Disseratio de haereisi valentiniana* appeared as an appendix to this work. This took the form of a catalogue of the depraved forms of this thought and thus offered a necessary complement and the completion of his work by enabling the exclusion of the depraved from the choice of useful philosophical forms.

Buddeus's book begins indeed with the question of Adam's wisdom.[42] He dismisses as uncertain speculations about the foundation of whole disciplines in Adam's time, but remains totally convinced by the idea of perfect knowledge. Later his conviction on this point weakened, and it seems that Gundling's criticism did not pass him by without leaving any trace.[43] Buddeus had referred to Locke and LeClerc in his *Institutiones theologiae moralis*, and, in his review of Buddeus's work, Gundling comments that this point at least leaves open the possibility of change: "It is still not certain, whether Adam did not doubt, and had to learn many thousands of things through experience."[44] But in 1702, when the new projects for the history of philosophy were being developed, Buddeus still viewed Jewish philosophy as a consequence of Adam's wisdom. Therefore, his selective writing of history is a direct result of his fundamental convictions about natural theology and moral simplicity. Yet both were impulses towards new reforms and developments. In as far as it proceeded electively, Buddeus's writing of history could be seen as part of the new critical revisions that official church history and history of philosophy had been undergoing since the time of Gottfried Arnold. This is particularly clear in the case of Buddeus's essays in the *Observationes selectae*, which removed from figures such as Guillaume Postel and Francesco Giorgio Veneto their forgotten and outlawed status. On the other hand, Buddeus, like Jacob Thomasius and Bayle before him, subjected theories that he himself rejected or viewed as dangerous to "historicization," in the sense that he portrayed them as originating from movements in which the current errors were already present. Clear examples of this are Hobbesian materialism and Spinoza's monistic atheism. Nevertheless it is possible here to recognize a strategy of avoidance frequently used by Christian apologetic to deal with any problems with which it otherwise could not cope. Such "historicization" is, however, also part of the era's commonly observed tendency to legitimate contemporary aims by drawing upon the past.[45]

In the years around 1700, Buddeus produced a large number of short works and dissertations on the history of moral philosophy. Gundling's *Historia philosophiae naturalis*, which appeared in 1705, was nothing other than an attempt to destroy Buddeus's understanding of history. This work is in many ways both unusual and a new beginning.[46] Here again Gundling follows the pattern which has already been seen in the particular case of Adam's wisdom, by gathering new critical research creatively into a unique whole. Part one of this work, the only part to appear, deals with the philosophy of the Egyptians, the Chaldeans, the Persians, the Arabs, the Indians, the Gauls and the Teutonics, the Hebrews and the Phoenicians. Drawing upon the chronologies of Marsham and Spencer, Gundling does not begin with the Hebrews; instead he prefers to avoid an order based upon chronological priorities or even upon *translationes studii*. Like Buddeus's introduction to Hebrew philosophy, which was meant to be followed by an introduction to other philosophical sects,[47] this "first" part of Gundling's work remained alone. Each seems to have completed only the part that was dearest to his heart. But while Buddeus's book achieves a constructive aim, the first part of Gundling's projected work is only a *pars destruens*. It is destructive in two senses. Firstly, he argues against any attempt to develop an a priori and "geometrical" method for moral philosophy, in the way that, for example, the young Christian Wolff had tried to do not so long before.[48] In conjunction with this, Gundling emphasizes the way in which the theory of the passions introduces opaqueness and "coloratura" which appear in the sphere of the *entia moralia* established by Pufendorf in the humanities.[49] For this reason, his preferred solution is a historically based moral philosophy, operating with hypotheses but still, in Locke's terminology, capable of being kept stable.[50] All this is still absolutely compatible with Buddeus's arguments. However, the book is, secondly, destructive in its historical exposition, which is intended to act as a systematic criterion for the *electio* of the theorems.

For Gundling, real philosophy begins only with the Greeks.[51] What he describes in this work is only philosophy's prehistory, or "barbaric" philosophy. Gundling deals with Hebrew philosophy purely and simply as a part of this "barbaric philosophy"; he does not concern himself with the distinction between and the contradiction of the *historia profana* and the *historia sacra*. He does so on the basis of a hermeneutic, shaped by his understanding of natural law, by which he argues that all human doctrines, whether those of heathen atheists or those of believers, are determined by the coloratura of their passions. Into this Pufendorfian, Hobbesian framework he introduces also Bayle's famous thesis that it is nothing other than prejudice to argue that atheists can only produce bad philosophy and can only lead bad lives. For, argues Gundling in addition, they, like all others, are influenced by the urge for self-preservation and the wish to maintain the status quo.[52]

From this perspective it seemed right to judge the philosophy of the past with taking account of its own context.[53] With a plethora of footnotes reminiscent of Bayle, Gundling prunes back the transmission of different philosophical traditions according to newer historical scepticism. In the light of this contextual process, a history like Buddeus's, which seeks to construct general lines which lead from the beginnings to the present day, can only be seen as a hermeneutic error and an inadmissible mixture. In the case of Buddeus, Gundling is fully aware of his reliance upon More and Cudworth. Buddeus had conquered Sparta anew, so to speak.[54] It may be assumed that Gundling was thinking of his colleague when he accused More of having confused Christian faith and Jewish philosophy: "ut nubem pro Iunone saepe arriperet, Pythagoricum pro Cabala, fidem Christianam pro Philosophia Iudaeorum."[55] Gundling has to reject this way of doing philosophy, for not only does it assume a state of original wisdom, but it draws from a number of different philosophical traditions to create a single mystical, Platonist, cabbalist theory. Moreover, Gundling observes the problems arising from closeness of the Cabbala to Spinoza's thought. Johann Georg Wachter had asserted the relationship between Spinoza's teachings and the Cabbala since the publication of his *Spinozmus im Jüdenthumb*, and Buddeus had already criticized this thesis in the first essay of the *Observationes selectae*.[56] But all Buddeus's efforts to distance himself from Wachter's thought were in vain; Gundling simply said: "Unum adhuc circa opinionem Cabalistarum monendum est, qui mundum hunc per emanationem a Deo conditum esse statuunt; videndumque an Spinozismi rei peragi possint, an minus."[57] He viewed Wachter's thesis as at least tenable, a view which was disastrous for Buddeus's whole conception, since Buddens was more interested in approving than denigrating the Cabbala as it originated in the Old Testament. What makes this argument even more forcible is the fact that Gundling's rejection of the More-Cudworth tradition has a theoretical basis upon which Buddeus, like Gundling, can and paradoxically usually does, call: its basis is the history of philosophy as written by Jakob Thomasius.[58]

## On the Way to a Critical History of Philosophy

At this point I must avoid the temptation of following the increasingly complex developments of the differences between Gundling and Buddeus. The name Jakob Thomasius stands with that of Bayle for the most important source of the concept of the history of philosophy formulated by thinkers of the early German enlightenment. Following Thomasius, they believed that the only adequate Christian understanding of the world was that it had been created by God from nothing; theories which assumed an original uncreated matter were seen as the precursors of gnosticism and of Spinozism and judged to be extremely dangerous. Buddeus had written a treatise *De Spinozismo ante Spinozam* in 1701, in which he sought the antique and medieval roots of the so-called "problem of Spinozism." At first sight Gundling's essays on the atheism of Plato, Parmenides and Hippocrates, written a few years later, continue this tradition.[59] However, he was in fact using the same approach to the same problem in such a way as to seriously undermine Buddeus's work, for Gundling complemented Buddeus's work by considering precisely those authors—but especially Plato— whom Buddeus had acquitted of the charge of atheism. The philological, historical basis for Gundling's interpretation are clear: a Neoplatonic, stoic interpretation of Plato and a Neoplatonic, cabbalist interpretation of Spinoza are so interwoven with each other in his thought that Plato and Spinoza are brought very close to one another.

That this stance intensified Gundling's differences with Buddeus was the result of a double shift in perspective. The assessment of Spinozism as the most immediate danger had two important effects in the thought, first of Bayle, but later and more clearly of Gundling. First, Spinoza's thought was now deemed to spring from the same platonist origins which Jakob Thomasius had identified as the source of the pantheism of the followers of Jacob Böhme. Second, the anti-atheist position developed by Cudworth against Hobbes became irrelevant to the discussion; indeed the implied relationship between Platonism and Spinozism meant that it had to be discarded. All this took place against the background of the corruption history of Christianity in which Platonism was seen as the mainstay of Christianity's dogmatic errors. Therefore, the controversy between Gundling and Buddeus (and later the Zurich theologian Johann Jakob Zimmermann) is roughly equivalent to that between Bayle and LeClerc: Bayle's anti-Spinozism sought to distinguish itself from LeClerc's defence of Cudworth and of his platonizing natural theology against the Hobbesians.

Gundling seemed at first to be continuing the line of Buddeus's *De Spinozimo ante Spinozam*, but the results make it clear that what he was actually doing was quite the opposite. He presents the Platonism of the Church Fathers as so closely integrated into the platonic tradition and to Plato himself that it is impossible to draw any further on Platonism. Buddeus's intended route forward towards a natural theology is thus cut off. These are the results. The motives for Gundling's work and the direction of his sometimes radical interpretations are more difficult to identify. Nevertheless, the closeness of his work to Souverain's *Platonisme devoilé* allow us to suspect that, going beyond the differences in interpretation, the moving force leading Gundling to his philosophical convictions could have been a scepticism about theological questions which was not far removed from an anti-trinitarianism. His work *Hobbes ab Atheismo liberatus* demonstrates that Gundling's background, like Bayle's, may be taken to have been a sceptical fideism.[60] Here Hobbes's comments about the corporeality of God are removed from the context of his polemic against incorporeal spirits and interpreted in the framework of a sceptical reserve about theological discussion. From this it can be seen that Gundling cannot be counted a member of the group of "heretic hunters" (*Ketzermacher*), but that he is concerned to protect from their theological critics authors such as Hobbes, who had previously been outlawed despite his importance for contemporary understandings of natural law. Gundling's discourse still contains strategic accusations of atheism, but these are reserved exclusively for antique authors "whom the present . . . charge no longer . . . harms."[61]

Let us draw all this to a conclusion. For the early Enlightenment in Germany it was necessary for scholars to work through the issues of the late seventeenth century once again before they could come to any sort of final conclusions about the way in which the history of philosophy was to be written or how this history was to be "disciplined." Gundling was instrumental in causing the reappearance in Germany of the debate between Hobbes and the Cambridge Platonists, between Bayle and LeClerc and others; once there it was integrated into the specific context of the discussion of natural law. As Jakob Friedrich Reimmann has said so well, Gundling was able to say clearly and to the point many a thing *quod alii mussitando indicarunt*, that is, that others only dared to mention quietly if at all.[62] Although Jakob Brucker was a student of Buddeus and not of Gundling, this did not stop him from drawing his own conclusions from the different positions held by Buddeus and Gundling in this debate. Moreover, Brucker was influenced by Christoph August Hernnmann who had devloped and clarified some of Gundling's ideas. In conclusion, therefore, I wish to try to identify the typical characteristics of the differences which have become clear in this study.

In order to prevent us from being misled by the language on the surface of the discussion, it seems appropriate to differentiate between four levels of the controversy. These levels only become clear from an analysis of the pragmatic content of the argument such as that which has been undertaken here.

1. The debates begin in the language of "accusation," or of its opposite, "rehabilitation." This is a result of the growth of the history of philosophy out

of Christian apologetic. The "rescue" of past thinkers who had been unjustly condemned was very important in Germany of the eighteenth century.[63] However, the controversies show clearly that opposition to this did not necessarily arise from conflict between orthodox heretic hunters on the one hand and liberal or radical Enlightenment thinkers on the other. It was often more the case that there were disagreements *within* the different groups of the enlightenment. Gundling slid into the role of the accuser of atheism, albeit with a certain amount of calculation. It was, however, possible to functionalize the language of the defender and of "rescue" above all other. The emphases of this language, popular in Protestant circles, could serve those who wished to find in the different forms support for their own direction; for no thinker in the early Enlightenment wanted to have anything to do with the orthodox heretic hunters.

2. The controversy centers on the interpretations of the thought of a number of different philosophers, but the choice of these philosophers seems also to belong to a superficial level of the debate. They are the pretext for a particular debate, rather than being an interest per se. Plato is taken to stand for questions of the Trinity, Hippocrates for the relationship between Christianity and natural science. Henry More can be drawn into the debate in order to ally him with Hobbes. The pretext is, therefore, not simply coincidental, for these positions can be recognized as a kind of shorthand which could, on the whole, be deciphered immediately by contemporaries. History was simply the favored field upon which the issues of the age were decided.

3. Behind the pretexts are concealed the real dangers which were seen as existing in contemporary problems such as, for example, Spinozism, scientific materialism, the loss of a solid foundation for Christian dogma, the potential aggression of confessional polemic, or the structures of authority in an established church. But the pretexts disguise also the possibilities which were seen in the development of a simpler Christianity, a *Gelehrtenrepublik* free from prejudice, or of a secular civil society.

4. The motives upon which these opinions are constructed do not appear explicitly in the discussion, but must be deduced from it. Nevertheless it is possible to find confirmation for certain suspicions in the search for coherence in these works. For instance, it is possible to argue convincingly that Gundling's thought centered upon his use of sceptical arguments through which he ensured a conclusive distinction between religious claims and those based upon the law of reason. The Hallensians' juridical propensities for a civil society were not only the product of their open antagonism towards ecclesiastical orthodoxy, but also of a latent internal distrust of reform theologians, whose conception of natural religion was the perfect vehicle for the restriction of the lawyers' impulses towards secularization. On the other hand, one of Buddeus's primary motives seems to have been the necessity of giving religion a rational, demonstrable basis. He saw any process of secularization that abandoned the fundamental Protestant traditions as disintegrating into ethical scepticism, the destruction of moral values and loss of all tradition.

Since these different levels may stand in very different relationships to one another, it is difficult, indeed virtually impossible, to speak here in terms of "radical," "conservative," or "modern." An appreciation of the differences between the Newtonian enlightenment and the radical enlightenment, often applied in the last few years to the situation of the early Enlightenment in England, may give an indication of what might be involved here. The "conservative enlightenment"[64] in the sense of the established Christian Newtonians, for whom science offered a support for religion and sociability a synonym for commerce, may be taken to include the north German physicotheologians and the "patriots" in the circle around Johann Albert Fabricius, and later also a broad section of popular philosophical movements. This description does not, however, apply to the situation in Halle. The "radical enlightenment," made up of politically critical, philosophically materialist or deist thinkers, did exist in Germany, but only as a marginal group made up of a few persecuted spirits and what was probably a larger number of extremist students. The positions which have been discussed here lie in the spectrum between these extremes. Both Buddeus and Gundling are part of movement for reform which finds a consensus in a certain degree of moderacy. The only disagreement is over the direction this reform should take. Their philosophy is certainly anti-scholastic, but this does not mean that Gundling cannot, for instance, employ scholastic, Aristotelian distinctions to serve his anti-Platonism. In the sensualist epistemology there is also a fluid transition between Aristotle and Locke. In the other corner, where Buddeus is to be found, it is possible to find coalitions of pietist theologians with positions sceptical of the power of the human intellect; here too the Platonist tradition is played off against the Aristotelian while at the same time Platonism is rejected as the ruin of Christianity. These fluctuating coalitions make it difficult to bring the picture into focus.

It might be argued that Gundling is in fact taking the "more radical" position through his defence of Hobbes, of freedom of thought and of Bayle's scepticism. But elements of the "conservative enlightenment" can also be found in his thought in that the coexistence of church and philosophy are not affected by the latent fideism of his position, or in as far as his anti-Platonism is at one with the conservatives' warning about the "enthusiasts," that is, the fanatical Platonists. In contrast, Buddeus is prepared to support some of the "modernizing" tendencies against the orthodox theologians, such as, for instance, his attempts to introduce the theorems of Christian Thomasius or Cudworth's natural theology of the Trinity. What seems to be constant is the fundamental decision as to whether the modern way is to be sought in the unification or in the separation of religion and reason. Whether the first position, that of Buddeus, then takes the more pietist direction of Poiret and Lange or a more rationalist road with LeClerc, and whether the second then tends more to anti-religious scepticism or to fideism is only a secondary consideration, one which is not an explicit theme of the discourse.

It is not sufficient to respond to the complexity of this situation with the comment that we are dealing with a period of transition. The epoch prior to the influential appearance of Wolff has its own signature and this must be read. Moreover, it has now been recognized that decisions with great importance for the enlightenment did not only begin to be made later, but were already being made in the time around 1700, the era of the *crise de la conscience européenne*.[65] One thing at least is clear: the German intellectuals were an integral part of the tapestry of international discussion. The German enlightenment did not go its own special way; it stood in the middle of the European debates of its time. But it was precisely the German enlightenment's unique appropriation of this wider discussion, and the consequent emergence in Germany of a particular approach to history, that made it possible for the Germans to lead the way in the writing of the history of philosophy. Many elements of this process are still unclear. Too little is known of the further effects of Buddeus's influential teaching in Jena, of the reception of LeClerc in Germany, of the coalitions between different trends in pietist historicizing, such as that of Gottfried Arnold and that of the followers of Christian Thomasius, of the mutual influences between the interpretation of the antique world or of the early church and early modern thought. It can however be concluded that the contours of this picture were fixed by the 1720s or 1730s, and that these were extremely influential well into the eighteenth century. Johan Lorenz Mosheim's treatise *De turbata per recentiores Platonicos ecclesia*, published in 1725, is the first attempt at a full picture of the interactions between Neoplatonism and early Christianity.[66] Although his position is close to that of Buddeus, Mosheim's "latitudinarian" theology can cope better with anti-Platonism.[67] In his work, Mosheim attempts to draw a differentiated historical picture which avoids the onesidedness to be found in the works of both Souverain and Baltus. Despite his efforts, the result has a clearly anti-Platonic tenor. Nevertheless, his approach was that adopted by Jacob Brucker in his *Historia critica Philosophiae*.[68] Reimmann's *Historia universalis atheismi et atheorum falso et merito suspectorum*,[69] published, like Mosheim's work, in 1725, offers what is effectively a summary of the debates about atheism and about the innocence of all the philosophers whose legitimacy had earlier been discussed with such passion. Mosheim's critical comments on Cudworth's intentions of establishing a *prisca sapientia*, found in his translation of and commentary to Cudworth's works published in 1733, are also symptomatic of the end of the phase of strict opposition. The need of assimilation had now been overtaken: the Platonic text and the anti-Platonic criticism have been united into a single whole.[70]

In the 1720s and 1730s the time seemed ripe to leave the pioneering work and the fights over direction found in a Gundling or a Buddeus and to go on to the search for a full picture. In this picture, the growth of the history of philosophy out of the apologetic discussions about the atheists of past and present, the "disciplining" of these discussions in the spirit of historical pyrrhonism and of the new natural law and their distancing of themselves from the myths of human origins had resulted in a first conclusion in which the interests of both

Gundling and Buddeus had been integrated. It would be wrong to ask whether it was Gundling or Buddeus who had passed the *Examen rigorosum . . . im Todtenreich*. It is much more the case that their productive arguments had sown the seeds from which the history of philosophy bloomed.

## Notes

1. This work is mentioned by Gundling's biographer, C. F. Hempel, in Gundling's *Vollständige Geschichte der Gelahrtheit* (Frankfurt und Leipzig 1736), vol. 5: *Umständliches Leben und Schrifften, Collegia, Studia, Inventa und eigene Meinungen*, 7006.

2. See L. Lindenberg, *Leben und Schriften David Faßmanns (1683–1744) mit besonderer Berÿcksichtigung seiner Totengespräche* (Berlin, 1937). Gundling himself had used this form for a dialogue between Montaigne and Archimedes: *Gundlingiana* 2 (Halle, 1715): 146–58. In 1728, Gottsched translated Fontenelle's *Nouveaux Dialogues*. In 1729 a *Besonders curieuses Gespräch In dem Reiche derer Todten, Zwischen Zweyen im Reiche der Lebendigen Hoch-Berÿhmten Männern / Christian Thomasio, . . . und August Hermann Francken* appeared in two parts. Its anonymous author may also have been the author of the *Examen rigorosum*. Other dialogues of the dead published at about the same time featured, for example, Buddeus and Leibniz discussing the question of theodicy. This seems to indicate the development of a text form in which conflicts and tensions which were latent or not explicitly adressed in the early Enlightenment could be featured as central themes, albeit at the level of the pamphlet. This aspect of such texts still requires investigation. Unfortunately I have not been able to locate a copy of the *Examen rigorosum*.

3. For Gundling (1671–1729), see the biography by Hempel (cf. note 1), and also: *Allgemeine Deutsche Biographie* (ADB) vol. 10 (1879), 129f.; *Neue Deutsche Biographie* (NDB) vol. 7 (Berlin, 1966), 318f.; N. Hammerstein: *Jus und Historie. Ein Beitrag zur Geschichte des historischen Denkens an deutschen Universitäten im späten 17. und im 18. Jahrhundert*, (Göttingen, 1972), 205–65; H. Rüping: *Die Naturrechtslehre des Christian Thomasius und ihre Fortbildung in der Thomasius-Schule* (Bonn 1968), passim; H. Klenner, "Eine fast vergessene Quelle deutscher Menschenrechts- und Rechtsideen: Nikolaus Hieronymus Gundling," *Dialektik* (1994/1): 123–30.

4. For Buddeus (1667–1729), see ADB, NDB; A.F. Stolzenburg, *Die Theologie des Joh. Franz Buddeus und des Chr. Matth. Pfaff. Ein Beitrag zur Geschichte der Aufklärung in Deutschland* (Berlin 1926, reprint Aalen, 1979), M. Wundt, *Die deutsche Schulphilosophie im Zeitalter der Aufklärung* (Tübingen, 1975, reprint Hildesheim 1992), 63–75; S. Masi: "Eclectismo e storia della filosofia in Johann Franz Buddeus," *Memoria della Academia delle Scienze di Torino II—Classe di Scienze Morali, Storiche e Filologiche*, Series 5, vol. 1 (Turin, 1977): 163–212; P. Zambelli, *La formazione filosofica di Antonio Genovesi* (Naples 1972), 366–417; W. Sparn, "Vernünftiges Christentum. Über die geschichtliche Aufgabe der theologischen Aufklärung im 18. Jahrhundert in Deutschland," in R. Vierhaus, ed., *Wissenschaften im Zeitalter der Aufklärung* (Göttingen, 1985), 18–57, esp. 26f.; F. Nüssel: *Bund und Versöhnung. Zur Begründung der Dogmatik bei Johann Franz Buddeus* (Göttingen, 1996).

5. Nevertheless, this is one of the first indications of the tension between Gundling and Buddeus: Buddeus explicitly employed eclecticism and published a three volume *Philosophia eclectica* between 1697 and 1703, while, although his philosophizing is not less eclectic in its practice, Gundling's work is notable for the absence of the term "ecclecticism." Compare M. Albrecht's comprehensive study: *Eklektik. Eine Begriffsgeschichte mit Hinweisen auf die Philosophie- und Wissenschaftsgeschichte* (Stuttgart, 1994), esp. 576ff.

6. Compare L. Braun, *Histoire de l'histoire de la philosophie* (Paris, 1974). To date the best introduction to research into the historiography of philosophy in early Enlightenment Germany is offered by M. Longo, "La storia della filosofia tra eclettismo e pietismo," in G. Santinello, ed., *Storia delle storie generali della filosofia*, vol. 2: *Dall' età cartesiana a Brucker* (Brescia, 1979), 329ff.

7. For the exemplary character of the works of Naudé in this trend, see M. Mulsow, "Appunti

sulla fortuna di Gabriel Naudé nella Germania del primo illuminismo," *Studi filosofici* 14–15 (1991-1992): 145–56.

8. See R. Chartier, *Cultural History: Between Practices and Representations*, trans. Lydia G. Cochrane (Cambridge, 1988), introduction.

9. Seneca, *De beneficiis*, II, 29, *Philosophische Schriften*, vol. 5, ed. M. Rosenbach, 199ff.

10. N.H. Gundling, "Reflexion über den Locum Senecae De benefic. Lib. II. Cap. XXIX", in ders., *Otia*, 1. Teil (Halle 1705), 158–216.

11. See J.F. Buddeus, *Elementa philosophiae instrumentalis* (Halle, 1703), ch. 2 ("De intellectus humani vitiis et imbecillitatibus"), 121.

12. "Reflexion," 202.

13. U. Huber, *Oratio de paedantismo* (1678), in C. Thomasius, *Introductio ad philosophiam aulicam*, (Leipzig 1688).

14. *Observationes selectae* Bd. I (Halle, 1700), Obs. 19. To the attribution, see C.H. Starcke, *Ad V. Cl. Vincentii Placcii Theatrum Anonymorum epimetron, Observationum Hallensium Latinarum Auctores quosdam detectos exhibens* (Lübeck, 1716); see also Gundling's comments in the *Historia philosophiae moralis* (Halle, 1706), 61. Pierre Bayle also argued against the fables of academic disciplines and books in the time of Adam: *Dictionnaire historique et critique* (Rotterdam, 1697, ²1701), s. v. "Adam" rem. D und K. In the German translation, Gottsched, with the complex German discussions in mind, accuses Bayle of not making enough out of this debate. To the question of prehistoric learning see the *Habilitationsschrift* by Helmut Zedelmaier, to appear shortly. I have benefited from many discussions with Helmut Zedelmaier on this topic.

15. Polycarp Lyser, *De trifolio verae religionis veteris testamenti Adamiticae, Abrahamiticae et Israeliticae iuxta unifolium religionis Lutheranae disputatio* (Wittenberg, 1664); J. Gisenius, *De vita Adami*, to give only two examples.

16. For this development, see H. Denzer, *Moralphilosophie und Naturrecht bei Samuel Pufendorf* (Munich, 1972); H. Medick, *Naturzustand und Naturgeschichte der bürgerlichen Gesellschaft* (Göttingen, 1981); F. Vollhardt: *Selbstliebe und Gesellgkeit. Untersuchungen zum Verhältnis von naturrechtlichem Denken und "schöner Literatur" im 18. Jahrhundert* (forthcoming).

17. Although this thesis already stood in the context of the new natural law it employed the theological formulated concept of the Fall. Thus it resists the clear division between 'pythagorean' and 'modern' natural law drawn by J.B. Schneewind (see the article by Schneewind in this volume). Its widespread acceptance also meant that the appearance of the concept of nature as the positive norm for the "natural," and with it the loss of the implication that the natural was somehow *per se* corrupt, were delayed by several decades.

18. Gundling, "Reflexion," 204: "Mir kommet es zum wenigsten so ungereimet nicht für / der ich die verlohrne Perfection in dem Willen des ersten Menschen setze / welcher heilig und unbefleckt gewesen; nicht aber in dem Verstand; der noch biß auf den heutigen Tag ist / wie der Verstand des Adams."

19. See C. Thomasius, *Ausübung der Sitten-Lehre* (Halle, 1696).

20. Gundling, "Reflexion," 205: "Die Socinianer haben ihn [Adam] einfältig, wie die alte Teutsche Tacitus beschrieben, die aus Dummheit nichts böses thun können."

21. C. Thomasius, "De scholis antediluvianis," 291. For the importance of the difference between *sapientia* and *philosophia* for the history of philosophy, see C. Blackwell, "The Definition of Philosophy and Science and the End of *Prisca Theologia* and *Prisca Sapientia*. Thomasius, Conring and Le Clerc and Gundling and Brucker" (in print).

22. Compare C. Thomasius, *Ausübung der Sitten-Lehre*, 87: "Denn ob wir wohl nicht läugnen / daß der Verstand / wenn er einmahl von dem Willen verderbet worden / nicht den Willen wiederumb antreiben / und weiter hinein führen solle; So haben wir doch allbereit oben dargethan / daß der Verstand ursprünglich von dem Willen verderbet werde / und daß der Wille ja so wohl seine eigene Vorurtheile habe als der Verstand des Menschen."

23. To the idea of Adam's wisdom, see also R. Häfner, "Die Weisheit des Ursprungs. Zur Überlieferung des Wissens in Herders Geschichtsphilosophie," in W. Malsch und W. Koepke, eds., *Herder Jahrbuch / Herder Yearbook 1994* (Stuttgart, 1994), 77–101. Häfner separates the strands of

tradition which assumed the *sagacitas hominum* (Seneca), humankind's practical skills, to be the basis of cultural development from those which see it as founded upon an original sapientia. This division corresponds well to the "Copernican revolution" from a history of corruption to the cumulative understanding of history indicated here (see below for further discussion of this point). It thus also becomes clear that it is no coincidence that Gundling draws upon Seneca's thought.

24. See also J.F. Reimmann, just a few years after Gundling, *Einleitung in die historiam literariam antediluvianam*, (Halle, 1709), 4: "... daβ die Logica der ersten Eltern, die sie von Gott empfingen, nicht in der künstlichen Ausgrübelung und Beschaulichkeit vieler unnützer Wahrheiten, sondern in dem vielfältigen Gebrauch und lebendigen Anwendung und thätigen Ausdruck einiger wenigen und nutzbaren Wahrheiten bestanden...."

25. J. Locke, *An Essay concerning Human Understanding*, III, ch. 6, sect. 44–51.

26. J. LeClerc, *Logica sive ars ratiocinandi* (London, 1692), 16f.: "Substantiarum singulorum ideae obscurissimae sunt, nec quicquam earum nominibus intelligimus, nisi subjecta nescio quae ignota, in quibus quaedam constanter coexistunt proprietates.... Idem judicium ferendum de reliquis substantiis, spiritualibus nempe, (an aliae plures sint hic non quaerimus) videbit quicunque earum ideas diligenter expendet, nec se sinet inanibus vocibus falli." For this argument see also Gundling's own Logic, *Via ad Veritatem* (Halle, 1713).

27. Gundling, "Reflexion," 203.

28. J. Locke, *Essay*, II, ch. 23, sect. 13: "... but how extravagant soever it be, I doubt whether we can imagine any thing about the knowledge of angels but after this manner, some way or other, in proportion to what we find and observe in ourselves."

29. J. LeClerc, *Genesis sive Mosis prophetae liber primus* (Amsterdam, 1693). Compare also the paraphrase with commentary in J.W. Feuerlein (Praes.) / V.H. Regenfus (Resp.), *Dissertatio philosophica de philosophia Adami putatitia* (Altdorf, 1715), 36: "Denique post lapsum per longissimam experientiam Protoplasto seminum, stirpium, radicum, herbarum, aliorumque rerum naturalium Phaenomena innumera innotuisse, eodemque quorundam effectuum causas ratiocinando indigasse, atque imprimis hanc notitiam ad praxin et felicitatem tam supremam quam subordinatam bene applicasse, ambabus largior manibus, integrum autem systema hypothesium hinc inferri posse ob rationes § 5. et 6. propositas nego atque pernego." For LeClerc's biblical exegesis, see H. Graf Reventlow, "Bibelexegese als Aufklärung. Die Bibel im Denken des Johannes Clericus (1657–1736)," in H. Graf Reventlow, W. Sparn, and J. Woodbridge, eds., *Historische Kritik und biblischer Kanon in der deutschen Aufklärung* (Wiesbaden, 1988), 1–20.

30. See J. Spencer, *De Legibus Hebraeorum Earumque Rationibus* (London, 1685); and the study by J. Gascoigne, "'The Wisdom of the Egyptians' and the Secularisation of History in the Age of Newton," in S. Gaukroger, ed., *The Uses of Antiquity. The Scientific Revolution and the Classical Tradition* (Dordrecht, 1991), 171–212. Spencer's ideas appear in the work of both Thomasius (see "De scholis antediluvianis," 301) and Gundling (*Historia philosophiae moralis*, 7, among others). Spencer's new formulation is closely connected to the new understanding of providence which appeared in the era of the scientific revolution. Here providence came to be seen as acting through secondary causes rather than through direct intervention, so that it did not contradict the laws of nature. History with all its byways then comes to be interpreted as the continuous formation of the people of God.

31. Gundling (in "Reflexion," 203) argues thus of the necessity of human development.

32. Hermann von der Hardt, *Epistola ad Paulum Martinum Noltenium ... de vocatis ab Adamo animalibus* (s. 1., 1705) (and compare Gundling: *Historia philosophiae moralis*, 61f.). The Helmstedt theologians were quick to condemn this view as heretical.

33. J.W. Feuerlein (Praes.) / V.H. Regenfus (Resp.), *Dissertatio philosophica de philosophia Adami putatitia* (Cf. note 29); J.W. Feuerlein (Praes.) / V.H. Regenfus (Resp.), *De Adami Logica, Metaphysica, Mathesi, Philosophia Practica, et Libris* (Altdorf, 1717). A number of Gundling's arguments and references are to be found in these works: for instance, in the first disputation his references to Seneca (16), to Hermann von der Hardt (25), and to LeClerc's commentary on Genesis (31ff.) together with his argument that names are inadequate as designations of ideas. It is interesting that Feuerlein mentions Reimmann with his *Einleitung in die historiam literariam antediluvianam*, but

does not mention the name of either Gundling or of Locke. Probably it was necessary in Altdorf to preserve a distance from Halle.

34. B. Pascal: *Pensées sur la réligion* (Amsterdam, 1688), 113ff.: "En un mot l'homme connoit, qu'il est miserable, il est donc miserable, puis qu'il le connoit. Mais il est bien grand, puis qu'il connoit, qu'il est miserable. S'il se vante, je l'abaisse: si s'abaisse je le vante...."

35. C. Taylor, *Sources of the Self. The Making of the Modern Identity* (Cambridge, 1989), Part 3, esp. 234ff.

36. See M. Mulsow, "Eclecticism or Skepticism? A Problem of the Early Enlightenment," to appear in: *Journal of the History of Ideas* (1997).

37. J. Lange, *Medicina mentis* (Halle, 1704); J.W. Zierold, *Einleitung zur gründlichen Kirchen-Historie mit der Historia philosophica verknüpfft, darinnen die Krafft des Creutzes Christi als der einige Grund des wahren Christenthumbs wider die Feinde des Creutzes, von Anfang der Welt biss auff unsere Zeit* (Leipzig and Stargard, 1700); see Longo, "La storia della filosofia," 350ff. For Lange, see B. Bianco, "Libertà e fatalismo. Sulla polemica tra Joachim Lange e Christian Wolff," *Verifiche* 15 (1986): 43–89.

38. See J.F. Reimmann, *Critisierender Geschichts-Kalender von der Logica* (Frankfurt, 1699); and originally, *Suidae Lexicon*, s. v. "Adam," ed. A. Adler, Teil I, (Leipzig, 1928), 43–46.

39. H.M. Barth, *Atheismus und Orthodoxie. Analysen und Modelle christlicher Apologetik im 17. Jahrhundert* (Göttingen, 1971).

40. Thus argues Gundling in *Der Philosophischen Discourse anderer und dritter . . . Theil* (Frankfurt and Leipzig, 1740), 702, about Buddeus's *Theologia moralis* of 1711: "Herr Buddeus hat auch in allen Capiteln seiner Theol. Moral zeigen wollen, was die revelation vor der Philosophia Morali voraus habe. Allein er excedieret darinnen dermasen, daβ seine Theologia moralis zur Philosophia morali wird, welche er nur mit Sprüchen aus der Schrift ausgezieret hat."

41. Compare [J.F. Buddeus], "De conciliatione Philosophorum inter se," in *Observationes selectae*, Bd. 3 (Halle, 1701), Obs. XIV, 258–80 (for the attribution, see once again Starcke, *Ad V. Cl. Vincentii Placcii Theatrum*, for Buddeus's program of attempting a view of the *historia integra* before one starts choosing from it an ecclectic. For his criticism of the Stoa, see Buddeus's "Exercitatio historico-philosophica . . . de erroribus stoicorum in philosophia morali," which had provided material for four series of disputations for Buddeus's students in Halle in the 1690s and which he republished in his *Analecta historiae philosophiae* (Halle, 1706), 97–203. For the influence of J. Thomasius and the problem of the difference between God and world, see this work, 169ff.

42. Buddeus, *Introductio ad historiam philosophiae Ebraeorum* (Halle, 1702), 1f.: "Ebraeorum philosophiam ab ipso Adamo derivare nullus dubito."

43. He could not hide this from the attentive C.A. Heumann. See Heumann's *Acta philosophorum*, 5. Stück (Halle, 1716), 755–809 and 7. Stück (Halle, 1716), 1–58, "Von der Philosophie der Patriarchen."

44. [Gundling], "J.F. Buddei . . . Institutiones . . . ," in *Neue Bibliothec*, 14. part, 275–307 (cf. p.288). Gundling, who wrote the review anonymously, does not mention LeClerc's name but speaks only of a "learned man."

45. See W. Sparn, "Formalis atheus? Die Krise der protestantischen Orthodoxie, gespiegelt in ihrer Auseinandersetzung mit Spinoza," in K. Gryˉnder and W. Schmidt-Biggemann, eds., *Spinoza in der Frühzeit seiner religiösen Wirkung* Heidelberg, 1984, 27–63.

46. This judgment was supported by Gundling's contemporaries. See, for instance, C.A. Heumann, *Acta philosophorum*, vol. I, part 6 (Halle, 1716), 1032–39, for his review of this book. It was also Heumann who was most decided in his use and development of Gundling's "disciplining" of the history of philosophy from the spirit of historical pyrrhonism. See his *Acta philosophorum*, part 3 (Halle, 1715), 179–236, "Von den Kennzeichen der falschen und unechten Philosophie" and his studies "Von der Philosophie der Patriarchen."

47. Buddeus did, however, publish an *Introductio ad philosophiam stoicam* in 1729.

48. C. Wolff, *Disputatio de Philosophia practica universali, methodo mathematica conscriptum*, Magisterpromotion (Leipzig, 1703).

49. *Historia philosophiae moralis*, Prooemium, §9.

50. Ibid., §9–13. Gundling refers to passages from Locke's *Essay*, Book 4, ch. 3, sect. 3–20, which deals with the limited nature of human knowledge, for which, however, relative moral demonstrations such as "Where there is no property, there is no injustice" are still possible.

51. A side effect of this decision to see true—not barbarian—philosophy as beginning only with the Greeks was that certain followers indentified their resentment of dominant French culture, common in Germany, with antique "barbaric" thought and with a projected nationalism began to believe Greek philosophy to be superior to that of the barbarians in just the same way as they held German scholars to be superior to French. The thought of Gabriel Wagner, a student of both Thomasius and Gundling, is just such an early example of what has been called by Martin Bernal the "aryan model" of antiquity (M. Bernal, *Black Athena*, vol. 1 [London, 1987]). See Wagner's works written under the pseudonym "Realis de Vienna," which will soon appear as volume 3 of the series *Philosophische Clandestina der deutschen Aufklärung*, ed. S. Wollgast. Gundling himself does not seem to have drawn such conclusions.

52. Gundling, *Historia philosophiae moralis*, Praefatio, 5–6.

53. Even before Gundling, Jean LeClerc's brother, Daniel LeClerc, had written a history of human knowledge which attempted to take into account the new gnoseology of John Locke and Jean LeClerc and which did not assume an original wisdom: *Histoire de la médicine* (Amsterdam, 1702). For D. LeClerc's view of Adam, see ch. 4, p. 7: " . . . il n'y a pas apparence que le premier [homme] de tous ait eu assez d'occasions pour pousser bien loin la Médicine, ou pour la réduire en Art." Gundling knew and used this work.

54. Gundling, *Historia philosophiae moralis*, 93.

55. Ibid., 96.

56. See J.G. Wachter, *De Spinozismus im Jüdenthumb* (Amsterdam, 1699); then, with the opposite omens (pro Spinoza), *Elucidarius cabalisticus*, (Rome, 1706). (Both texts have been reprinted with an introduction by W. Schröder in the series *Freidenker der Europäischen Aufklärung*, vol. 1,1 and 1,2 (Stuttgart, 1994 and 1995); [J.F. Buddeus], "Defensio Cabbalae Ebraeorum contra auctores quosdam modernos," in *Observationes selectae*, vol. 1, (Halle, 1700), Obs. I.

57. Gundling, *Historia philosophiae moralis*, 97.

58. For J. Thomasius (1622–1684), see G. Aceti, "Jakob Thomasius e il pensiero filosofico-giuridico di Goffredo Guglielmo Leibniz," in *Jus* N.S. 8 (1957): 259–318; U.G. Leinsle, *Reformversuche protestantischer Metaphysik im Zeitalter des Rationalismus* (Augsburg, 1988), 146–49; P. Petersen, *Geschichte der aristotelischen Philosophie im protestantischen Deutschland* (Leipzig, 1921), reprint Stuttgart, 1964); G. Santinello: "Jakob Thomasius (1622–1684): *Schediasma historicum*," in G. Santinello, ed., *Models of the History of Philosophy*, Vol. I: *From its Origins in the Renaissance to the "Historia Philosophica"* (Dordrecht, 1993), 409–42. R. Häfner, "Jacob Thomasius und die Geschichte der Häresien," in F. Vollhardt, ed., *Christian Thomasius*, (forthcoming).

59. Gundling, "Hippokrates atheos," in his *Otia* (Halle, 1706), 82ff.; "Gedancken über Parmenidis Philosophie," in Gundling, *Gundlingiana* 14 (Halle, 1717), 372ff.; "Plato atheos," in Gundling, *Neue Bibliothec*, 31. Stück (Halle, 1713), 1–31; "Von Platonis Atheisterey," in *Gundlingiana* 32 (Halle, 1724), 103–46; "Velitatio prior de atheismo Platonis," in *Gundlingiana* 43 (Halle, 1729), 187–280; "Velitatio posterior de atheismo Platonis," in *Gundlingiana* 44 (Halle, 1729), 281–360.

60. Gundling "Hobbes ab Atheismo liberatus" (1706), available to me in Gundling, *Observationes selectae* (Halle, ²1737), Bd. 1, 37–77; "Von Thomas Hobbesii Atheistery," in *Gundlingiana* 14 (Halle, 1717), 303–39.

61. Gundling, "Erste Reflexion über Herrn D. Trillers Hippocratum atheismi accusatum," in *Gundlingiana* 22 (Halle, 1719), 87–186, cf. 97.

62. For the debate about Plato's atheism, see Reimmann, *Historia universalis atheismi et atheorum falso et merito suspectorum* (Hildesheim, 1725, reprint Stuttgart-Bad Canstatt, 1992 [ed. W. Schröder]), 161.

63. For an initial introduction to this question see M. Mulsow, "Appunti" (Cf. note 7).

64. J.G.A. Pocock, *Edward Gibbon in History: The Tanner Lectures on Human Value* (Cambridge, 1990), 291–364, particularly 338ff.; and compare his "Post Puritan England and the Problem of the Enlightenment" in *Culture and Politics*, ed. P. Zagorin (Los Angeles, 1987). Pocock

follows the characterization of the "Newtonian enlightenment" offered by J.R. Jacobs and M. Jacobs. See J.R. Jacobs: *Robert Boyle and the English Revolution* (New York, 1977); M.C. Jacobs, *The Newtonians and the English Revolution* (Ithaca, NY, 1978); M.C. Jacobs, *The Radical Enlightenment. Pantheists, Freemasons, and Republicans* (London, 1981).

65. See P. Hazard, *La crise de la conscience européenne 1680–1715* (Paris, 1935).

66. J.L. Mosheim, *De turbata per recentiores Platonicos ecclesia*, diss. (Helmstedt, 1725); extended version in *Dissertationes ad historiam ecclesiasticam pertinentes*, vol. 1, 85–216.

67. For Mosheim, see M. Mulsow, R. Häfner, F. Neumann, and H. Zedelmaier, eds., *Johann Lorenz Mosheim 1693–1755. Theologie im Spannungsfeld von Philologie, Historie und Philosophie* (forthcoming).

68. J. Brucker, *Historia critica philosophiae* (Leipzig, 1742–44).

69. Reimmann, *Historia universalis atheismi* (cf. note 62).

70. See *Rudolphi Cudworthi Systema intellectuale huius universi*, trans. J.L. Mosheim (Jena, 1733).

# No Discipline, No History
## *The Case of Moral Philosophy*

### J. B. Schneewind

How are we to understand the history and the historiography of moral philosophy? As a number of recent works in ethics show, philosophers think that more than our knowledge of the past is at issue. In these works, interpretations of the history of ethics play an important part in the construction of an acceptable view about morality. MacIntyre, Williams, Donagan, Taylor, Irwin and Annas offer various views about whether moral philosophy has made progress or not. They ask whether we are now doing better at the subject than our predecessors did, and they offer answers—quite different answers, to be sure, but that is no surprise in philosophy. Historians are apt to be uncomfortable with this question. To ask it is to suppose not only that there is some common enterprise engaging earlier and later moral philosophers, but also shared standards by which we may judge improvement or decline. Historians tend to think that such assumptions are not their business, and that in any case they make for "Whiggish" or "triumphalist" history. They are thus likely to be uncomfortable with histories of moral philosophy written with such philosophical assumptions; philosophers are likely to think histories written on any others are irrelevant.

In this paper I shall raise some questions about the supposition that there is enough significant continuity in the concerns of moral philosophers to warrant discussions of progress and regress in the discipline. I should like also to indicate why it is not possible to consider the history of moral philosophy seriously without some views that, properly speaking, can be adjudicated, if at all, only within the discipline itself. It seems obvious that if there is no single discipline of moral philosophy there cannot be a history of it; it may seem less obvious that unless you are pretty well versed in, and have views about, moral philosophy—unless you have acquired the discipline—you cannot properly study whatever history it may have. The history and the discipline, I shall argue, are inseparable.

### I

The version of the history of moral philosophy that is most commonly accepted today goes back at least as far as Xenophon. He tells us that Socrates broke with his predecessors by attending to a new set of issues. He did not dispute, as they did, about the cosmos and the nature of things in general. He asked instead about human affairs.[1] Cicero elaborates on the point. Socrates, he says, "was the first to call philosophy down from the heavens and set her in the cities of men ... and compel her to ask questions about life and morals and things good and evil."[2] In the eighteenth century Thomas Reid taught his pupils that Socrates "has always been reckoned the Father of Moral Philosophy."[3] In the opening paragraph of *Utilitarianism* (1861), John Stuart Mill calls upon this tradition. "From the dawn of philosophy," he says, "the question concerning the *summum bonum* or ... the foundation of morality, has been accounted the main problem of speculative thought ... And, after more than two thousand years, the same discussions continue, philosophers are still ranged under the same contending banners, and neither thinkers nor mankind at large seem nearer to being unanimous on the subject than when the youth Socrates listened to the old Protagoras, and asserted ... the theory of utilitarianism against the popular morality of the so-called Sophist."

Mill offers a theory to account for the fact that the question of the first principle of morality remains open after so many centuries of inquiry. In all the sciences, he says, much information and many low-level theorems come to light long before the most basic principles are discovered. Mankind learns many more or less general truths from experience; only later does careful analysis enable us to extricate the fundamental concepts and principles of a science from the mass of details. Moral beliefs are like others. Hence it is not surprising that common sense should possess some sound *beliefs* about moral rules even though we will not have secure *knowledge* about morality until we discover its true foundations.[4]

It is still standard to say that moral philosophy began with Socrates and has been carried on continuously ever since. Thus Bernard Williams begins his important study *Ethics and the Limits of Philosophy* (1985) as follows: "It is not a trivial question, Socrates said: what we are talking about is how one should live. Or so Plato reports him, in one of the first books written about this subject. ... The aims of moral philosophy ... are bound up with the fate of Socrates's question ... " (1). Although we have not reached agreement about the basis of morality, the Socrates story draws on a clear picture of the tasks that moral philosophers should undertake. We are trying to answer the question Socrates raised: how to live. People have always had opinions on the matter, but it is very hard to get an indubitable answer based on an undeniable foundation. It is so hard that skeptics ask us to doubt that there is an answer, or even a real question. Perhaps, as Mill says, the difficulty exists in all disciplines. Or perhaps, as others think, there are special problems about morality that make the task of developing its theory harder than the tasks facing physics. These problems may account for the fact that we seem not even to have made any generally accepted progress toward the answer, much less found it. Still, the issues are there, and we should continue working on them. If we study earlier moral philosophy, it is because we may gain some insights from our predecessors, or learn at least to avoid their errors.

## II

Because the Socrates story is simply taken for granted today it is important to be aware that it is not the only possible narrative of the history of the subject. For many centuries an alternative view of that history was widely held. Like the Socrates story, it carries with it a distinctive view of the tasks of the discipline. The underlying thesis of the alternate history is that the basic truths of morality are not the last to be discovered. They have been known as long as humans have been living with one another. Whatever moral philosophy is, therefore, it is not a search for hitherto unknown scientific knowledge.

The alternative narrative takes two forms, one religious and one secular. The religious version is the older. It gives importance to a question most of us would not naturally ask. We will be inside the story once we see why we might ask it. The question is: Was Pythagoras a Jew?

The question arises from two assumptions. One is that the biblical narrative provides the unquestionable framework within which all human history must be located. Bossuet's *Discourse on Universal History* (1681) is perhaps the greatest modern monument built on this assumption. As Santinello's study of Renaissance histories of philosophies shows, it was long common to assume that all wisdom comes from God. One major task for historians, therefore, was to explain its presence in cultures not directly descended from the Jews.[5] Those who undertook these enterprises believed that philosophy is an important human activity, which must have a providentially assigned role. They had a special problem about morality and its relation to moral philosophy. The truth about morality was revealed very early in human history, and it has not changed. William Law, arguing against Mandeville, gives us a clear eighteenth-century statement of this point:

> When Noah's Family came out of the Ark, we presume, they were as well educated in the Principles of Virtue and moral Wisdom, as any People were ever since; ...
> There was therefore a Time, when all the People in the World were well versed in moral Virtue ....
> He therefore that gives a *later* account of the Origin of moral Virtue, gives a *false* account of it.[6]

Belief that the Noachite revelation was the origin of moral knowledge itself would make it natural to ask why we have moral philosophy anyway. It would also lead us to wonder about how the Greeks could have been the ones to start it.

The answer to the first question lies in human sinfulness. Our nature was damaged by the fall. It not only dimmed our faculties, lessening our ability to become aware of God's commands and understand them. It also unleashed the passions. Evildoers, driven by their lusts, seek to avoid the pangs of conscience, so they blind themselves to its clear dictates. They also strive to veil and confuse the moral thoughts of those whom they wish to entangle in their wicked schemes.[7] Bad reasoning is one of their basic tools. Now reason is one of God's gifts to humanity. Among other things it enables us to hold on to at least some of the moral knowledge we need, once revelation has ceased. If reason makes moral philosophy possible, pride leads men to try to outdo one another in inventing schemes and systems of morality, and morality itself gets lost in their struggles. Since the causes of the misuse of reason and of bad philosophy are now ingrained in our nature, there will be no final triumph of good philosophy until after the last judgment. But the battle must be kept up. Moral philosophy is to be understood as one more arena for the struggle between sin and virtue.

As to the Greeks, it may be mysterious *why* God chose them to be the first to philosophize. We can, however, find out *how*, lacking the Noachite and the Mosaic revelations and being as corrupt as the rest of mankind, they could have done as well as they did with morality (how well they did being, again, a subject of debate).

The first part of the answer is due to a frequently cited remark attributed to Aristotle. In *Magna Moralia* 1.1 he says that Pythagoras was the first who attempted to treat of virtue. Thomas Stanley, the first English historian of philosophy to write in the vernacular, repeats the claim, citing this source.[8] Given Aristotle's standing as the first historian of ancient thought, it seems that one could hardly ask for more impressive testimony. We can discover the importance of Pythagoras's priority from a parenthetical remark that Scipion Dupleix inserts in his assertion of it. In his *L'Ethique ou Philosophie Morale* of 1603 he says that although Socrates is praised for his discussion of the rules delivered by moral philosophy, he was not the first in the field: "it is certain that Pythagoras himself, whom the Greeks took for a philosopher of their nation (although St. Ireneus assures us that he was Hebrew and had read the books of Moses) had worthily treated of morality" before Socrates did.[9] Here as elsewhere Dupleix was unoriginal. Ficino, for example, thought he recalled that St. Ambrose "showed that Pythagoras was born of a Jewish father;" and there were others.[10] Thus the problem of transmission is solved. If Pythagoras was the one who initiated moral philosophy among the Greeks, and he was a Jew, it is clear how the Greeks managed to get the subject going.

Not everyone thought Pythagoras was actually Jewish; but there were second-best stories. It was a commonplace that the Greeks got much from the Jews.[11] John Selden, who traced our grasp of natural law back to the Noachite commandments, devoted long pages of his *De Jure Naturali et Gentium* of 1640 to analyzing the testimony of Jewish and Christian writers about the Jewish influence on Pythagoras. He preferred the Greek authorities to the Jewish, as having, he thought, less of a vested doctrinal interest in proving such a debt to the Jews. His conclusion is that the weight of the evidence makes it clear that Pythagoras,

> the primary teacher of Greek theology and the first to be called a philosopher, to whom some also attribute the first doctrine in Greece concerning the immortality of the soul . . . and others wish to credit the first disputations about the virtues, that is, the principles of moral philosophy . . . consulted and heard the Hebrews.[12]

Selden thinks it quite possible that Pythagoras was taught by no less a figure than the prophet Ezekiel.

Henry More is also explicit about the importance of Greek philosophy's debt to the Jews:

> Now that Pythagoras drew his knowledge from the Hebrew fountains, is what all writers, sacred and prophane, do testify and aver. That Plato took from him the principal part of that knowledge, touching God, the soul's immortality and the conduct of life and good manners, has been doubted by no man. And that it went from him, into the schools of Aristotle, and so derived and diffused almost into the whole world, is in like manner attested by all.[13]

We have here the germ of a history of moral philosophy. I do not know how old it is.[14] But I think that some version or other of the Pythagoras story, as I shall call it, must have been assumed, however indistinctly, by a great many philosophers. There is a large amount of room for maneuver within this kind of historical schema. Even the religious version leaves a role for reason while not making revelation superfluous.[15] Locke and Clarke in England, and Crusius in Germany, all concerned to defend the view that morality *at present* is not dependent on revelation, are still determined to keep revelation historically essential. They replace Pythagoras's Noachite revelation with Christ's, as that through which alone we became able to know the full truth about morality. It seems, Locke says, that

> 'tis too hard a task for unassisted reason to establish morality in all its parts upon its true foundation . . . We see how unsuccessful in this the attempts of philosophers were before our Saviour's time . . . And if, since that, the Christian philosophers have much outdone them, yet we may observe that the first knowledge of the truths they have added [is] owing to revelation.[16]

Now that Christ has revealed the truth, we can see for ourselves the reasonableness of his teaching, and can even turn our knowledge into a demonstrative science. A truth is reasonable and philosophical, Crusius says in 1744, when it can be proven by valid arguments from rational starting points. It does not matter where we first learned it. "The duties that the Christian religion imposes on us are grounded in reason. Because our knowledge of them was dimmed by our corruption [*Verderben*] they had to be repeated. . . . we learn the extent of human corruption from the fact that without divine revelation we would not have grasped the most important and most fully-grounded rational truths."[17]

## III

Neither Locke nor Clarke nor Crusius say anything at all about Noah or the idea that Pythagoras had a Jewish connection. Yet in holding that the Greek philosophers were never able to get very far in figuring out what morality requires, they share with the Pythagoras story the belief that reason without revelation could not discover morality. The Pythagoras story's explanation of the role of moral philosophy is implicit in their work as well. With it in mind, we can see, for instance, that Clarke's standard description of Hobbes as "the wicked Mr. Hobbes" is not just an incidental expression of personal revulsion. But none of them gives any account of the history that leads up to their own moral theories.[18]

Jean Barbeyrac is of great interest because he gives an only partly secularized version of the history embodying the basic philosophical assumption at work in the religous Pythagoras story.[19] Following Pufendorf, Barbeyrac assumes that the basic truths of morality are always readily accessible to human reason. They must therefore have been known in the earliest ages, so that no revelation of morality was necessary. But Barbeyrac also believes, with the religious Pythagoras story, that human sinfulness leads men to try to evade the demands of morality and to use reason in the effort. It is worth looking at some of the details as they are spelled out in his 1706 *Historical and Critical Account of the Science of Morality*.[20]

Since the first inhabitants of the world lived in the "Eastern Countries," he says, it must have been among them that there originated "the most general Notions of Morality, and the other Sciences . . . The Greeks," he adds, "for all their Vanity, were forc'd to own themselves Debtors for these notices, to those, they call'd Barbarians." As for Pythagoras, he travelled in the East, among the Egyptians, Persians, and Chaldeans, and brought back "many of his Notions" as well as his symbolic and enigmatic way of teaching (Barb., 37,45,47). Barbeyrac also cites the important sentence of Aristotle, on which the whole tradition hangs (Barb., 50). If he does not explicitly say that Pythagoras was a Jew or studied the books of Moses, he clearly has something like it in mind.

The problem that gives the structure to his *Account* shows his divergence from the purely revelational account of the origins of moral knowledge given in the older Pythagoras story.[21] In his first two chapters he tells us that the principles of morality are so simple that they are within the reach of everyone. His adherence to Pufendorf's natural law view of morality requires him to add that in thinking of morality we must also be thinking of God, but he finds no problem with this because he thinks it easy to acquire the natural knowledge that God exists and actively governs our lives (Barb., 1–2). Morality is not only plain and simple, it is also, as Locke showed, demonstrable (Barb., 4–5). Why, then, has the science of morality remained so backward? And what use are philosophers?

Here Barbeyrac falls back on the older history. Sin provides the answers. Barbeyrac cites Hierocles—significantly, he cites a comment on Pythagoras— to say that sinful people can have clear ideas about many things while still being blind to morals (Barb., 6). Not only self-interest, but long-standing custom or tradition can also conceal moral truth from us, as can prejudices acquired in early education. We might think that priests would at least teach morality properly, but it has not been so. Pagan priests had such misguided ideas of the divinity that they could not hope to get morality right. The Jewish priests were too busy with ceremonial and civil affairs to teach adequately "the revelation of which they were the depositaries." Their carnal prejudices, moreover, kept them tied to the letter of the law. Although Christ re-established morals in all their purity, there were false teachers even in the time of the Apostles who corrupted his doctrine. Barbeyrac goes on at great length about the decay that followed and about the awful morals of the Christian Fathers. He includes St. Augustine among those condemned: did not Augustine write in defence of persecution? Barbeyrac was a Huguenot, and he could hardly excuse Augustine for this failing (Barb., 24).

## IV

Although Barbeyrac cites him as an authority for Pythagoras's priority as student of virtue, Aristotle—if indeed it was Aristotle who wrote the *Magna Moralia*—does not make a very strong claim about it.[22] The whole of what he says on the matter is this: "Pythagoras first attempted to speak about virtue, but not successfully; for by reducing the virtues to numbers he submitted the virtues to a treatment which was not proper to them" (1182a12–14). Socrates and Plato are also mentioned as having attempted to understand virtue, but with similar lack of success; and we are left with the distinct impression that Aristotle sees himself as the first to succeed in moral philosophy. When Barbeyrac cites the passage in his text, he omits the words "but not successfully," although he plainly knew they were there. Pythagoras was in any case something of an embarrassment. Little of his writing survives, and that little obscure. He was the subject of fantastic stories. Stanley unquestioningly repeats some of the stories contained in Diogenes Laertius and in the (alleged) Aristotelian fragments—that he showed his golden thigh in public, was in Croton and Metapontum at the very same time, and convinced an ox to stop eating beans forever by whispering in its ear.[23] The lack of clarity in Pythagoras's writings and the other mists that shroud him may have made him ideally suited for construal as the link between God's revelation to the Jews and the ability of the Greeks to have and philosophize about a proper morality. But by the end of the seventeenth century he was evidently beginning to seem a broken reed.

The Pythagoras story, however, still kept its hold. Adam Glafey, a German historian of natural law writing some thirty years after Barbeyrac, refuses to report on the thought of the Eastern nations because they left no adequate written accounts. But he takes Pythagoras as the first of the Greek thinkers to give serious attention to morality, and after discussing his philosophy remarks that "we can see in general from this short summary of Pythagorean morality that, just as this man borrowed much from the Jews and the Egyptians, so also the succeeding Greek philosophers themselves made use of his doctrine."[24]

Vico found it necessary to challenge a number of stories about Pythagoras, among them the one about his having learned from the Jews. Quite aside from the difficulty of accepting the tales of Pythagoras's numerous travels, there is the strong probability that like priests everywhere, the Jewish priests kept their mysteries secret. Vico holds that it was "by grace of a most sublime human science" that Plato and Pythagoras "exalted themselves to some extent to the knowledge of the divine truths which the Hebrews had been taught by the true God."[25] He thus denies the essential presupposition of the Pythagoras story, that knowledge of morality could only have been acquired at first from a divine revelation.

Vico was not widely read; and it was not until the end of the eighteenth century that the Pythagoras story was critically examined and dismissed. In 1786 the German scholar Christoph Meiners published a history of the sciences in which he devoted much space to a critical examination of the alleged Pythagorean writings, dismissing almost all of them as unreliable.[26] An effective positive replacement for the Pythagoras story was not published until 1822, in what I think should be considered the first comprehensive *modern* history of moral philosophy, Carl Friedrich Stäudlin's *Geschichte der Moralphilosophie*.[27]

Stäudlin opens with a brief remark suggesting that morality arises from the interaction between the native powers and dispositions of the human mind and our situation in the world. Its origins lie so far back in antiquity that there is no use speculating about them. There was morality everywhere before there was philosophizing about it, and there were unsystematic and poetic articulations before anything rational appeared.[28] We are as naturally moved to reflect on our own powers as on the world in which we act, and that reflection, carried far enough, is philosophy. Moral philosophy begins with the Greeks: pre-eminently with Socrates.[29] Stäudlin gives Pythagoras a chapter; but in it he expresses great admiration for the work done by Meiners (whom on other points he attacks) enabling us to dismiss all the old claims about his importance. Allowing that Pythagoras had some interesting thoughts about morals, he himself is not willing to concede that there is any live issue about a Jewish connection. Yet he notes several recent writers who do, and the amount of effort he devotes to getting rid of the Pythagoras story suggests that it is still a live option.[30]

What makes Stäudlin's work modern is not mainly its dismissal of the Pythagoras story and its kin. That, after all, is a scholarly position that might alter. Thus there is a more recent version of the Pythagoras story—surely not intended as such—according to which Zeno, the founder of Stoicism, was himself the son of a Jew. Giovanni Reale says that both Zeno and Chrysippus were Jewish and hypothesizes that the Stoic notion of *kathekonta* reflects Zeno's effort to bring Jewish moral categories into Greek philosophy.[31] What makes Stäudlin's work modern is essentially its attitude toward error. He treats error in moral philosophy as like error in any science, no more due to wicked desires or self-aggrandizing tendencies than blunders in mathematics. Error comes not from original sin, but from the great difficulty of the subject. The function of moral philosophy is not to defend God's revelation from sinful and perverse reasoners. Like Kant, Stäudlin holds that it expresses the human tendency to reflect on our own powers and dispositions. The Greeks, Stäudlin thinks, exhausted almost all the possibilities and explored almost all the blind allies. Only rarely does a new insight, such as Kant's, enable us to advance. With such insights moral philosophy may increase our grasp of moral principle from time to time, or correct honest mistakes made along the way, and so contribute to the progress of morality as well as of moral philosophy.

Stäudlin was a Kantian; and Kant would have agreed with much of his approach. But he would have added that reflection also has an important moral function. He thinks that we have all always known the basic principle of morality. Because of our tendency to selfishness, however, a natural dialectic arises in which we try to convince ourselves that prudential reason is the only practical reason there is.[32] The philosophical reflection that shows the reality of pure practical reason therefore has its own practical importance. Kant has developed his own version of the secularized Pythagoras story.

V

Both the Socrates story and the Pythagoras story (in its secular as well as its religious versions) illustrate the interconnections among our conceptions of the aim or task of moral philosophy, the proper understanding of its history, and the nature of morality. The two grand narratives are similar in holding that moral philosophy has essentially a single task, though each assigns it a different one. But the assumption that there is one single aim that is essential to moral philosophy gives rise to difficulties for both views.

One difficulty lies in formulating the aim. Perhaps it is plausible to hold that we and Socrates are asking the same question if the central issue is described as Williams describes it. Yet we might wonder whether identifying the question of moral philosophy as "How should one live?" is useful for those interested in the history of the subject. The Socratic question, so stated, is extremely general. To take it as locating "the aims of moral philosophy" we must surround it with a number of unspoken assumptions. For instance, we must not take it to be a question about how one should live with respect to health, or income, or eternal well-being. Are we then to take it as a general question about how we should live in order to be happy? We have only to think of Kant's ethics to see that this will not identify an inquiry central to all moral philosophy.

The single-aim view seems to rest on a theory about the essences of philosophical disciplines which is itself contestable. If we look historically at what moral philosophers have said they were trying to do, we do not come up with a single aim uniting them all. Compare, for instance, Aristotle's claim that moral philosophy should improve the lives of those who study it with Sidgwick's belief that "a desire to edify has impeded the real progress of ethical science."[33] Recall the Stoic aim of finding the way to personal tranquillity; Hobbes's aim of stabilizing a society put in danger by religious fanaticism; Bentham's aim of locating a principle to show everyone the need for major political, social and moral reform; Parfit's aim of developing a new, wholly secular, science of morality.[34] Unless we leave the statement of the aim quite vague, it will be difficult to find one on which these thinkers agree. If we are more definite, then it seems that we will be required to say that anyone not sharing the favored aim is not really doing moral philosophy. Whatever the single aim assigned to the enterprise, we would be forced to deny the status of moral philosopher to many thinkers usually included in the category.

Those holding a Pythagoras story version of moral philosophy's single task face some additional difficulties. They must assume that the moral knowledge which is always to be defended can be identified in some way that does not presuppose the truth of any specific theory, and that it is always and everywhere essentially the same. Yet it is implausible to claim that Greek morality, the morality of the Decalog, and the liberal morality of modern Western democracies are in essence identical. The claim can be made out, if at all, only by proposing as "the essence" of these moralities some interpretation of them, probably in philosophical terms, which was not available to some or all of those whose moralities are at issue.

These objections to single-aim views about moral philosophy are themselves both historical. The historian will have a further problem with the outlook. It implies that since we and past moral philosophers share aims and goals, the best way to understand the work of our predecessors is to look at them in the light of our own view of the truth about morality. Even allowing, as some philosophers do, that our own views may not be the last word, it is still tempting, on a single-aim approach, to suppose that ours is the best word yet, and that therefore no other standpoint is needed for examining what has gone before.

The historian will complain that insistence on describing the views of past thinkers in our own terminology forces us into anachronism. If we are interested in what our predecessors were doing and thinking, we must try to understand them in terms they themselves had available. It is obvious that Hume could not even have conceived the aim of "anticipating Bentham." But it is just as misleading to describe him as "trying to develop a rule-utilitarian theory of justice." Although he discovered some of the important differences between the morality of actions within social practices and the morality of independent actions, the idea of utilitarianism as well as the distinction between "act" and "rule" versions of it are much later inventions. We may have good reason for thinking of his theory in terms like these, but we are not, in so doing, giving an historical account of it. Worse, we may be overlooking its historical distinctiveness by forcing it into our own molds.

# VI

We cannot, it seems, write a history of moral philosophy without having some philosophical idea of the aims of the discipline; and we cannot have a well-grounded idea of its aims without having some awareness of its origins and history. The difficulties for the historian arising from this conclusion are not wholly avoidable. But they are less acute if we give up thinking of moral philosophy as having some single essential aim and suppose instead that philosophers at different times were trying to solve different problems.

As historians, we can work with a very general concept of morality, taking it vaguely and imprecisely as the norms or values or virtues or principles of behavior that seem to be present in every known society. We will study those who try to reflect philosophically on the matters thus described. No doubt our idea of what counts as "philosophical reflection" will be marked by our present conception. But we will not try to impose much more uniformity on past efforts than is carried by these two reference points. We will not need to decide whether common sense morality, ancient or modern, is mere opinion or genuine knowledge. We will not, in particular, suppose that everyone who thought about morality in a way we consider philosophical was trying to solve the same problem or answer the same questions. We will think instead that the aims of moral philosophy—the problems that moral philosophers thought required reflection—are at least as likely to have changed as to have remained constant through history.

Why might there be such alterations in the questions or problems that set the differing aims of philosophical reflection about morality? One answer is that there have been times of upheaval when the norms involved in our common life have been called into question by social, religious, and political changes. The need to blend Christian belief with an inherited culture coming from Greece and Rome is one such case. The problems arising from the disintegration of even the appearance of a unified Christendom was another. Perhaps Parfit's concern to work out a wholly secular morality is another. Perhaps the apparent fact that there is no hope for agreement on conceptions of the good presents another such juncture. The history of moral philosophy, we may think, itself provides important clues to the eras at which the stresses on widely accepted norms and values became overwhelming and change was necessary. If philosophers do little to bring about the strains, they sometimes provide means to diagnose or even to cope with them.

If we take this approach we will be led naturally to ask some kinds of question about the history of moral philosophy that we may overlook if we think the discipline centers on only one question. On the single-aim assumption we will suppose we always know what moral philosophers were trying to do. They were trying to solve the essential problem. Without this assumption, we will need to ask what past philosophers were doing in putting forward the arguments and conclusions and conceptual schemes they favored. We will ask about the point or purpose of using these arguments. The answers will have to be historical. Holding that the answers to such questions may vary from time to time, we will ask just how the thinkers we study differ from earlier thinkers and from those of their contemporaries whose work they knew. What our subjects refused to ask or assert will matter as much to us as their positive claims. Knowing the former will enable us, as knowing the latter alone will not, to understand what their aims for moral philosophy were. To know what they refused to include we must know what they might have included, and did not. Here only historical information—not rational reconstruction of arguments in the best modern terms—will tell us what we need to know.

One benefit of this approach is that it gives us a way of checking on our interpretations or readings of past moral philosophy. There is historical evidence about the vocabularies available to our predecessors, and about the issues that mattered to them and to their publics. We may lack documentary evidence about a philosopher's own specific intentions in publishing a given book. But we can assume that he meant to be understood by a living audience, and not just by posterity; and what writers as well as readers could have understood is set to a large extent—not wholly—by the language they already possessed. Even innovative terms and concepts require some sort of introduction via existing notions. To learn what resources were available to a philosopher, we must look outside his writings, and outside of philosophy. If we do not check our accounts of a past philosopher in this manner, we are in serious danger of mistaking our own fantasies about what he "must have meant" for what he really did mean.

Single-aim theorists may reply that on this view there is no continuing subject of moral philosophy whose history we can try to write. But to say this is to oversimplify. Continuities are quite compatible with the discontinuities that arise from changing problems and aims. It seems highly probable that all societies complex enough to generate philosophical reflection must handle certain problems of social and personal relations. Views about the fair or proper distribution of the necessities of life, or about the relative praise- or blame-worthiness of individuals, seem always to arise in such societies. Study of different ways of structuring such views is a constant theme that gives moral philosophy some of its identity amidst its differences.

Some arguments and insights about what makes for coherent views of morality may carry over from one situation to another. They provide further elements of continuity in the work of moral philosophers. One illustration must suffice. When Cudworth said that 'good' could not be defined as "whatever God wills," he turned against Descartes and Hobbes the same kind of argument that Plato's Socrates sketched against Euthyphro. G.E. Moore later presented other arguments against the definability of "good." Cudworth was trying to preserve the possibility of a loving relation between God and man that could not have concerned either Socrates or Plato. A century later Diderot appealed to the same argument precisely because it "detaches morality from religion."[35] Moore had still other aims in view. One could write a useful history concentrating simply on the question of definability.[36] But to do so would be to ignore historically crucial differences in the uses to which the point was put. I do not wish to minimize the importance of portable arguments. They do indeed provide a major set of linkages between past and present moral philosophies. But they do nothing to support the claims of the single-aim historian. Praxiteles and Brancusi both used chisels, no doubt, but we do not learn much about their art from noticing the fact.

Single-aim philosophers will undoubtedly feel that more significance must be assigned to these portable arguments. They will say that such arguments represent what moral philosophy is all about—the discovery of the truth about morality. Plato and Cudworth and Moore all saw the same thing, even if they described somewhat differently what they saw and put their discovery to different uses. They did not discover a mere tool for carrying out some external aim. They themselves say that they are in search of the truth about morality itself, and it is quite possible that they found an important part of it. Progress in moral philosophy, as in science, involves replacing false and one-sided theories with true and comprehensive ones about the designated subject matter of the discipline. History is useful only when philosophical assessment of the arguments of past philosophers helps us with our present projects.

Histories of moral philosophy can of course be written on such assumptions; and at the very least it is true that assessment of arguments given in the past is indispensable. The historian needs to know what led to the alteration or abandonment of various views. Since failure to achieve coherence or to produce valid supporting arguments may explain it in some cases, the historian who is not sensitive to such matters will write defective history. If the single-aim view is asserting only that knowledge of the discipline is prerequisite to writing its history, one cannot object to it. But the single-aim view leaves unexplained a great deal that the historian will naturally wish to consider. Why do some theories emerge and flourish and then disappear? Why do some recur? Why is there so little convergence, what does moral philosophy as a practice or discipline do in and for the societies in which it is supported? It is more useful for the historian to turn away from the single-aim view and adopt a variable-aim approach instead.

## VII

If we take the variable-aim view of the subject, we will not be strongly inclined to make much of the question of progress or regress. We will look at the enterprise of rationally examining norms and virtues as one of the tools that various societies have used to cope with different problems they faced in shaping or preserving or extending a common understanding of the terms on which their members could live with one another. We will not think of moral philosophy as standing apart from and above the moral discourse of a society. We will take it as being simply one voice in the discussion of moral issues. In moral philosophy we will hear the voice that asks us to stand back from current issues and look at them in the most general terms we can call upon—or invent. Its hope is that by so doing we can reformulate the problems in more manageable ways. The very stance seems to make it natural to use an atemporal mode of discourse, but the rhetoric of moral philosophy need not conceal the fact that those who use it are located in their own times as well as in a timeless web of abstractions.

It is not hard to understand how questions that were of great importance at one time may lose their hold at another. The conditions giving the questions urgency may have altered. Or new and more pressing problems may have emerged. The abandonment of one question and the move to consider a new one may itself be a major kind of progress in moral philosophy. Perhaps only the assessment of questions can keep moral philosophy from the sterility and irrelevance that we sometimes call "scholasticism."

## Notes

1. Xenophon, *Socratic Memorabilia*, I.11–12.

2. Cicero, *Tusculan Disputations*, V.iv.10–11.

3. Thomas Reid, *Practical Ethics*, ed. Knud Haakonssen (Princeton: Princeton University Press, 1990), 110.

4. John Stuart Mill, *Utilitarianism*, I. §§1–2.

5. See *Models of the History of Philosophy*, ed. Giovanni Santinello, vol. 1: *From Its Origins in the Renaissance to the "Historica Philosphica"* by Francesco Botten et al., trans. C.W.T. Blackwell et al. (Dordrecht: Kluwer Academic Publishers, 1993), 21, 26, 28, and especially the discussion of Thomas Burnet, 330ff. See also Peter Harrison, *"Religion" and the Religions in the English Enlightenment* (Cambridge: Cambridge University Press, 1990), ch. 5, for discussion of the "single source" theory of religion. D. P. Walker, *The Ancient Theology* (Cornell, 1972), is another important study of this kind of view.

6. William Law, *Remarks upon . . . the Fable of the Bees*, in *Works* (London, 1762, 1892), 2:7.

7. "There has ever been an uninterrupted succession of men, who, seduc'd by a secret desire to shake off the troublesome yoke of duty; and to indulge themselves in the gratification, if not of their sensual and gross Desires, yet at least of their more delicate and refined Inclinations, have employed all the Faculties of their Souls, in extinguishing the Evidence of those Truths, which were most clear . . . in order to involve in their Ruin all certainty of the Rules of Virtue." From Jean Barbeyrac, *An Historical and Critical Account of the Science of Morality*, trans. Carew (London, 1729) §III, p.5.

8. Thomas Stanley, *The History of Philosophy*, (1655–62) (London, 1721), 395. The passage from *Magna Moralia*, 1182a12–14, is cited in full below at the opening of section 4.

9. Scipione Dupleix, *L'Ethique, ou Philosophie Morale* (Paris, 1603, 1632), 4. Dr. Sebastian Brock informs me that to the best of our knowledge St. Irenaeus said no such thing.

10. I owe these last references to the excellent book by S.K. Heninger, *Touches of Sweet Harmony* (San Marino: The Huntington Library, 1974), 201–2. Heninger lists (229, n.5) half a dozen studies from the seventeenth and eighteenth centuries that contain bibliographies on Pythagoras's debt to Moses.

11. Herodotus and other ancients attested to Greek debts to Eastern thought generally, and Isocrates held that Pythagoras in particular had brought into Greek the philosophy he learned from the Egyptians. See W.K.C. Guthrie, *A History of Greek Philosophy*, vol. 1 (Cambridge: Cambridge University Press, 1962), 160, 163. Guthrie devotes nearly two hundred pages to reviewing the difficulties of studying Pythagoras and Pythagoreanism and summarizing the results of modern scholarship.

Heninger's fifth chapter, 256–84, gives a full and fascinating account of various views of what specifically the moral philosophy of Pythagoras, or of the Pythagoreans, was supposed to be, and the many ways in which Pythagorean views were given Christian legitimacy and propagated widely.

But although he has earlier noted Pythagoras's alleged debts to the Jews for his moral thought, he does not explore the bearing of claims about the debts on the historiography of moral philosophy.

12. John Selden, *De Jure Naturali et Gentium iuxta disciplinam Ebraeorum, Collected Works* vol. 1, col. 89. The examination of testimonies occurs in cols. 82–85 and elsewhere. I am deeply indebted to Michael Seidler for having put Selden's passages concerning Pythagoras into quotable English. Selden offers an explanation of why "we do not find many vestiges of Hebrew doctrine in the writings of the Greek philosophers–indeed, that nothing at all occurs there which sufficiently retains the pure and unadulterated nature of its Hebrew origin." The various Greek sects themselves commingled so much, and splintered the old teachings so greatly, that the result is everywhere a hodge-podge. But, he adds, no one doubts that in Platonic as well as Pythagorean doctrine there are teachings derived from the Hebrews (col. 91).

13. Henry More, *Enchiridion Ethicum*, English translation of 1690 (London), 267. For More and the "ancient theology" see Peter Harrison, cited in n.5 above, 133–35. He does not discuss the Pythagoras story about moral philosophy; the ancient theology was concerned less with moral matters than with such doctrinal concerns as trinitarianism.

14. Josephus, in *Against Apion*, trans. H.St.J. Thackeray, Loeb Classical Library, (Cambridge: Harvard Univ. Press, 1926, 1966), claims the Greeks learned much of their science and law from the East, and specifically from the Jews; he mentions Pythagoras in this connection but does not explicitly claim that he was the originator of moral philosophy. See I.13–14, I.165, II.168, where the translator suggests that the dependence of Greek on Jewish thought was first suggested by Aristobulus. Eusebius in the *Prepration for the Gospel* gives a famous description of Plato as Moses Atticizing; but he does not tie Pythagoras to the origins of moral philosophy.

15. More himself says that the eternal son as the *Logos*, or human reason, as well as revelation, can enlighten us about morals.

16. Locke, *Reasonableness*, MPMK I.194–95.

17. C.A. Crusius, *Anweisung vernünftig zu Leben* (1744), reprint ed. G. Tonelli (Hildesheim: Olms, 1969), "Vorrede," fol. b4 and following.

18. John Locke, *The Reasonableness of Christianity* (1695), ed. I.T. Ramsey (Stanford: Stanford University Press, 1958), β241, pp. 60–61. Samuel Clarke, *The Unchangeable Obligations of Natural Religion* (London, 1705), VII.1. For Clarke it is the "wicked Mr. Hobbs" whose evil philosophy makes his own virtuous philosophical activity necessary. Reid does not say that the heathens could not have discovered the principles of morals; but he does say that revelation allowed Christians to surpass the heathen in matters of natural religion. Reid, *Practical Ethics*, 108–9.

19. Richard Tuck rightly finds his history important for its part in the propagation of Grotian natural law theory, but does not discuss the historiography as such. See Tuck, *Natural Rights Theories* (Cambridge: Cambridge University Press, 1970).

20. The English translation of Barbyrac's 1706 French work serves as a lengthy introduction to the fourth edition of Kennett's English translation of Barbeyrac's French translation of Samuel Pufendorf's *Of the Law of Nature and of Nations* (London, 1729). Barbeyrac's *Account* is hereafter cited in the text, with the abbreviation "Barb."

21. I am greatly indebted to Dr. Jennifer Herdt for her comments on an earlier version of this paper, in which I failed, in the present section, to notice Barbeyrac's adherence to what I am calling the "secular" variant of the Pythagoras story. Her remarks led me not only to correct this error but to rethink the whole paper, which is, I hope, improved as a result.

22. Guthrie presumably thinks that the *Magna Moralia* is spurious, since he does not list its remark about Pythagoras and the study of virtue in his review of sources concerning Pythagoras. Heninger also thinks that the *Magna Moralia* is spurious (277 n.2)—a thesis John Cooper challenges in "The *Magna Moralia* and Aristotle's Moral Philosophy," *Amer. J. of Philology* 94.4 (Winter 1973), 327–49. For my purposes it does not, of course, matter whether the attribution to Aristotle is correct; it suffices that in the sixteenth and seventeenth centuries it was thought to be so.

23. Stanley, *The History of Philosophy*, 360–61. For Aristotle, see Fragments 190–91, in Barnes, ed., *Complete Works of Aristotle* (Princeton: Princeton University Press, 1984) vol. 2, 2441. Stanley

partly follows Diogenes Laertius, *Lives*, VIII.11, who includes other wonders. The "Golden Verses" once taken as evidence about his moral doctrine are a much later composition attributed wrongly to him. See Heniger, who gives Stanley's translation of them (260–61) as well as a rich note on the controversies about them (278–79 n.18).

24. Adam Friedrich Glafey, *Vollständige Geschichte des Rechts der Vernunft* (Leipzig, 1739, reprint, Aalen: Scientia Verlag, 1965), 26–28; on Selden, 23–25.

25. Giamabattista Vico, *The New Science*, trans. Bergin and Fisch (Ithaca: Cornell University Press, 1948), 43. This is the third edition of 1744, §94–95. Leon Pompa, *Vico: Selected Writings* (Cambridge: Cambridge University Press, 1982), gives comparable passages from the first edition, 1725; see §§36, 39, 86.

26. I have not been able to consult Meiners's work; I rely on Lucien Braun, *Histoire de l'Histoire de la Philosophie* (Paris, Editions Ophrys, 1973), 173–77. By the end of the eighteenth century another German scholar produced a brief history of ethics in which Pythagoras is mentioned along with Aristotle's claim about him, lacking, again, the phrase "but not successfully." But he says nothing about any link between a first revelation of moral truth to the Jews and its elaboration by the Greeks. See Johann Christoph Hoffbauer, *Anfangsgründe der Moralphilosophie und insbesondere der Sittenlehre, nebst einer allgemeinen Geschichte derselben* (Halle: Kümmel, 1798), 295–96.

27. Adam Smith surveys part of the history of moral philosophy in Part VII of *The Theory of Moral Sentiments*. Although showing what I here describe as a modern attitude, it makes no effort to be comprehensive.

28. Vico in *New Science*³ says that human thought begins in particulars, not in theorizing, and that it was a mistake to think that universal laws were the most ancient form of the direction of action: 498–501.

29. Carl Friedrich Stäudlin, *Geschichte der Moralphilosophie* (Hannover, 1822), 22. Referred to hereinafter as Stäudlin.

30. Stäudlin, 1–3; on Pythagoras, 32–59. On the several writers who still take seriously the thought that moral philosophy originated among the Jews, or Egyptians, etc., see 19n.

31. See Reale's *The Systems of the Hellenistic Age*, trans. John Catan (Albany: SUNY Press, 1985), 209, 216, 280–81. He refers to Max Pohlenz as his authority for Zeno's Jewishness. See Pohlenz, *Die Stoa* (Göttingen, 1948–49), 1.22, 24–25, 28, and the evidence, rather scant and with an anti-Semitic tone, 2.14n. For doubts about the thesis, see Brent Shaw, "The Divine Economy: Stoicism as Ideology," *Latomus* 44 (1985), 20 n.8.

32. *Grundlegung zur Metaphsik der Sitten*, Kant, *Gesammelte Schriften* (Berlin), 4:405.

33. Aristotle, *Nicomachean Ethics*, 1179a35–b4; Henry Sidgwick, *Methods of Ethics*, 7th ed. (1930), vi.

34. *Reasons and Persons* (Oxford, 1984), 453.

35. Diderot makes this remark in one of his contributions to Raynal's *Histoire des Deux Indes* (1772 and later editions). I cite from Denis Diderot, *Political Writings*, ed. John Hope Mason and Robert Wokler (Cambridge: Cambridge University Press, 1992), 211.

36. See Arthur Prior, *Logic and the Basis of Ethics* (Oxford: Oxford University Press, 1949).

# VICO AND THE
# BARBARISM OF REFLECTION

## Donald Phillip Verene

In his autobiography published in 1728, Vico presents himself as a lone figure in corrupt times. After describing his near death from a fall at the age of 7, which fractured his cranium and which involved a slow recovery, Vico says that he "grew up, from then on, with a melancholy and acrid nature which necessarily belongs to ingenious and profound men, who through ingenuity flash like lightning in acuity, through reflection take no pleasure in witticism and falsity."[1] In the first part of this statement, Vico is echoing Cicero's gloss on Aristotle's view that those who are ingenious and profound have a melancholic temperament.[2] Vico carries this further, and contrasts himself as part of the class of those who are ingenious and profound with those who employ reflection only to achieve witticism and falsity. When Vico goes to Vatolla as tutor to the children of the Rocca family, he finds himself "wedded to the corrupt style of poetry" (i.e., *barocchismo*). To cure himself of this corrupt style of modern poetry he devised a special method of reading the masters of the Tuscan tongue (Boccaccio, Dante, and Petrarch) against the Latin masters (Cicero, Virgil, and Horace), pairing them off with each other and reading them in a cycle, on successive days. In this way he completed the process of becoming an autodidact.[3]

When he returns from his nine years of residency at Vatolla to reside permanently in Naples, Vico says he "found the physics of Descartes at the height of its renown among the established men of letters," and that Aristotle's physics had become a laughingstock. The best thinkers of the sixteenth century were not revered. Medicine had "declined into scepticism" and law had suffered a similar fate. Plato was seen as important only for the quotation of an occasional passage. The study of rhetoric, poetry, history, and classical languages and culture generally was no longer pursued. In other words, all those fields that Descartes had found inessential for the search for truth in the *Discourse* were eschewed. Vico felt fortunate that he had sworn allegiance to no teacher and that he could continue to be guided by his good genius.[4]

In his continuation of the autobiography, written in 1731, Vico sums up the conditions of his whole career by saying: "Among the caitiff semi-learned or pseudo-learned, the more shameless called him a fool, or in somewhat more courteous terms they said that he was obscure or eccentric and had odd ideas."[5] Among those who called him a fool would be the author of the false book notice on the *Scienza nuova*, that was sent to and printed in the Leipzig *Acta* for

August 1727, in which Vico is described as an "abbé" of the Vico family and which proceeds maliciously to misdescribe his views and the general contents of the work. To this notice Vico wrote the reply, *Vici Vindiciae*.[6] The author of the notice, whom Vico calls an "unknown vagabond," would be a prime example of one taking pleasure in witticism and falsity.

In his letter of 1726 to Abbé Esperti in Rome, in which he relates the very limited reception that the *Scienza nuova prima* received, Vico says such a work cannot expect universal applause because it goes against the corrupt spirit of the times.[7] He plays upon a line of Tacitus, "*corrumpere et corrumpi seculum vocatur*," in which Tacitus criticizes Roman customs in contraposition to those of the barbaric but virtuous Germans (*Germania*, 19). Vico employs this in the *Scienza nuova seconda*. He says that Roman emperors, when they wish to give reasons for the ordnances they issue, claim to have been guided by the "sect of their times" (*sètte dei tempi*). Vico says that "the customs of the age are the school of princes, to use the term *seculum* (age) applied by Tacitus to the decayed sect of his own times, where he says, *Corrumpere et corrumpi seculum vocatur*—They call it the spirit of the age to seduce and be seduced—or, as we would now say, the fashion" (979).[8]

Although this quotation comes from a description by Tacitus of the Germanic customs and attitudes toward marriage, fidelity, and the sexes, the larger meaning of an age in which "to corrupt and to be corrupted" is the mode is not to be lost on the reader. Vico holds that when the intellectual virtues are corrupted into the fashion of witticism and falsity in thought this is accompanied by a parallel corruption in social life, in which the moral virtues are corrupted into flattery, soft embraces, and plots against intimates and friends. Vico portrays this corruption of social life in the most impassioned passage of his *Scienza nuova seconda* (1730 and 1744). He says that there are two types of barbarism, a barbarism of sense and a barbarism of reflection. The barbarism of sense typifies human life in the first stage of the "ideal eternal history traversed in time by every nation in its rise, development, maturity, decline, and fall" (245). The barbarism of reflection typifies human life at the point of the nation's decline and final fall. Vico says:

> But if the peoples are rotting in that ultimate civil disease and cannot agree on a monarch from within, and are not conquered and preserved by better nations from without, then providence for their extreme ill has its extreme remedy at hand. For such peoples, like so many beasts, have fallen into the custom of each man thinking only of his own private interests and have reached the extreme of delicacy, or better of pride, in which like wild animals they bristle and lash out at the slightest displeasure. Thus no matter how great the throng and press of their bodies, they live like wild beasts in a deep solitude of spirit and will, scarcely any two being able to agree since each follows his own pleasure or caprice. By reason of all this, providence decrees that, through obstinate factions and desperate civil wars, they shall turn their cities into forests and the forests into dens and lairs of men. In this way, through long centuries of barbarism, rust will consume

the misbegotten subtleties of malicious ingenuities [*ingegni maliziosi*] that have turned them into beasts made more inhuman by the barbarism of reflection [*la barbarie della riflessione*] than the first men had been made by the barbarism of sense [*la barbarie del senso*]. For the latter displayed a generous savagery [*un fierezza generosa*], against which one could defend oneself or take flight or be on one's guard; but the former, with a vile savagery [*un fierezza vile*], under soft words and embraces, plots against the life and fortune of friends and intimates. Hence peoples who have reached this point of reflective malice [*riflessiva malizia*], when they receive this last remedy of providence and are thereby stunned and brutalized, are sensible no longer of comforts, delicacies, pleasures, and pomp, but only of the sheer necessities of life. And the few survivors in the midst of an abundance of the things necessary for life naturally become sociable and, returning to the primitive simplicity of the first world of peoples, are again religious, truthful, and faithful. Thus providence brings back among them the piety, faith, and truth which are the natural foundations of justice as well as the graces and beauties of the eternal order of God (1106).

Vico sees himself as living in an age of the barbarism of reflection. It is an age we share with Vico. Within the "great city of the nations" (1107) that is all of human history, in the sense of the gentile peoples, a divine or providential order prevails in which each nation passes through a threefold cycle that begins in an age of gods (in which all of nature and the functions of communal life are formed in terms of gods), develops through an age of heroes (in which all human virtues and meanings of social institutions are formed in the figures of the heroes) and ends in an age of humans (in which the world is made intelligible in wholly secular and logical terms).

The pattern of gods, heroes, and humans is a pattern that is lived out over and over in the courses and recourses of the nations. In Vico's conception of Western history, Homer marks the end of the age of gods and heroes in the first *corso*. The *Iliad* and the *Odyssey* summarize these previous stages, after which comes the age of philosophy and reflection. Upon the fall of the ancient world comes the *ricorso* of "returned barbarism."[9] The first two ages of this *ricorso* are summarized in Dante's *Divine Comedy*. Vico calls Dante "the Tuscan Homer" (786, 817). After Dante comes the third age of the recourse of Western history. The philosophers arrive once again, and this time their thought is built upon the recollection of the philosophy of the earlier *corso*. By Vico's time, the fruitful powers of Renaissance philosophy have given way to the full powers of reflective thought. The quarrel between the ancients and moderns has begun—the quarrel between the *studia humanitatis*, that emphasizes the importance of rhetoric and *verbum* as the basis of human knowledge and civil wisdom—and the science of the *moderni*, that emphasizes sensation and experiment as well as mathematical method as the standard of human knowledge.

What does Vico mean by the "barbarism of reflection"? And what is its significance as a description of modern knowledge and the modern condition?[10]

Vico produces the term "barbarism of reflection" in this dramatic passage, in the middle of his compact conclusion to the *Scienza nuova*. It is a passage that places Vico's entire project in relation to his own age.[11] Vico's new science is not a form of the barbarism of reflection itself, but stands against this as the condition under which the new science is formed. Vico's work is a counter to such a state of mind and society. Reflection is a modern term. There is no genuine counterpart for it in Greek. *Reflectere* occurs as a term relating to mental action, in classical Latin, for example, in Cicero and in Vergil, but it has the sense of reversing one's opinion or turning one's mind in the opposite direction. Reflection as a term, distinguished from sensation, has roots in scholastic thought, but not as a major philosophical term.[12] The notion of reflection, in the sense of mirroring of images between objects or events in the physical world is, of course, known to the Greeks. Such mirroring can be found in Plato, and Aristotle makes reference to this phenomenon in a number of places.[13] Reflection or refraction is part of modern optics. Descartes makes use of the term in this way in his correspondence and in his scientific discussions.[14] This is in accord with the basic Latin meaning for *reflecto*, "to turn back," "bend back," "to divert."

The origin of reflection as a philosophical or psychological term describing the internal operation of the mind for modern thought is Locke's *An Essay concerning Human Understanding* (1690).[15] Locke, in describing the origin or our ideas, distinguishes between those from external sensible objects and those from the initial operations of our minds. He says: "These two are fountains of knowledge, from whence all the ideas we have, or can naturally have, do spring" (I.2.2.). He says one source of what is in the mind is due to what the senses convey to it from external objects (e.g., yellow, white, heat, cold, soft, hard, bitter, sweet): "this great source of most of the ideas we have, depending wholly upon our senses, and derived by them to the understanding, I call SENSATION" (II.1.3.). The other source of ideas is the experience our own mind has of its operations that cannot be had from without (e.g., perception, thinking, doubting, believing, reasoning, knowing, willing). Locke says "This source of ideas every man has wholly in himself; and though it be not sense, as having nothing to do with external objects, yet it is very like it, and might properly enough be called *internal sense*. But as I call the other Sensation, so I call this REFLECTION, the ideas it affords being such only as the mind gets by reflecting on its own operations within itself" (II.1.4.). The problems of whether Locke's conception of reflection is to be understood intellectually or empirically are well known, and such problems of the internal consistency and viability of Locke's epistemology are not the issue here.

Although the term reflection exists in use in various senses in English before Locke, his *Essay* is the source for the use of reflection as a philosophical term having the above meaning. In French the case is similar, in the sense that *réflexion* is derived from the Latin *reflectere*, and it has similar variations in meaning to the term in English. In French, however, the etymology of reflection in the philosophical sense of the mind's action on itself is traceable to Descartes's use of the term in the fifth part of the *Discours de la méthode*. Here

Descartes reaffirms the essentials of his proof of God and the soul as the mind having tested its ideas of them through its doubts, and Descartes uses the phrase, "avoir fait assez de réflexion."[16]

I believe that Vico intends the terms "barbarism of sense" (*barbarie del senso*) and "barbarism of reflection" (*barbarie della riflessione*) to refer to Locke's distinction between sensation and reflection in the *Essay*. He may also intend them to have some resonance with Descartes, for although *réflexion* is not a major term in Descartes's philosophy, the act of reflection itself is. Descartes forms his case in the *Discourse* and in the *Meditations* in terms of a general distinction between sense and thought, and, as mentioned above, he pins his proofs of the nature of God and the soul on the mind's successful action of grasping the certainty of its own operations. In *Meditation* I, Descartes speaks first of the status of his sensation of external things, then of his sensation of bodily state (his example of dreaming) and then of the possibility of doubting the results of the pure operations of his own mind, apart from sensations. His genus-species definition of himself as *cogito* is similar to Locke's inventory of operations that characterize reflection as "internal sense." My interest here, as with Locke above, is not with the problems inherent in Descartes's position, nor with its relationships to and distance from Locke's position. For Vico, Descartes and Locke represent the basis of modern philosophy.

Vico understands all of modern thought as revolving around two alternatives—the Stoic and the Epicurean, the latter of which he also often associates with a sceptical position. There are those who believe in a version of the deaf necessity (*sorda necessità*) of the Stoics and those who embrace one version or another of the blind chance (*cieca fortuna*) of the Epicureans. As he states in his autobiography, both of these metaphysics involve a moral philosophy of solitarities.[17] In the passage quoted above, these are examples of that solitude of spirit and will (*solitudine d'animi e di voleri*) that characterize the modern age of barbarism of reflection.[18] The Stoic ideal is a doctrine of the independence of the individual from the social whole, and the Epicurean advocates the peace of the soul through withdrawal into the garden. Both turn their backs on those conceptions of civil wisdom of Plato and Aristotle, and advocate a solitary life that is false to the true nature of the human as social. Vico's principle of providence as the ideal eternal pattern in human history is intended as the basis for a metaphysics that goes between the horns of the dilemma of necessity and chance of the Stoic and the Epicurean metaphysics. Vico's ideal of "wisdom speaking" (*la sapienza che parla*) that he takes from the ancients, especially Cicero, and from the Humanist tradition, is intended to go between the horns of the Stoic individual who claims to be a whole society unto himself and the Epicurean who withdraws to his own ideal society of the garden, within the larger whole. The rhetorical notion of the connection of *sapientia* with *eloquentia* has as its classical third term *prudentia*—the ideal of thinking, speaking, and acting well as the ideal of human life and human greatness.[19]

Vico repeatedly identifies Descartes with the Stoic position—and indeed, Descartes so identified himself. He identifies Locke with the Epicurean and sceptical position because he regards Locke as deriving everything, including God, from the corporeal. This view of Locke's conception of God was not an uncommon one in Vico's time.[20] Vico—in a short chapter, "Reprehension of the Metaphysics of René Descartes, Benedict Spinoza, and John Locke," that he drafted as an addition to the *Scienza nuova seconda*—says that Locke "is compelled to offer a God all body operating by chance."[21] Although Locke believed that he had established God as immaterial, and thus as unknowable and not in conflict with established religion, his thought was subject to the general charge of materialism, in Vico's time. Locke was suspected of having an expanded version of Spinoza's pantheism and holding that there was only one substance and that it was material. Since, on Locke's view, man cannot know with certainty the nature of substance, materiality and immaterial substance may be the same.

That Locke is a materialist and has an untenable conception of God and the soul is the general direction of the polemic taken by Paolo Mattia Doria in his *Difesa delle metafisica degli antichi filosofi contro Giovanni Locke ed alcuni moderni autori* (1732).[22] In his "Reprehension," Vico draws a contrast between these three modern thinkers, who base all metaphysics in a doctrine of substance and metaphysical thinking in the principle of supposition (*supposizione*) and the thought of the ancients, especially Plato and Aristotle, who, he claims, derive their metaphysics from Being as it is originally captured in the images of the theological poets.

*Riflessione*, as a philosophical term, enters Italian from the work of Locke.[23] Its general use in Italian in various senses parallels that of English and French, and is derived from the Latin original. The *Essay* was translated into Latin in 1701 and into French in 1735. Vico most likely employed the Latin edition, *De intellectu humano*.[24] Moreover, as Vico was a member of various Academies that met for discussions in Naples, he was no doubt part of many discussions of Locke's views, as well as of those of Gassendi and the Cartesians.

Gustavo Costa, in an essay, "Vico e Locke," has explored a number of connections between the Latin translation of Locke's *Essay* and Vico's *Scienza nuova*, especially between Locke's description of sensation and his use of the mind of children.[25] Costa holds that, what Locke claims about sensation, Vico reformulates in historical terms regarding the primordial mentality of the first men. Costa sees a parallel between Vico's description of how the first men stretch words to make their imaginative metaphysics by forming the world in terms of their own passions and sensations (405), and Locke's discussion of the "Abuse of Words" (Bk. III, ch. 10). What is described by Locke in logical and epistemological terms, Vico puts in rhetorical and historical terms. It is not Costa's claim, nor mine, that Vico bases his *Scienza nuova* on Locke's *Essay*, but only that there are at certain points Lockean elements and apparent borrowings and transformations of Locke's views within the *Scienza nuova*.

Vico originally wrote the *Scienza nuova* in negative form (*Scienza nuova in forma negativa*), as a criticism of the standing philosophies of his time. He arrived at the *Scienza nuova* as a complete recasting of the text into positive form, which must have entailed the reformulation of the positive elements of such philosophies into his own account. Seen in this way, we might expect to find that Vico's text is a kind of palimpsest, which, when gazed at closely, can be seen to have the texts of others beneath it, including those he is ultimately against. The more closely we gaze at Vico's text, the more we may see beneath its outlines.

In a great act of *ingenium*, Vico saw that the notion of sensation that was part of the rise of modern empiricism, if understood in historical terms, could become a key to understanding the mind and the metaphysics of the first men. They are the perfect empiricists, the men of sensation, who think with their bodies and are all body, living in a world of bodies. In a corresponding act of *ingenium* Vico saw that the notion of reflection held the key to understanding the modern world, that which arrives within the life of any nation in its period of modernity, its third age. Sensation and reflection were both part of the mind, but they were developed by the mind within the process of history. The notion of reflection as the mind's attention to its own operations may have been in the air, and in the discussions of the moderns who met in the various academies of Naples. But through Locke, and probably also as a feature of Descartes's rationalist procedure of thought, Vico saw prophetically the significance of reflection as the basis of modern philosophy and, in fact, of modern life, for with reflection also comes a form of language and speech that dominates society. There is a reflective way of thinking and there is a reflective way of speaking and of acting. These are replacements for the humanist ideals of wisdom, eloquence, and prudence, as I wish to show now.

What is the significance of the barbarism of reflection for our conception of modern knowledge and modern life? Throughout his works—from at least the *Study Methods*, through the attack on Descartes in the *Ancient Wisdom*, and within the autobiography and the two versions of the *New Science*, as well as elsewhere, in his letters and orations—Vico is claiming that something is wrong with modern thought and with modern education. He states this directly in terms of human education in the *Study Methods*, in which he asks for a balance between the approach of the moderns and that of the ancients, and advocates that students be first educated in the art of topics—in memory, imagination, metaphor, and in plane geometry, because it employs figures—and should only later be introduced to the abstract methods of scientific reasoning, analytic geometry, and metaphysics. To do otherwise is to make the adult mind sterile, unable to find the beginning points of thought and unable to grasp the ways of thinking and speaking that are necessary to civil wisdom.[26] As he says in the *Ancient Wisdom*, anyone who would make a speech or live a life by means of the geometrical method engages in a form of rational madness.[27]

I think that most of Vico's criticism of modern philosophy, modern education, and the conception of modern knowledge generally, is caught up in his term, "barbarism of reflection." It is no accident that he springs this term on the reader just at the end of the *Scienza nuova*. Vico thinks like a poet; this means that he is oriented to give forth his ultimate point in the ultimate lines. His final line of the work is: "that this Science carries with it the study of piety, and that he who is not pious cannot be truly wise" (1112). This thought, as I wish to show, is closely connected with his conception of the barbarism of reflection. The barbarism of reflection is, above all, the loss of the ancient idea of piety. But to make this clear we must consider briefly how much Vico was right about the dominance of reflection as the preoccupation of modern thought and how reflection is connected with the idea of the understanding as the basis of human knowledge.[28]

It is Kant who establishes the firm connection between reflection and the nature of the Understanding. Kant, in the *Critique of Pure Reason*, criticizes both Locke and Liebniz on the basis of his conception of "transcendental reflection." Kant states: "The act by which I confront the comparison of representations with the cognitive faculty to which it belongs, and by means of which I distinguish whether it is as belonging to the pure understanding or to sensible intuition that they are to be compared with each other, I call *transcendental reflection*."[29] Both Locke and Leibniz, in Kant's view, lack a transcendental conception of reflection by which they would be able to sort out what belongs rightfully to the pure understanding and what is given in appearance by sensibility. Locke would reduce all objects to what appears in the sensibility and Leibniz would attempt to obtain the inner nature of things by comparing all objects to the understanding. Transcendental reflection, then, is the proper operation of the understanding of the knowing subject delineating the conditions of its own knowing so that its powers to sense the object are kept in relation to its powers to form logically what is sensed. The subject reflects on its own operations—sensible and conceptual—and grasps the conditions under which it has an object that it knows. My aim here is not the interpretation of the complexities of Kant's "amphiboly of the concepts of reflection" (*Amphibolie der Reflexionsbegriffe*) but on the prominent place reflection is given in Kant's conception of knowledge. Reflection becomes the inner form of the understanding. Transcendental reflection is a synonym for critique itself. Critique or criticism is the reflection of the knower on the very conditions of the possibility of the object known. The project of philosophy is thus defined as the encounter of the mind with its own operations, to be able to say with certainty how the phenomenal object can be known by the knower.

It is Hegel who recognizes the error of basing philosophy in reflection. He says that ancient metaphysics believed that thought could achieve a true knowledge of things, "Aber der *reflektierende* Verstand bemächtigte sich der Philosophie" (But *reflective* understanding took possession of philosophy)."[30] Hegel says that "philosophy is essentially reflective" has become a slogan.[31] The understanding (*Verstand*) of critical philosophy can only give us a table of contents of experience in which sense perceptions fill up the concepts of thought with content. In Hegel's view only Reason (*Vernunft*), which is by nature dialectical, can produce the inner form of what is actual in thought. For Hegel the true is always the whole, and this whole is achieved first by consciousness and then thought coming to know its own nature by grasping all of the forms of its own activity in a total account. After Vico, Hegel is the one philosopher to identify reflection as contrary to the proper search for philosophical truth.[32]

Reflection is barbarous as a philosophical doctrine because it substitutes the idea of self-reflection for the ancient project of self-knowledge. By self-knowledge I mean the dictum attributed to various of the seven Sages, of *gnothi seauton*—Know thyself—and the goal of Socratic thinking—that the purpose of philosophy is to examine life and the nature of the *polis*. Philosophy is to take place in a civic fashion in the *agora*, not in the chamber or *poêle* that is established as the solitary model of Descartes. Ancient philosophy begins in wonder, and the object of wonder is Being, not substance—the Being that the poets originally form as Jove and the pantheon of gods. Substance does not invoke wonder. Modern philosophy begins in the closet, in the philosopher's private thoughts based in doubt and suspicion that result in a quest for certainty. As Vico points out against Descartes, in the *Ancient Wisdom*, to arrive at the proof that we can with certainty know what we think we know is a different project than the quest for truth. The quest for truth is to know *per causas*, to know the cause of things. This cannot be achieved by thought reflecting on its own operations. Only critical perspective can result when thought is carried to the level of transcendental reflection. The principles that thought must necessarily employ can be established, but a knowledge of the causes of the object known are left behind, a knowledge of things in themselves. Kant with his notion of transcendental reflection simply carries on the Cartesian project of critical philosophy and the quest for certainty.

Reflection is the act of knowing coming to know itself, that is, thought turns back upon itself and assesses how it knows what it knows. This at first glance can appear to be self-knowledge, because it is an action of the knower on the knower's own operations as opposed to sensation, that is, the notion of the knower's relation to what is in some sense external or is appearance. By the point that reflection has developed in modern philosophy, reflection and its companion, the understanding, appear to be the very basis of the world of the self. This is a cruel joke thought has played on itself, because the self has been left behind, as little more than the Kantian "I think." The ancient conception of self-knowledge is based on the notion of wisdom, and wisdom, as Cicero defines it and as Vico accepts it, is a "knowledge of things human and divine" (*Tusc.* IV.26.57; *De off.* II.2.5), the distinction that is tied to the original definition of philosophy by Pythagoras and present in Socrates' speech in the *Apology* (23A).

In the modern conception of reflection, the self as that reality that falls between the human and the divine is lost. The ancient conception of self-knowledge has a natural bond with topical philosophy as opposed to critical philosophy. For the self is the *topos* from which the speech on any topic can and must be brought forth. The self's being is what it is in terms of its apprehension of the divine Being, the other reality that Vico's first men experience in the phenomenon of Jove, and although originally only felt with the senses, later, at the hands of the ancient philosophers, it becomes a guidepost of thought. The faculties upon which self-knowledge depends are memory, imagination, and ingenuity, and these are also the faculties upon which rhetoric and eloquence depend.

Vico's first public statement is his inaugural oration of 1699, which has the title: "That Self-knowledge is the Greatest Incentive to the Study of the Cycle of Fields of Knowledge in the Shortest Possible Time." And his last public statement, the oration he gave on philosophy and eloquence to the Academy of Oziosi in 1737, was an endorsement of the Socratic approach to philosophy and the need for wisdom to be joined to language as the perfecter of human nature. At the end of his autobiography Vico compares himself to the ideal of Socrates. In his oration "On the Heroic Mind," Vico exhorts the students to give their minds and their studies a heroic shape. It is not possible in a barbaric age actually to be a hero, to be the living embodiment of virtue. Heroes are not present in our world, not once the heroic age has passed in the course of any nation. The actual presence of virtue in the deeds and being of the hero is lost, replaced by moral philosophy.[33] In the age dominated by reflection, we can aspire to be more than critical thinkers and orient our studies and learning to accord with the highest in our natures. Education, for Vico, is based in the ideal of self-knowledge. In the *Scienza nuova*, Vico refers to the dictum of the seven Sages, "Know thyself" (414), and describes how this ideal is first present in the fables of Aesop and later in the reasoning of Socrates.

As Vico states in his passage on the two barbarisms, quoted in full at the outset of this essay, the barbarism of reflection is not simply the result of abstract thinking that is cut off from the reality of the self and the self's relation to the divine—it is a condition of society as well as a condition of thought. Vico speaks of the barbarism of reflection as containing "a vile savagery [that] under soft words and embraces, plots against the life and fortune of friends and intimates." This line echoes Dante's sins of the *lonza*, the lowest region of the Inferno, in which the social fabric of humanity is eroded by fraud, by treacheries against guests and hosts, friends and associates, and relatives. Reflection and the understanding are not sources for moral philosophy, as is the ideal of self-knowledge. Once thought gives up its relationship to wisdom as a knowledge of things human and divine and becomes critique, that is, an activity of reflection on its own operations of the understanding—piety is no longer relevant. Language can become simply the instrument of desire, and language becomes a form of praxis—not a form of *phronesis* or prudence—but a practice that is connected to wisdom. Thus rhetoric is not held together with dialectic in a bond, as strophe to antistrophe (as Aristotle defines it in the first sentence of the *Rhetoric*), but rhetoric became strictly an instrument for using speech to play upon the passions and to fulfill desire. Reflective thinking can offer us a means but not an end. It cannot be the product of virtue because there is no critical basis for virtue. Virtue requires the self to be in relation to an order that is more than itself. This is one way to see why Kant in his ethics insists that persons must be treated as ends and not means. He holds his philosophical heart in the right place, but in the end can offer no real basis for so treating persons, because he has given up the doctrine of virtue when he gave up the doctrine of self-knowledge for the critical philosophy of reflection and the forms of the understanding.

The correlate to reflection, in modern life, is the belief in the reality of the individual, coupled with the paradox of the uniformity of the social mass. This is the view that the individual can guide his life by introspection, coupled with the desire for applied ethics and decision procedure. Reflection makes the individual's personality the source of his existence, and at the same time (because of its emphasis on the purity of the universal form), reflection encourages the individual to face civil life as an arena of problems, methods, and programs. The line of philosophy from Locke and Descartes to Kant has removed judgment from its connection with virtue and placed it in connection with form. Thus there are forms of judgment, and judgment itself is essentially the establishment of form.

The term "barbarism of reflection" puts together a Latin word, reflection, with a Greek word, barbarism. Barbarism by some accounts is thought to have its origin in onomatopoiea, to imitate the seeming babbling of foreign tongues. Later, in the late Middle Ages and the Renaissance, barbarism is asssociated with a conception of language as barbarous, in its incorporation of foreign words and expressions. Barbarism is originally an incomprehensible speech that later becomes a dishonest or suspect form of speech, a speech that abandons honesty and true eloquence. It is a synonym for a false and affected way of the expression of thoughts. Barbarous speech violates the Socratic ideal that Vico endorses in his oration to the Academy of Oziosi, where he says: "'Right thinking is the first principle and source of writing,' because there is no eloquence without truth and dignity; of these two parts wisdom is composed. 'Socratic writings will direct you in the choice of subjects,' that is, the study of morals, which principally informs the wisdom of man, to which more than in the other parts of philosophy Socrates divinely applied himself, whence of him it was said: 'Socrates recalled moral philosophy from the heavens.'"[34]

In his autobiography, Vico describes the thesis of his oration on the *Study Methods*, saying that its purpose is to weigh and balance the advantages of the ancient and modern conceptions of knowledge. He says that "by adding only a Plato, for example, to what we possess beyond the ancients, we should have a complete university of today; to the end that all divine and human wisdom should everywhere reign with one spirit and cohere in all parts, so that the sciences lend each other a helping hand and none is a hinderance to any other."[35]

In the *Study Methods*, Vico says that the "whole is the flower of wisdom." Self-knowledge, unlike reflective knowledge, is the quest not for certainty but for the whole. Vico's answer to this, in terms of philosophy, is his attempt to join philosophy to philology. Philosophy naturally tends to the formation of abstract sentences and to saying what ought to be the case. Vico's method of the new science is to "have philosophy undertake to examine philology," understanding philology to be the doctrine of all the things "that depend upon human choice; for example, all the histories of languages, customs, and deeds of peoples in war and peace" (7). By following such a method of thinking through the whole of the human world, Vico hopes to have a philosophy that can act against the barbarism of reflection. This would be a philosophy based in heroic mind or heroic thinking, a form of thinking that joins wisdom and piety.

Vico identifies philosophy as we understand it as the creation of the third age of any nation's life and thought with reflection. When, at the hands of Socrates, philosophy enters the stage of the first course of Western history, after Homer's summation of the forms of *fantasia* of the ages of gods and of heroes, it arrives in the form of self-knowledge, but this eventually dissolves into the critical thinking of the schools of the Stoics, Epicureans, and the sceptics. Philosophy finally dissolves itself into logic and solitude and becomes an activity of abstract language, separate from civil life. Following Dante, the Tuscan Homer, philosophy recreates itself in the Renaissance, with the ideal of self-knowledge in the orations on moral philosophy of Pico della Mirandola, Ficino, Vives, and others. But this moves toward the specific doctrine of critical philosophy forged by the moderns at the hands of Descartes and Locke, and, beyond Vico's time, with Condillac, the Enlightenment, and Kant, to the philosophy of the understanding and reflection. At various points throughout his *New Science*, Vico uses reflection as a term identified with philosophy and with philosophical reasoning, and he does this in a positive way, or seemingly positive. But when the philosophers arrive in any nation's course of ideal eternal history, it is the beginning of the end. The reader of the *New Science* should never forget this. Vico has a cyclic, not progressive, conception of history. Each new recourse is based upon the memory of what has gone before, to the extent that it can be remembered, but the recourse is not a kind of spiral or advance. History like all human events operates in cycles. What we find in history is not novelty but repetition, at least if we have an eye good enough to see the analogies between things, if we have the combined powers of memory, *fantasia*, and *ingenium*.

What does Vico offer us, as readers of his *New Science*, in the face of the third age of barbarism? Vico's counter to reflection as the basis of philosophy is meditation and narration, the two acts upon which he claims rests the proof that governs his new science. The combination of meditation and narration, as Vico understands them, is not open to the understanding. Here Vico has in mind Descartes, who gives the reader a *feigned* narration of the order of his thoughts, and consequently, a feigned meditation.[36] Vico in describing his proof to the reader, says that the reader is to narrate and meditate the new science to himself, by that principle that it "had, has, and will have to be" (349). In saying this, Vico is quoting the power assigned to the Muses, the daughters of Memory who govern the arts of humanity.[37]

Vico's new science is a museum, in which all the arts of humanity are brought together and put on orderly display. It is the autobiography of humanity for the reader to pass through the museum in its natural order of development of course and recourse, stopping at each place to meditate on the nature of what is seen. In this way the reader educates himself in the whole; it is a complete wisdom that is absorbed. It is self-knowledge because the reader who remakes the new science through his powers of memory, *fantasia*, and *ingenium* confronts not only the world of human things but also the divine, because what is seen is not simply history but the circular, providential order of all events. Thus wisdom is achieved, accompanied by piety. Moreover, two other things are achieved in Vico's proof, that cannot be achieved by reflective understanding: eloquence and prudence. The museum exists only as an oration in words. Like the poets, the philosopher produces in words his knowledge of things human and divine. Thus Vico's new science is itself not an essay but an oration, in which all things human are spoken of according to their natural and providential order of causes, their causes human and divine.

What is the nature of this oration? It is an oration on moral philosophy. Its point is to display a doctrine of providence, the divine order of history. In this sense, the new science is intended to convey the basis for practical wisdom. In Latin, *providentia* and *prudentia* are synonyms. Vico's doctrine of prudence is based, not on the traditional conception of the study of the actions of great men in history, but on the study of the actions of history itself. To understand the providential order of events and how they are related to authority and to human choice is to discover the model of prudence, or the wisdom of human actions. It is a guide for life in barbaric times—something that reflection cannot supply.

In summary, one way to comprehend the truth that Vico has attempted to convey against the barbarism of reflection, is to bring to mind the image of Raffaello's *School of Athens*, in the center of which stands Plato, who holds the *Timaeus* in his hand, and beside him is Aristotle, who holds in his hand the *Ethics*. Arranged througout the rest of the painting are figures who founded or mastered all the other fields of human knowledge. In the *Scienza nuova*, Vico has taken the most crucial feature of the *Timaeus*, the Demiurge, and moved it from the process of nature onto the processes of history, and recast it in the role of Providence. He has then joined this with the *Ethics*, from which he has taken the notion of human nature acting in accordance with virtue, and the *phronesis* that is required to accomplish this. This is to take the Socratic ideal of self-knowledge, the ideal that is as old as philosophy itself, and as old as Western moral thought itself, and to place it within the dialectic between Plato and Aristotle.

It is an inviting speech, full of fortitude to face the flattery and soft embrace of the reflective language of the understanding. But the philosopher who would cross its threshold can expect no better or worse than Vico got, which was: "Among the caitiff semi-learned or pseudo-learned, the more shameless called him a fool, or in somewhat more courteous terms they said that he was obscure or eccentric and had odd ideas." And he might later have said, with Rousseau, quoting Ovid: "Here I am the barbarian because no one understands me."

## Notes

1. See my *The New Art of Autobiography: An Essay on the "Life of Giambattista Vico Written by Himself"* (Oxford: Clarendon Press, 1991), 161 (my trans.). For the Italian text, see *Vita di Giambattista Vico scritta da se medesimo* in Giambattista Vico, *Opere*, ed. Andrea Battistini, 2 vols. (Milan: Mondadori, 1990), I:5.

2. "Aristotle indeed affirms, all ingenious men to be melancholic" (*Aristoteles quidem ait, omnes ingeniosos melancholicos esse*) (Cicero, *Tusc.* I.xxxiii.80). Cf. "Why is it that all men who have become outstanding in philosophy, statesmanship, poetry or the arts are melancholic?" (Aristotle, *Prob.* xxx.1).

3. Giambattista Vico, *Vita, Opere*, I:13. The English translation is *The Autobiography of Giambattista Vico*, trans. Max Harold Fisch and Thomas Goddard Bergin (Ithaca, NY: Cornell University Press, 1983; orig. pub. 1944), 120.

4. *Vita, Opere*, I:24–25; *Autobiography*, 132–33.

5. *Vita, Opere*, I:84; *Autobiography*, 199–200.

6. *Vita, Opere*, I:73–75; *Autobiography*, 187–89. The source of this malicious book notice was most likely Vico's colleague in Naples, his constant *tormentatore* Nicola Capasso, but Vico may have believed it to be the Neapolitan historian Pietro Giannone, who was in exile in Vienna.

7. Vico, "All'Abate Esperti in Roma," *Opere*, I:322–25.

8. *Principi di scienza nuova d'intorno alla comune natura delle nazioni* (1744), *Opere* I; *The New Science of Giambattista Vico* (unabridged translation of the 3rd ed., 1744, trans. Thomas Goddard Bergin and Max Harold Fisch [Ithaca, NY: Cornell University Press, 1984]).

9. Vico speaks of the "returned barbarian times" (*tempi barbari ritornati*). He means by this the returned beginnings of the *ricorso* of Western history, generally speaking, the Middle Ages. In this he is following Petrarch's phrase of the "barbarism of the Middle Ages." Vico says that the heroic Latin poets of the returned barbarian times were like those of the first barbarian times in their power to tell only true stories: "since barbarians lack reflection which when ill used, is the mother of falsehood. . . . And, in virtue of this same nature of barbarism, which for lack of reflection does not know how to feign. . . . " (817; see also 816, 516, 708). The language of this returned barbarism is not the pure language of the barbarism of sense of the first *corso* per se, but it is a language not capable of reflective abstraction and hence more honest than the language that is to succeed it, once modern society develops within the *ricorso* of Western culture.

10. Beyond a few remarks in standard commentaries, Vico's "barbarism of reflection" has not become a topic in Vico literature. There is an excellent recent essay by Alain Pons, "Vico et la 'barbarie de la réflexion,'" in *La Pensée politique* (revue annuelle), ed. Marcel Gauchet, Pierre Manent et Pierre Ronsanvallon (Paris: Gallimard, 1964), 178–97. Pons focuses on putting together a systematic statement of what Vico says about the barbarism of reflection and does not explore its implications for a criticism of modernity, as I wish to do. I do not wish to go over completely the same ground that Pons does, but the reader would find Pons's essay very useful as a counterpart to my inquiry. A second essay exists on the subject, by Stephen Taylor Holmes, "The Barbarism of Reflection," in *Vico: Past and Present*, ed. Giorgio Tagliacozzo (Atlantic Highlands, NJ: Humanities Press, 1981), 2:213–22. I find little to agree with in Holmes's essay, which is a quick argument

of his own (with no footnotes to Vico's texts) that claims that only "*scientific reflection*," in Vico's view, "is able to stem the rebarbarizing tide of reflection itself." And he concludes: "Thus, the question of how Vico thought reflection capable of turning on itself remains yet unanswered" (222). If Holmes is wrong in his presentation of Vico's position (and I think he is), then the paradox of Holmes's conclusion is just one of his own making, not Vico's. There is also my chapter on "Wisdom and Barbarism," in my *Vico's Science of Imagination* (Ithaca, NY: Cornell University Press, 1981; reissued 1991).

11. In the paragraph formation introduced by Nicolini in the standard Laterza edition of the *Scienza nuova*, vol. 4 of *Opere di G. B. Vico* (Bari: Laterza, 1911–16), and reproduced in subsequent editions of this edition, and in many other editions by other editors, the rhetorical structure of Vico's "Conclusione dell'opera" becomes distorted. When we realize that, in the original editions, both published in Naples during his lifetime, in 1730 by Felice Mosca and in 1744 by the Stamperia Muziana, Vico has divided his conclusion into four paragraphs or parts (that are glossed over by Nicolini's logical divisions into sixteen paragraphs), we see that this follows the classical division of legal forensics as stated by Quintilian in book 4 of the *Institutio oratoria*. Paragraph 1 is Vico's *exordium* (in which Quintilian says we can stir up the judges); Paragraph 2 is the *statement of facts* (Vico's account of what really happens in human affairs in history). Paragraph 3 corresponds to the *proofs* "whether we are confirming our own assertions or refuting those of our opponents"; Paragraph 4 is the *peroration*, "whether we have to refresh the memory of the judge by a brief recapitulation of the facts, to do what is far more effective, stir his emotions." Vico confirms his own assertions in Paragraph 3, and in Paragraph 4 he both recapitulates and stirs the emotions of the reader (the judge) with feelings for the ancients and for the divine (his references to piety and wisdom), once again pagan and Christian.

12. In later medieval philosophy, in the psychology of Roger Bacon, reflection can be found used in the sense of the soul's knowledge of itself, for example, the power of reflecting on the contents of consciousness is more fully described in the passage: "'In anima . . . reflexio potest fieri vel conversio intellectus supra speciem absolute, non considerando cuius rei sit illa species vel ymago, et sic fit pura apprehensio speciei et non memoria, vel potest fieri reflexio supra illam considerando cuius rei sit, et conferendo ad rem cuius est, et sic fit cum apprehensione memoria.' *Quaest. Met., xi, p. 88;25.*" E. D. Sharp, *Franciscan Philosophy at Oxford in the Thirteenth Century* (Oxford: Oxford University Press, 1964; orig. pub. 1930), 165. *The Revised Medieval Latin Word-List* (Oxford, 1965) indicates that *reflexion* in the sense of self-knowledge is not earlier than the thirteenth century.

13. E.g., in Plato, *Rep*. 6.510a; *Soph*. 266b; *Tim*. 71b; in Aristotle, *Meteorology*, bk. 1, chs. 5 8, bk. 2, ch. 9, and bk. 3, ch. 4; *De anima*, bk. 2, ch. 8, and bk. 3, ch. 12; *Sense and Sensibilia*, ch. 2.

14. A search of *Oeuvres de Descartes*, ed. Adam and Tannery, 12 vols. and suppl. (Paris: Cerf, 1897–1913), shows Descartes to use *réflexion* in the sense of a mathematical and physical phenomenon in his correspondence with other thinkers, e.g., Mersenne (vols. 1–5), and in his *Cogitationes privatae*, and in the piece "Descartes et Beeckman" (vol. 10), and also, more obviously, in the essay on *Dioptrique* appended to the *Discours*.

15. John Locke, *An Essay concerning Human Understanding*, ed. Alexander Campbell Fraser, 2 vols. (Oxford: Clarendon Press, 1894). Cited above by book, chapter, and paragraph number.

16. *Oeuvres de Descartes*, VI:41. In French, the further important use of reflection is Condillac's reworking of Locke's views that aim at giving reflection a materialistic basis, which appeared after Vico's death: Condillac, *Essai sur l'origine des connaissances humaines* (1746), *Oeuvres philosophiques de Condillac*, ed. Georges Le Roy, 3 vols. (Paris: Presses Universitaires de France, 1947–51), 1:21–23 (I.2.v.).

17. Vico, *Vita*, *Opere*, I:15; *Autobiography*, 122.

18. The line reads: "e sì, nella celebrità [a Latinism, *celebritas*, i.e., a crowd] o folla de' corpi, vissero come bestie immani in una somma solitudine d'animi e di voleri" (1106). In Oration VI, Vico says: "in corporum frequentia summa est solitudo animorium" (*Le orazioni inaugurali I–VI*, ed. Gian Galeazzo Visconti [Bologna: Mulino, 1982], p. 195 ("assemblies of men may appear to be societies, but the truth is that isolation of spirits is greatest where many bodies come together")).

*On Humanistic Education: Six Inaugural Orations 1699–1707*, trans. Giorgio Pinton and Arthur W. Shippee (Ithaca, NY: Cornell University Press, 1993), 129. The source is Cicero: "And who does not believe that those are more alone who, though in the crowded forum, have no one with whom they care to talk, than those who, when no one else is present, either commune with themselves or, as we may say, participate in a gathering of most learned, men finding delight in their discoveries and writings" (*De re pub.* I.xviii.28). See Battistini, ed., Vico, *Opere*, II:1751. The paradox of solitude runs throughout Vico's life and work.

19. The theme of the interconnections of *sapientia*, *eloquentia*, and *prudentia* runs throughout Vico's *Inaugural Orations*; see my introduction to *On Humanistic Education*. On the ideal of *la sapienza che parla* see *Vita*, *Opere* I:84; *Autobiography*, 199, and Vico's oration, "Le accademie e i rapporti tra la filosofia e l'eloquenza," *Opere*, I:408 ("The Academies and the Relation between Philosophy and Eloquence," trans. Donald Phillip Verene, in Giambattista Vico, *On the Study Methods of Our Time*, trans. Elio Gianturco (Ithaca, NY: Cornell University Press, 1990), 89. Cf. Cicero: "For eloquence is nothing else but wisdom delivering copious utterance" (*De part. orat.* xxii.79.).

20. The best study of this is John W. Yolton, *John Locke and the Way of Ideas* (Oxford: Oxford University Press, 1953). Yolton writes: "The charge of materialism against Locke was commonly made by his contemporaries on the ground of his suggestion that God might be able to add to matter a power of thinking" (pp. 145–46).

21. See my "Giambattista Vico's 'Reprehension of the Metaphysics of René Descartes, Benedict Spinoza, and John Locke': An Addition to the *New Science* (Translation and Commentary)," *New Vico Studies* 8 (1990): 2–18.

22. Vico says that Doria was the most intelligent man he had ever met, and the only one with whom he could discuss metaphysics (*Vita*, *Opere*, I:29; *Autobiography*, 138).

23. See Salvatore Battaglia, *Grande Dizionario della Lingua Italiana*, vol. 16 (Turin: UTET, 1992), 267, entry for *riflessione*. But prior to Locke is the work of Paolo Sarpi (1552–1623): "niun conoscente conosce di conoscere se non facendovi riflessioni" (Sarpi, *Scritti filosofici e teologici*, ed. R. Amerio [Bari: Laterza, 1951], 130). In his work on *L'arte di ben pensare* (the original text of which is lost), Sarpi anticipated Locke. "In this he distinguished between sensation and reflection, and referred our ideas of sound, colour, taste, and smell to reflection. In it he also spoke of human language, of words, of genus and species, of faith and truth, and error, of the law of association, of demonstration and probability, and to a very large extent he covered the same ground that Locke did about a hundred years later in his celebrated 'Essay on the Understanding.'" Alexander Robertson, *Fra Paolo Sarpi*, 2nd ed. (London: Sampson Low, Marston, 1984), 53.

24. John Locke, *De intellectu humano*, trans. Ezekiel Burridge (London, 1701) was reviewed in the *Acta eruditorum* in 1702 and reprinted in Leipzig in 1709; in Amsterdam in 1729.

25. Gustavo Costa, "Vico e Locke," *Giornale critico della filosofia italiana* 3 (1970): 344–61; See also Costa, *Vico e L'Europa* (Naples: Guerini e Associati, 1996), ch. 2.

26. Vico, *Study Methods*, sec. 3.

27. Vico, *De antiquissima Italorum sapientia*, *Opere filosofiche*, ed. Paolo Cristofolini (Florence: Sansoni, 1971); *On the Most Ancient Wisdom of the Italians*, trans. L. M. Palmer (Ithaca, NY: Cornell University Press, 1988).

28. Vico appears at various points throughout the *New Science* to speak in an approving way about reflection, e.g., "The human mind is naturally inclined by the senses to see itself externally in the body, and only with great difficulty does it come to understand itself by means of reflection" (236; cf. 218–19). But the powers of reflection are always a mixed blessing for Vico (as is Philosophy itself, for its arrival in the third age of the ideal eternal history marks the beginning of the end of *corso* or *ricorso*). A good example of this ambiguity is: "philosophy should make the virtues understood in their idea, and by dint of reflection thereon" (1101), yet Vico adds: "but as the popular states became corrupt, so also did the philosophies. They descended into scepticism" (1102).

29. Kant, *Kritik der reinen Vernunft* (Hamburg: Meiner, 1956); *Critique of Pure Reason*, trans. Norman Kemp Smith (London: Macmillan, 1958), A262; B317. Kant's term for reflection here is *Überlegung*, which he is using as the equivalent for the Latin *reflexio*. He makes this clear in the first

sentence of the "Amphiboly of Concepts of Reflection": "Die überlegung (*reflexio*) hat es nicht mit den Gegenständen selbst zu tun . . ." (A260; B316).

30. Hegel, *Wissenschaft der Logik*, 2 vols. (Hamburg: Meiner), I:26.

31. What Hegel identifies as a "slogan" in his day has become the dominant concern of philosophy as it is practiced in ours, from Husserl to Sartre to Dilthey to Gadamer. The best single guide to this development, that I have found, is the entry on *reflexión* in vol. 4 of Jose Ferrater Mora, *Diccionario de filosofia* (Madrid: Alianza, 1980), 2807–11. Also R. Hébert, "Introduction à l'histoire du concept de réflexion: Position d'une recherche et matériaux bibliographiques," *Philosophiques* 2 (1975): 131–53.

32. See my "Two Sources of Philosophical Memory: Vico versus Hegel," in *Philosophical Imagination and Cultural Memory*, ed. Patricia Cook (Durham, NC: Duke University Press, 1993), 40–61.

33. Vico, "De mente heroica," *Opere*, II:367–401.

34. Vico, *Opere*, I:408; *Study Methods*, 89. See Horace, *Epistola ad Pisones* (Ars Poetica), 309–11 and Cicero, *Tusc*. V.4.10.

35. Vico, *Vita*, *Opere*, I:36; *Autobiography*, 146.

36. Vico, *Vita*, *Opere*, I:7; *Autobiography*, 113.

37. See my "The New Art of Narration: Vico and the Muses," *New Vico Studies* 1 (1983): 21–38.

38. "Barbarus hic ego sum quia non intelligor illis" (Ovid, *Tristia*, v.x.37). Affixed to Rousseau's title page of the *First Discourse* (*Discours qui a remporté le prix a L'Académie de Dijon. En l'année 1750*).

*Part III*
*Human Sciences*

# An Antiquary between Philology and History

## Peiresc and the Samaritans[1]

Peter N. Miller

Once upon a time, when the world of learning was smaller and its prospects were grander, antiquaries prowled the landscape collecting, describing, comparing, ordering, and re-ordering all that could be known of the world's history. Their questions and practices have since been lost to posterity with the subsequent partition of that homeland where philology, philosophy, anthropology, and archeology once met and mingled. The antiquary worked with antiquities, what Bacon called "history defaced, or remnants of history which have casually escaped the shipwreck of time." Indeed, it is Bacon who has left us one of the most evocative and perspicuous accounts of this practice.

> *Antiquities*, or remnants of histories, are (as was said) like the spars of a shipwreck: when, though the memory of things be decayed and almost lost, yet acute and industrious persons, by a certain perseverance and scrupulous diligence, contrive out of genealogies, annals, titles, monuments, coins, proper names, and styles, etymologies of words, proverbs, traditions, archives, and instruments as well public as private, fragments of histories scattered about in books not historical,—contrive, I say, from all these things or some of them, to recover somewhat from the deluge of time; a work laborious indeed, but agreeable to men, and joined with a kind of reverence; and well worthy to supersede the fabulous accounts of the origins of nations; and to be substituted for fictions of that kind.[2]

New interest in the history of early modern scholarship and in the history of art and archeology has served to focus attention on antiquarianism.[3] The classic work on the subject by Arnaldo Momigliano[4] has begun to be revisited and the present essay, a sketch of a particular antiquary's interest in a circumscribed subject that turns out to have far-ranging implications, is a contribution to this deepening engagement.

Why Peiresc? Nicolas-Claude Fabri de Peiresc (1580–1637)[5] was one of the most famous Europeans of his generation, hailed by Momigliano as "that archetype of all antiquarians" and celebrated in *Tristram Shandy* as an "indefatigable labourer ... out of love for the sciences." In the *Polyhistor*, Daniel Morhof singled out Gassendi's *Vita Peireskii* as the examplary scholarly life and Guez de Balzac merged person and practice when he identified Peiresc, in an echo of Bacon, as himself "a piece of the shipwreck of antiquity and relic of the Golden

Age."[6] Looking carefully at how he worked can help us understand the antiquaries' place in the history of scholarship. Theirs is a story that branches off from the main line that runs from Scaliger and Casaubon to Bentley and veers towards the foundation of that edifice to be built, later, by Gibbon, Burckhardt, and Huizinga. It is by studying texts and objects with equal seriousness, and seeking to augment their quantity by catalyzing a wide-ranging intellectual network that extended from England to Ethiopia, and which included the planning of scholarly expeditions, that Peiresc presents a striking example of an intellectual practice that stands poised between philology and cultural history.

Like many of the antiquaries, Peiresc was also a polymath; his studies of antiquities took place alongside dissections and telescope-aided observation of, among other items, the Medicean planets and the first nebula, in the constellation Orion, which he discovered. Contemporaries recognized that what was being constructed was "*votre Encyclopedie.*"[7] No man, claimed Gassendi in his biography of Peiresc, "was more desirous then he, to run through the famous Encyclpoedia, or whole Circle of Arts" (*celebre illud liberalium disciplinarum coronamentum*). Jean-Jacques Bouchard, too, in his funeral oration, praised Peiresc's letters as so crammed with all sorts of learning that he might "have been said to have gone through the whole Encyclopedia or perfect Orbe of all Learning and liberal Arts" (*universum omnium doctrinarum et liberalium disciplinarum orbem*).[8] Much of the interest in people like Peiresc is derived from an ongoing effort to understand better early modern encyclopedism, including of course, the volume in which this appears. Our encyclopedia, however, looks very different, and as it began to crystallize in the later seventeenth century people like Peiresc ceased to fit—and then ceased to matter.

Why the Samaritans? They re-emerged on to the scholarly map in the seventeenth century for the first time since late antiquity because they offered an alternative version of Judaism to an age obsessed by the beginnings of Christianity.[9] But they also represented a link in the transmission of culture from East to West since their alphabet was shared by both ancient Jews and Greeks. Scaliger was the first to perceive that these two stories coincided in the history of the Samaritans and Peiresc seized on the implications of this for understanding the relationship between Biblical and classical history. With this step three important developments in the early modern history of scholarship hove into view. First, the antiquaries' study of the ancient Near East marks the extension to the extra-European world of the recognition that past and present were discontinuous that was the fruit of Renaissance historical thought. Second, the way in which the Bible's account of the ancient Levant could now be fit into the received history of the classical world succeeded, finally, in making the Bible into history. It was this very success in making the sacred *historical* that was to render it vulnerable to all the skepticisms that beset the study of the human past. Third, the antiquaries' researches provided a means of forging a common narrative that could integrate the classical and the extra-classical, or non-European, worlds with all of the obvious implications for what counted as the *oecumene* and its natural forms of morality, religion, and society.

References to Joseph Scaliger in Peiresc's work mark this trail from philology to the broader study of culture. It was, typically, in Italy that the twenty-year-old from Aix first came into contact with Scaliger, a French exile in Leiden. Equally typically, this contact was epistolary. The letters they exchanged prior to Scaliger's death in 1609 show Peiresc continually striking the pose of client, protesting his willingness and desire to serve Scaliger's interests. These included the acquisition of Hebrew books and coins and information concerning della Scala family history. In addition, he took upon himself the task of seeing to the recovery of Scaliger's newly-acquired Samaritan Pentateuch that was lost with the foundering of the *St. Victor*. "I will employ all my friends in Marseilles who trade with the Levant to endeavor to recover it. All that I desire in this world is to have occasions to render such service."[10]

What was the intellectual legacy of Scaliger for Peiresc? As Anthony Grafton has shown, Scaliger applied philological methods to texts and artifacts of the non-classical East that were communicated to him by both scholarly travellers and well-informed natives.[11] It was precisely this approach that led him to the breakthrough in historical chronology constituted by *De emendatione temporum* (1583) and *Thesaurus temporum* (1606). Chronology itself, as he envisioned it, was a discipline whose essence was synthetic: time was the same for the Babylonians, Chinese, Egyptians, Jews, Greeks, and Christians.[12] Since their local narratives ought, then, to fit together, chronology could be projected as the foundation for a universal history—or a new encyclopedia.[13] Scaliger classified etymology as pseudo-science,[14] but recognized that a historically-informed comparative linguistics could provide rich results.[15]

Scaliger offered a model that a young admirer like Peiresc could emulate. Although lacking the technical skills and intuition that enabled Scaliger to work through the chronological material, Peiresc pursued the insight that classical philology together with oriental studies could yield a new history of civilization. As Momigliano has argued, Gibbon's *Decline and Fall* was the eventual fruit of this insight.[16] Peiresc's synthetic approach to questions of ancient metrology and comparative linguistics pushed at the limit between classical and non-classical history and fed on a constant flow of new materials relayed from residents of the extra-European world and travellers he kitted out and dispatched with shopping lists. What La Popelinière, in his famous letter to Scaliger of 4 January 1604, theorized as the next and necessary step towards "the perfection of history," namely scholarly travel, Peiresc took as a given.[17] Peiresc also understood the relationship between chronology and universal history. In a letter of 1632 he politely refused to loan out his copy of the *Thesaurus temporum* because it was covered with marginal comments "for my use in diverse places."[18] Peiresc was, as Grafton has noted, an "imitator" whose interests and approach, even if broader and more diffuse than Scaliger's, nevertheless constituted the "true continuation" of his work.[19]

Perhaps nowhere is this clearer than in Samaritan studies. Scaliger had been drawn to them because of their calendar;[20] this, in turn, drew him into a much wider investigation of Second Temple and rabbinic Judaism.[21] In response to a letter accompanying their calendar in 1584, Scaliger addressed a series of questions about their rituals to leaders of the Samaritan community. Their replies of 1590 never reached him, but they were eventually recovered by Peiresc in 1629 and sent to Paris to help Jean Morin with his work on the Samaritan Pentateuch.[22] Copies of the Latin translation in Peiresc's hand are annotated in his customary fashion of underscoring passages of interest.[23] He paid closest attention to geography (the Samaritan temple on Mt. Gerizim), rituals (observance of Sabbath, Passover, and Circumcision) and, in particular, the institution of the priesthood and the tradition which linked the first high priest, Aaron, with the current one, Eleazar, the letter writer himself. In addition, Peiresc underlined those questions addressed by the Samaritans to Scaliger and which were precisely the sort that Peiresc included in the instructions that he drew up for travellers to alien lands: What language do you speak? Where do you live? Who is your ruler? What is your law? Who are your priests?[24] In Scaliger's questions, the Samaritans' answers, and Peiresc's annotations it is already clear that the scholarly study of the Bible could be shaped by the antiquary's practice: texts were illuminated by a context that could be literary, material, or living, and in the best case, as with the Samaritans, all three.

But Scaliger also grew interested in the Samaritans because they conserved the ancient Hebrew alphabet that was shared with the Phoenicians. In a much reworked passage in the *Thesaurus temporum* Scaliger argued for the Phoenician derivation of the Ionic alphabet. Anthony Grafton has noted the importance attached to this argument by its author and a friendly reader, Isaac Casaubon, who commented simply: "Digressio de literis Ionicis, admirandae eruditionis."[25] In a late letter to Richard Thomson (23 September 1607) Scaliger extended this same argument backward in time, observing that "Phoenician letters" were used in Canaan at the time of Abraham and served as the script of the ancient Jews; after the alphabet shift they remained in use solely among Samaritans.[26] Hence the implied claim that study of the Samaritan language could shed much light on the crucial Phoenician link between Biblical and classical history. In Jacques Leschassier's (1550–1625) memoire preserved in Peiresc's oriental register (see below) both passages in which Scaliger makes the Phoenician argument are recorded. Indeed, this same theme dominates Leschassier's letter to Peiresc of 10 May 1610.[27] Peiresc was, then, clearly aware of the importance contemporaries were beginning to attach to the Samaritans in the construction of a new world history.

Peiresc's interest in the Samaritans was fired by word received from Girolamo Aleandro in Rome in the late spring of 1628 of plans to publish their Pentateuch in the Paris Polyglot Bible.[28] Aleandro had received the news in a letter from the prospective editor and translator, Jean Morin of the Oratory, who had just published a new edition of the Septuagint with a preface stressing the utility of the Samaritan Pentateuch for biblical scholarship. Morin had written to inquire about the existence of ancient shekel coins bearing Samaritan inscriptions. In reply, Aleandro mentioned that two additional Samaritan Pentateuchs could be found in Rome, one obtained by Scipione Cobelluzzi, Cardinal of Sta. Susanna, and the other by Pietro della Valle, the famed aristocratic traveller.

Scaliger dominated Peiresc's involvement in *cose Samaritane*, and his posthumous authority was especially relied upon in the early stages when Peiresc needed to motivate others to work on his behalf. On the heels of Aleandro's letter, Peiresc informed Pierre Dupuy that there existed in Rome a Samaritan text "which I would esteem much more than all the rest. It would be worth undertaking an edition containing both. The late M. della Scala would have desired to see this with an extreme passion."[29] A week later, on 27 May, Peiresc wrote to Aleandro acknowledging his desire to help accelerate the appearance of Morin's edition of the Pentateuch. Peiresc agreed that if it were possible to include della Valle's text in "the true Samaritan language"—Samaritan-Aramaic as opposed to Hebrew in Samaritan characters—this ought to be done without depriving him of the original. What had captured his imagination was word that della Valle possessed a Bible that was written in "Egyptian" with the Arabic version on the facing page "which," Peiresc adds, "I esteem a treasure, among the richest and most noble of all antiquity."[30] Peiresc's repeated conjunction of Samaritan and "Egyptian" (really Copic) reflects his working hypothesis that comparative linguistics held the key to discovering, and rigorously establishing, the connection between the ancient eastern Mediterranean societies.

Peiresc's letters of Autumn 1628 to his chief intellectual contacts in the Barberini court, Lucas Holstenius and Aleandro, are full with questions and theories about the relationship between Greek, Coptic, and Samaritan. Holstenius is asked to examine the "Egyptian" fragments and advise him on the "language" and "characters" in which they were written.[31] In a letter to Holstenius of 10 November, Peiresc applauds his effort to familiarize himself with the oriental languages "from which derive the most notable origins of antiquity." In particular, he suggests that Holstenius make the acquaintance of della Valle and view his collection of manuscript books retrieved from the Levant and "especially the Samaritan and Egyptian."[32] Writing to Paris on the 22nd, Peiresc notes the arrival of letters from Rome, including one from della Valle, whose Samaritan and Egyptian books "are exquisite pieces that the late Mr. de l'Escale would have found perfectly to his taste."[33]

Finally writing directly to the famed traveller Pietro della Valle, Peiresc acknowledges his "great pleasure" in learning about these books. He agrees on the necessity of supplying a Latin translation, which he thought Morin could fashion. Peiresc does, however, admit that "it would be difficult to persuade me that one can rely on the diligence" of Morin for such a task, given his complete absence of familiarity with the Levant or with Samaritans. Peiresc's comments here shed some light on how he believed rare languages ought to be learned. Travel and immersion were one option, while living with a group of native language speakers in Europe was another. Both of these could be accomplished through the initiative and philanthropy of a "Great Prince." Morin had done neither and so his learning was less than the best. Nevertheless, Peiresc believed that it was necessary to keep him involved, so long as he worked with dispatch.[34]

The great Scaliger, Peiresc wrote, who so carefully studied oriental languages "had had such a great desire to penetrate into the Samaritan traditions" in order to read a computus that he spared no effort to acquire Samaritan texts. How did Scaliger learn the language? Since a Psalter was all he possessed, he read it alongside the Latin so as to master the vocabulary and form a grammar "which he showed me several times." Scaliger's unfulfilled desire was to acquire a Pentateuch which would substantially further his knowledge of the language. Peiresc declared that it was "for love of him for this purpose only" that he himself had written to the Levant to locate and purchase such a volume. Though his subsequent success was thwarted by shipwreck, della Valle's acquisition would "give greater ease to the *letterati* and practitioners of oriental languages who could extract from it a grammar . . . [and] . . . finish the work that Scaliger had only begun."[35]

In another letter to della Valle, this of October 1630, Peiresc turned to the question of the link between pronunciation and provenance. While Peiresc's newly-acquired Samaritan Targum, or paraphrase, came from Damascus, he recalled that Aleandro had mentioned that della Valle's was brought back from Persia. He speculated that differences between them in spelling and pronunciation reflected cultural factors, whether a harshness more common in the eastern Samaritan dialect, or the influence of Coptic on the Samaritan spoken in Egypt. Peiresc was wrong both in the particular—della Valle's Pentateuch was also purchased in Damascus—and in the general—there is no acknowledged differentiation between eastern and western Samaritan. But there was already, thanks to Scaliger, an awareness that there were Aramaic dialects that varied from Jerusalem to Antioch to Babylon, and one sees Peiresc applying something of this distinction as well as contemporary platitudes about the influence of climate on speech.[36] Peiresc was later to invite Morin to view his collection of medals with Phoenician and Punic legends. He had just received an inscription recovered from off the African coast written in Punic characters which merited closer inspection and "principally those which could have some relationship to the shape of some of the ancient Hebrew characters, or the modern Samaritan."[37]

Peiresc's support for Samuel Petit reflects this same intellectual commitment to the value of the Samaritans as a bridge between the Classical and Biblical worlds. In a letter to Pierre Dupuy of July 1629 Peiresc introduced Petit, a Protestant minister from Nîmes, who had shown him a small work he had written on the Samaritan computus which drew on material unknown to

Scaliger.[38] In a later letter to Lucas Holstenius, Peiresc described Petit as a translator of Punic who was trying to establish the rules governing the relationship between it and oriental languages.[39] Peiresc tried to find for Petit a position in the group of scholars working on the Paris Polyglot and praised him to his friends.[40]

How seriously did Peiresc treat the study of Samaritan? Was he, for example, able to read it?[41] His file on oriental languages in the Bibliothèque Nationale preserves a short Samaritan grammar entitled *Lashon Shamraita: Lingua Samaritica*, written by one Christopher Crinesius, professor of public theology at Altdorf and author of a Syriac grammar and several works on comparative Semitics. The text is more of a brief history of the origins of the language than a proper grammar. Its entry into Peiresc's collection can be precisely dated. In a letter of 20 March 1629, Jacques Dupuy, after acknowledging receipt of della Valle's book on Persia, notes that the book entitled *Lingua Samaritana* was no longer available and that he believed there were never more than 4 or 5 exemplars printed. He was sending Peiresc the copy he had obtained.[42] In the return letter of 14 April, Peiresc observes that he "did not find in this discourse on the Samaritan language what I expected, at least from the author's contribution."[43]

In the early letters to della Valle, Peiresc wished most of all to understand the relationship of the Samaritan language to Hebrew, Syriac, and Aramaic. Was it dependent on one or the other of these or, rather, a pastiche in which all "participated"?[44] In a subsequent letter, Peiresc emphasized the importance of philological collation. "And the comparison would make it easier to choose what would be more appropriate and more conforming to the Hebrew text." After requesting a sample from Deuteronomy, Peiresc repeated that his criterion for choosing amongst the variants was that it be "the most proportionate and most conforming to the most ancient Hebrew."[45]

Peiresc did indeed compare the passage with that in his own Samaritan Targum. A memoire preserved in the Bibliothèque Nationale records his comments (see appendix 1). The document shows Peiresc's ability to navigate the Samaritan alphabet and exactly how a seventeenth-century antiquary set about the task of "comparison." It also illustrates how Peiresc's particular multilingualism—fluent in French, Italian, and Provençal—enabled him to perceive immediately how language changed over time within families across dialect lines. Hence, differences between Samaritan texts from different locations could be assimilated to those between Gascon and Provençal.

A fascinating exchange of letters with Denys de Sailly, Prior of the Charterhouse in Aix, revisits the family relationship between languages now called Semitic in a model of the antiquary's comparative method. De Sailly's first letter (3 July) was prompted by Peiresc's gift of Morin's *Exercitationes Biblicae* (1631). It was precisely Morin's historical argument that the ancient Hebrew script abandoned by the Jews in the time of Ezra was retained by the Samaritans that elicited de Sailly's response. How could the Jews have exchanged the letters with which God himself wrote the Ten Commandments for those of the Assyrians, their idolatrous enemies? Morin had argued that the existence of ancient Judean coins bearing inscriptions resembling the modern Samaritan demonstrated that at a certain point the alphabet was shared. De Sailly's reponse—underlined by Peiresc as part of his filing system along with the earlier mention of R. P. Morin and "Characteres desquelz s'est servi Esdras"—is that this showed that the Jews employed two alphabets, one for sacred writing and the other for profane.[46] This "Egyptianizing" argument, since it followed from the contemporary view of the hieroglyphs as sacred letters, could accommodate the appearance of the Assyrian script while minimizing the importance of the Samaritans as a privileged source solely because they used this alphabet. De Sailly asked to borrow Pieresc's copy of Simon de Muis's work (a bitter critic of Morin's thesis) and to have his opinion on these matters.[47]

Peiresc's long response of 6 July, a clean copy of which he retained under the filing title "De Samaritanorum characteribus" (appendix 2) indicating its importance as his statement on the subject, takes as its point of departure the nature of change in a culture—here its alphabet—over time. It was, he thought, no more difficult to imagine the Jews abandoning the script in which God wrote on the tablets than Moses shattering them by his own initiative. Second, and much more central to Peiresc's answer, since comparison of the ancient and modern "Chaldaic and Syriac" scripts showed signs of change why could the same process not have affected Hebrew? His own collection of ancient and more recent Hebrew manuscripts revealed just such a variation. Moreover, reflection on the historical development of the European vernacular scripts offered a third proof of the ease with which alphabets could alter in very short periods of time. In any event, conclusive argument depended upon the presentation of contemporary evidence.[48]

Scaliger had sought out Jews to learn Hebrew, Maronites to learn Syriac, and Samaritans to learn Samaritan. A generation later, Peiresc could rely on printed grammars like Crinesius's; his initiative was in recognizing the broad intellectual implications of Scaliger's scattered scholarly intuitions and turning them into research projects. If the difference between Scaliger's interest in the Samaritans and that of his teacher Guillaume Postel[49] marks one transition in the history of humanist orientalism, that between Scaliger and his "disciple" Peiresc marks another. This is perfectly illustrated in a *memoire* preserved in Peiresc's volume of oriental manuscripts in the Bibliothèque Nationale. It has been the subject of a fascinating article by J. G. Fraser, and while both the specific authorship and dating of the text remain somewhat uncertain, the import is clear.[50] Scaliger's interests in Semitic epigraphy, the Samaritan Pentateuch, and religious practices are reflected in excerpts from his writings. Separate headings are then annotated to reflect the current state of evidence. This material ranges chronologically from Jerome through Scaliger to Claude Duret (*Thresor de l'histoire des langues de cest univers*, 1613). The final heading, listing "Books to be Recovered" seems exactly the sort of practical extrapolation typical of

Peiresc. As Fraser notes, this last list closely resembles the memorandum prepared for Theophile Minuti on the eve of his expedition in 1629.

While fascinated by the historical connections that the comparative study of objects and languages facilitated, Peiresc frequently put his tentative conclusions to the test: would further research bear them out? Not content to wait and depend on what others brought back to Europe for scholarly scrutiny he established his own independent network of diplomats, merchants, and missionaries who were given background briefings and lists of questions before departing, and were often debriefed in Provence upon their return. These provided a steady stream of raw material for his philological observations. In a long letter to Dupuy of 7 November 1629, for example, Peiresc announced the impending arrival in Marseilles of a Samaritan grammar and a Pentateuch in three columns, all in Samaritan script, but each in a different language, Hebrew, Samaritan-Aramaic, and an "ancient vulgar" that some judged to be Arabic and others Syriac.[51] The Samaritan "grammar" was actually a nearly complete dictionary in ten notebooks. Each word was defined in three different languages, leading Peiresc to surmise that the dictionary was meant to accompany the Samaritan triglot and that its languages were Hebrew, Samaritan, and either Syriac or "old Arabic," about which, Peiresc declared, he "understood nothing or almost nothing." There were, in addition, some seven or eight other notebooks each containing fragments of a grammar.[52]

The great Scaliger, Peiresc hastened to add, would have been able to draw some valuable observations from even these fragments, since he himself had gone through the "agony of fabricating a sort of grammar in this language." Peiresc suggested inquiring of his executor, Daniel Heinsius, if Scaliger had left any unpublished materials which could be of use to Morin in drawing "something more solid" from these remaining fragments. As for himself, Peiresc believed that these two versions of the Pentateuch would greatly assist in the study of the Samaritan language and the content of its literature. In the meantime, Peiresc would have his agents continue their search for Samaritan materials. He concluded, characteristically, that "it was necessary to see if he could succeed in either the one or the other and aid the public in every way possible."[53]

Among all those who did Peiresc's bidding in the East, Theophile Minuti, the Minim monk, was given the most detailed instruction. He was furnished by Peiresc with a list of contacts in Constantinople, Aleppo, Jerusalem, and Cairo, and with a series of memoranda designed to guide acquisitions. In the "Memoires sur les medailles et pierres precieuses Gravées, qui [se] peuvent reschercher et recouvrer en Levant," Peiresc stressed that he was interested in Greek coins "but above all those which are found written in characters resembling the Samaritan, of whatever sort of metal." In the event that any were found, sketches were immediately to be made "so that these could serve as instruction, at least for those who are doing research." Moreover, "since the Samaritans are of greater curiosity than the others," if their owners refused to sell, then Minuti was to press for permission to have lead or plaster casts made of the medals. Peiresc urged Minuti to look for engraved gems, especially those with inscriptions in Greek, Latin, and Samaritan. Whatever quantity could be acquired at a price "bien moderé" was to be purchased, but, he added, "principally those in which one could recognize Samaritan characters."[54] That six of the eight paragraphs concerning procurement gave priority to Samaritan things is, surely, a reflection on Peiresc's thinking in the year 1629, the year of his deepening involvement with della Valle and Le Jay.

A contemporary document, the "MEMOIRE POUR LES INDES" prepared for the trip of Ferrand Nuñes and Manuel da Costa Casseretz to Goa, includes the same sort of instruction. Medals with Greek, Latin, or Arabic inscriptions were to be purchased, but special care was to be taken "above all with those where there are Samaritan characters, or those which resemble them."[55] Copper coins in Greek or Samaritan "or those closely resembling" Samaritan were specifically mentioned. Just as the conjunction of Samaritan and "Egyptian" in the letters to della Valle reflects Peiresc's effort to place Egypt in the history of oriental languages, that of Samaritan and Greek points towards Peiresc's view of Samaritan as a fossilized form of Phoenician that could help explain the origins of the Greek alphabet in the ancient Levant. A comprehensive inquiry pursued along these lines would yield a history of the encounter between the Levant, Egypt, and Greece that was the subject of so much contemporary romance and scholarship. A fragmentary memo that seems to date from this period addresses the question of language families and their historical development: "Envoyer un eschantillon des trois Langues, et des Prieres, et des Epistres, pour faire determiner ce qui est du vray Language de chascune soit Hebreu, Syriaque, Arabe, ou Cophte, ou Samaritain."[56]

Minuti was also charged with the acquisition of books. This is detailed in a fragmentary autograph note preserved at the Bibliothèque Méjanes in Aix, "Les livres des Samaritaines qu'on desire avoir du Levant" (see appendix 3). At the top of the list was the Pentateuch in Samaritan Hebrew, which Peiresc describes as "touts divers des carateres Hebreus vulgaires." In terms of priorities, Peiresc was, as usual, exactingly clear. Alongside the first entry was a cross, and at the bottom of the page Peiresc explained that "One desires principally the first of these books which is crossed. And for the others, if they could be had easily then they should be acquired, but if not, one will be content with the first."[57]

On the verso, the "Memoire concernant les livres Sa[maritains] qu'on desire avoir du Levant" contextualizes what had been presented schematically. It begins by describing the Samaritans as a sect of Jews found mostly in Palestine in the area around Mt. Garizim who had preserved many books in Hebrew and Samaritan. "Of which I desire to have as many as could easily be recovered," Peiresc writes, "but principally the five books of Moses." At this point, more than half the text on each line is lost, but enough remains to indicate that Peiresc went on to mention the presence of Samaritan communities in Egypt and the advisability of inquiring there for the range of books listed on the reverse.

Fortunately, still another memoire prepared for Minuti illustrates Peiresc's knowledge of the Samaritan diaspora in Egypt, entitled "JUIFS, SAMARITANS, JUIFS de la Columbe au CAYRE." "In Cairo," it begins, "all the Jews are constrained to live in a single quarter that is not far from that of the French. There are three sorts, namely, those who adore the dove, who are the Samaritains, and who never exceeded the number of nine persons." In the margin Peiresc has noted that the Cappucin Gilles de Loche reported "that there are not 12 families of Samaritains in the entire Levant." The note describes the other types of Jews as "ordinary" and those called "Carrains" [Karaites], who have "more than 60 Synagogues in Cairo and hold the Pentateuch alone" sacred.[58]

These memoranda that Peiresc retained offer a glimpse of his intellectual practice. They show how he sought to apply book learning to experience and thereby create a deeper and more secure foundation for knowledge. The scholarly travel that he organized and the collection that he amassed reflect the seriousness with which he pursued, in the world, questions raised in ancient and modern texts. In the context of this quest for better texts and more documentation of a wider sort, new questions were being asked about the peoples of the Levant. Peiresc's goal, as it was that of Scaliger and Selden, and would be of Montfaucon and Creuzer, was to provide a documentable account of the origins of European "civilization."

# Appendix 1

(Paris, Bibl. nat. MS. Nouv. acq. fr. nouv. ff. 23–24)

[23r]
SAMARITANORUM
Dialectus

Notes sur le specimen du Samaritain
du sr. Pietro della Valle de Rome envoyé
par le P. Morin de Paris[1]

ex vers.1.capt.xv.Exod[.]

Les poincts de distinction ne sont poinct aprez la troisiesme parolle [sic] משׁה ains seulement aprez la onziesme למימר.[2]
la quattriesme parole le ב est en la place ובני et non un ר [.]
En la huictiesme le ד est fort distinctement exprimé par un dalet.

ex vers.27 & 28.cap.xii.Deuteron.

La lettre Aleph en la quattrisme parole ne semble poinct abusivement inserée, ואדמה car elle est pareillement repetee en la neufviesme ואדם[3] pour dire *sanguis* selon le languaige *Syriaque*. En la parole antepenultiesme du 28me, באזת, on a fort bien recogneu qu'il y debvoit avoir faulte ou equivoque du Coppiste, qui avoit obmis quelque lettre, ou n'en avoit pas sceu bien distinguer les figures. *Car au lieu de la letre Aleph* א *quil y met pour la seconde, dans mon M.S. il ya deux lettres* ע *ayin et chet* ח *qui faict* בחעות[4] BAHAZUTH, ou IN CONSPECTU. Ce qui prent bien le sens, tant de l' Hebreu IN OCULIS, que de l'Arabe, ענד, qui faict CORAM.

La troisiesme parole du 28^me verset ne se trouve poinct veritablement dans le texte Hebraique des Juifs. mais elle est dans l'hebraique des Samaritains, comme dans les 70. et dans leurs deux versions vulgaires, tant Arabique que Syriaque ou Samaritaine. Il est vray qu'il ya cette differance, Que dans mon M.S. ou Syriaque ou Samaritain, *il y a une lettre Tau de plus* qu'en l'exemplaire du Sr Pietro della Valle de Rome aprez la premiere lettre vau, Et s'y lict ותעבד pour dire, ET FACIES, comme en la precedante et deuxsiesme parolle, il insera un autre Tau, aprez le Vau, pour dire ET AUDIES. Et ce pour mieux exprimer l'Hebreu des Samaritains qui use des mots ושמעת ועשית.

La neufiesme n'est pas non plus dans l'Hebreu des Juifs, ne dans les Septante, mais elle n'est pas obmise en mon MS. En toutes les trois langues pour dire HODIE. Il est vray quil y a cette differance que *au lieu que dans les MS. de Rome il finit par la lettre He* ה *dans le mien y a une lettre Nun* יומן.

[23v]

En la seconde parolle du 27^me verset, dans mon MS. *y a une lettre de plus qui est un ˙ devant la derniere lettre*, qui change la signification du singulier au plurier, et respond beaucoup mieux au texte Hebraique tant des Juifs et Samaritains que des Septante, qui signifie HOLOCAUSTA, au lieu que celuy de Rome voudroit dire HOLOCAUSTUM, au singulier qui ne se trouve nulle part. Et possible n'est ce qu'une obmission du Coppiste.

*Les poincts de distinction* sont mal aprez la sixiesme parole du 27^me verset, au lieu qu'il doibt estre aprez la huictiesme comme il est dans mon M.S. [o]u il y a aussy bien à propos un autre poinct *de distinction aprez la XV*me [Domini dei tui] au lieu que *dans le M.S. de Rome elle est mise hors de place* aprez la XVII mot *Jusques aprez le premier mot du 28me verset, ce qui monstre que si le MS. de Rome est ainsi punctué, il fault qu'il vienne d'une main bien mal adroice, pour ne dire fort incorrecte ou ignorante.*

Le *premier mot du 28me* verset est de trois lettres en mon MS. שור [5] *celle du mitan desfaillant* en celuy de Rome, qui en rend le sens ou interpretation un peu plus difficile, et moings asseurée si le temps.

En la sixiesme parolle [sic] dudit 28^me verset מליה. *La redupplication des deux premieres lettres* ממלל, n'est poinct en mon M.S. et rend *la prononciation plus doulce*, et plus convenable à l'Hebreu qui veut dire [verba] que celuy de Rome[.] Mais cela se peult neantmoings tollerer bien que rude, comme un *Chaldaisme* ou *Cophtisme*, et pourroit faire inferer que cette version de Rome, eust esté à l'usaige de peuples habitez plus avant en l'AEgypte. *Et de faict Sr Pietro della Valle disoit l'avoir trouvé en Perse; et le nostre est de la Palestine.*[6]

La dixiesme parole dudit 28^me verset à en teste dans le MS. de Rome, un ל qui n'est poinct au mien, et semble *surabondante*, selon l'usaige des langues plus corrumpues, et moins pures.

[24r]

En la onziesme[7] *le MS. de Rome entrelasse un Tau entre les deux Jots, qui change de l'actif au passif*, bien que l'hebreu n'ayt point de tel passif en usage, Et *ne met pas un Dalet* qui est en mon MS. au commencement du mot et qui faict mieux la liaison du discours. En cette sorte דיהטב. Ut [bonum sit] tibi.

En la XVI^me de mon MS. *y a un lamet de plus* au commencement לעלם, qui faict difference comme qui dizoit [usque aethernum] au lieu de dire [usque in aethernum] et qui peult neantmoings passer pour surabondante.

En la XIX^me mon MS. à le mot דשפר [quod pulchrum est], dont *les lettres sont transposées* en celuy de Rome en cette sorte דפשר, qui est possible une aequivoque du Coppiste, autrement le sens n'y seroit pas bien intelligible si ce n'est que cette transposition ou soit ~~du Dia~~ de la corruption du dialecte, plus barbare, comme quand les Gascons disent CRABE pour CABRE, qui ne signifient pas moings l'un que l'autre, entre les Gascons, et Provençaulx chascun chez soy.

## Notes to Appendix 1

1. All Samaritan characters have been transcribed into the modern ("Assyrian") Hebrew script.
2. Also after the sixth word in the printed version of della Valle's text in the sixth volume of the Paris Polyglot Bible (1645[1631]). All references will be to this edition.
3. In the printed version: רד.
4. In printed version: בחזות.
5. In printed version: שר.
6. This last sentence is crossed out with vertical lines.
7. In printed version: twelfth word.

## Appendix 2

(Paris, Bibl. nat. Ms. Lat., fols. 79–79)
'De Samaritanorum Characteribus'

[f9]

Monsieur mon R.P.

Je vous suis trop redevable de l'honneur de votre souvenir et de la participation qu'il vous plaist me faire de voz bonnes et devotes prieres donc j'ay bien ressenty les effectz, en sortant de la grande maladie qui m'avoir accueilly dont je vous remercie trez humblement. Et vous envoye le livre de Mons. de Muys que vous me demandéz, ayant esté bien ayse que vous avez trouvé de l'entretien agreable en celluy du P. Morin. Quant à l'antiquité des Caracteres Samaritains, ce n'est pas une petite question, ne qui se puisse facilement traiter, et conclure dans une lettre missive, seulement vous diray-je, que quant il n'y auroit aultre Inconveniant que celluy qu'il vous a pleu de toucher sur le scrupule que pouvoit faire Esdras, d'abbandonner l'ancienne facon de l'escripture mosaique pour sa dignité, puis qu'il sembloit que dieu l'eust sanctiffiée en escripvant les tables de la loy, Je n'y trouverois pas tant d'Incompatibilité si on supposoit comme il se pourroit faire qu'elle eust esté lors comme prophanée puis que Moyse mesme n'avoir pas faict de difficulté de rompre et fracasser les tables de la loy, qu'il venoit recepvoir de la main de dieu, par une juste indignation contre le peuple d'Israel, qui s'en estoit rendu Indigne. Et quant à l'autre difficulté que vous faites sur la difference du Caractere moderne, tant Syriaque comme Caldaique, je vous diray que je n'estime pas que lesdits Caracteres modernes tant Caldaique comme Syriaque soient guieres anciens, ne possible guieres conformes, à ceux qui pouvoient estre du temps d'Esdras, non plus que les Caracteres, dont se servent les nations Italienne, Françoise, Espagnolle, & autres de l'Europe, pour escripre en langage tant latin que vulgaire, ne sont guieres conformes a ceux dont se servoient les Romains, avant la decadence de leur Empire, car les Caracteres majuscules, dont on s'est servy pour [79v] les frontispices des livres, depuis environ un siecle en cà, a esté emprunté & Imité du temps de nos peres seulement. Sur les marbres et Inscriptions anciennes, ou les vrayes figures et proportions du Caractere latin s'estoient conservées, Car la forme d'escripre en langue latine, qui sont conservée par traditive de pere à filz, n'a pas esté Inventée tout en un Coup soudainement pour passer d'une extremité à l'autre, à scavoir du beau Caractere majuscule & quarré, à celluy qui est arrondy que l'on apelle auiourdhuy dans les Imprimeries le Caractere Romain ou Italique, ou à celluy que l'on appelle dans les escrolles des Escripvains de Paris la lettre financiere, mais cela s'est abastardy petit a petit et par degréz ainsi qu'il se peut veriffier par les marbres mesmes, sur lequelz on voit bien de la difference de l'escripture de ceux qui sont gravéz de quelques siecles plus tard les uns que les autres. Et se recongnoist encores mieux dans les livres manuscriptz dans lesquelz l'escripture changeoit de mode à chasque siecle, tout de mesme comme le language vulgaire & comme les habillementz voire la diversité des nations à produict une grande diversité de Changementz comme il se voit par la Comparaison des Caracteres modernes tant Italique et Francois que Allemand. Or J'estime que la mesme chose est arrivée non seulement aux Caracteres Caldaiques et Syriaques modernes, ainsy que je l'ay recongnu par la Comparaison de deux manuscriptz que j'ay en langue Syriaque, dont l'un est plus ancien que l'autre de deux ou trois cens ans, mais aussy aux Caracteres Hebraiques modernes, dont je n'estime pas que la forme, ay esté arrestée en la facon qu'elle est de plus grande antiquité, que celle du temps de Mazorets ayant mesme des fragmentz de vieux livres hebraiques dont le Caractere à beaucoup de difference d'avec celluy qui est le plus en usaige, et le Caractere mesme des Rabins n'est pas tousjours Conforme a soy mesme non plus que l'autre, J'ay mesme des vieux manuscriptz ou se trouvent des allegations en [80r] language hebraique dont le Caractere est si different de Celluy des Mazoretz qui n'est presque pas recongoissable. C'est pourquoy je ne tiens pas qu'il faille trouver estrange que la difference soit sy grande du Caractere Hebraique au Samaritain, n'estimant pas mesmes que les Samaritains dans le scrupule et superstition qu'ilz ont eu pour cela ayent peu conserver si religieusement la figure du Caractere Mosaique quilz ne l'ayent de beaucoup alterée, sinon en tout, au moins en plusieurs Caracteres de leur Alphabet et surtout en la lettre Tau, dont Il semble quilz ayent affecté d'abolir la forme quelle avoir d'une Croix, en haine du Christianisme, aussy bien que les Juifz, ne se pouvant point revoquer en doubte que les Juifz, n'ayent retenu le mesme Caractere, que l'on appelle aujourdhuy Samaritain, fort long temps aprez Esdras, Et d'estimer quilz en eussent deux si differentz entr'eulx, comme se trouvent aujourdhuy celluy que l'on appelle Samaritain & celluy que l'on appelle Hebraique, c'est ce que je ne me scaurois persuader sans voir d'autres preuves plus precises et plus concluantes que tout ce que j'en ay peu voir à presnt, dans les livres du temps, si vous en avéz d'autres que cela, vous m'obligerez bien fort de m'en faire part et encores plus de me commander Monsieur, Comme votre tres humble & tres obeissant serviteur de Peiresc

A Boysgency ce 6 Juillet 1632

à Monsieur le R.P. Dom denis de Sailly prieur de la Chartreuse d'Aix a Aix

## Appendix 3
## (Aix-en-Provence, bibl. Méjanes, Ms. 1168 unfoliated)

[recto]
SAMARITAINS

Memoire concernant les livres Sam[aritains]
qu'on desire avoir du Levant[e]

Il y a en la Palestine tout plein de SAMARITAINS, qui est
une secte de Juifs diverse des Juifs ordinaires.
Les principaux prebstres de leur Loy se tiennent
au MONT GARIZIM qui est prez de la Ville de Caesarée Hi[ . . .
Et se font appeler d'un nom en leur Langue qui sign[ifie]
en Langue Francoise (Les Dependants du Mont B[enedictus])
Cez prebstres ont plusieurs livres, tant en langu[e Samaritaine]
Que en Langue Hebraique, escripts neantmoings en [caractere?]
Samaritain. Desquels on desire avoir tous . . .
pourront commodement recouvrer, Mais par [ticulierement?] . . .
les cinq livres de Moise.    Ils on [E] . . .
residents au Grand Caire et en autres gro . . .
lesquels dependent tous de ceux de ce de la . . .
ont quelques ungs desdits livres, mais n . . .
abondance. Et ceux de la Palaestine en . . .
du Caire et autres lieux, des Almanachs . . .
toutes les années en leur Langue et leur . . .
leurs autres livres selon qu'ils en [ont] besoing . . .
De sorte que si ceux qui ghab [itent] . . .
difficiles à despartir de l . . .
avoir plus tost par . . .
du Caire ou autres villes . . .
dudit Mont Garizim.

[verso]

Les livres des Samaritains qu'on desire
avoir du Levant

[#]1 Les cinq livres de Moyse en langue Hebraique escripts en
Caracteres/ Samaritains qui sont touts divers des Caracteres Hebreus Vulgaires.

[2] Les mesmes cinq livres de Moyse, traduicts en Langue/
Samaritaine, escripts en mesme Caractere Samaritain./
Le livre qu'ils appellent IOSUE qui est une chronique/
de leur Histoire depuis le deceds de Moise jusques à cent ans aprez Jesus Christ./

Une Grammaire en langue Samaritaine, qu'ils appellent/
leur Alphabet./
Une petit sommaire de leur chronique depuis Adam qui avoit esté continué jusques à
l'an de Christ 1584/
Leur Almanach, qu'ils renouvellent touts les ans/
ou Computation des jours de leur année &c./
Les autres livres qui se pourront trouver escripts en caractere/
desdits Samaritains./

On desire principalement le premier des susdits livres qui est croixsé, & pour les autres si on les peult avoir commodement on en serà bien aise, sinon, on se contenterà du premier.

. . . . entre les mss. apporteez par M. de Sancy, le Pentateuche

## Notes to Appendix 3

1. . . . [ce]-uz que lon/ . . . [des]ire voir/ principalement/ sur tous les autres
2. These are bracketed by Peiresc with the comment "Pour ceux icy on . . . / qu'aultant qu'il../ commodement et à . . . "

## Notes

1. A version of this essay was presented at a panel of the Renaissance Society of America in April, 1995, in New York. I wish to thank Agnés Bresson, Tom Cerbu, Anthony Grafton, and Ingrid Rowland for reading earlier drafts and generously sharing with me their knowledge of early modern antiquarian culture. I am extremely grateful to Agnés Bresson for her invaluable assistance in preparing the appendices for publication. I acknowledge the support provided this project by the American Philosophical Society.

2. Francis Bacon, "De Augmentis Scientiarum," *The Philosophical Works of Francis Bacon*, ed. John M. Robertson (Freeport, NY, 1970), 2.6.433.

3. Some recent important works include: Anthony Grafton, *Defenders of the Text: The Traditions of Scholarship in an Age of Science 1450–1800* (Cambridge, MA, 1991) and *Joseph Scaliger. A Study in the History of Classical Scholarship*, (Oxford, 1983, 1993), 2 vols.; Francis Haskell, *History and Its Images. Art and the Interpretation of the Past*, (New Haven and London, 1993); Alain Schnapp, *La Conquête du passé. Aux origines de l'archéologie*, (Paris, 1993); Francesco Solinas, ed. *Cassiano dal Pozzo. Atti del Seminario Internazionale di Studi. Napoli, 18–19 dicembre 1987*, (Rome, 1987); Bruno Neveu, *Erudition et religion aux XVII$^e$etXVIII$^e$*.

4. Arnaldo Momigliano, "Ancient History and the Antiquarian," *Journal of the Warburg and Courtauld Institutes* 13 (1950): 285–315; "L'eredità della filologia antica e il metodo storic o *siècles* (Paris, 1994); and *Documentary Culture. Florence and Rome from Grand-Duke Ferdinand I to Pope Alexander VII*, eds. E. Cropper, F. Perini and F. Solinas, (Bologna, 1991)," *Rivista Storica Italiana* 70 (1958): 442–58; and *The Classical Foundations of Modern Historiography* (Berkeley, Los Angeles, and London, 1990), esp. 54–79. Revisitings include, among other projects now in train, the memorial volume published by the Warburg Institute, *Ancient History and the Antiquarian. Essays in Honor of Arnaldo Momigliano* (London, 1995) and Mark Phillips, "Reconsiderations on History and Antiquarianism: Arnaldo Momigliano and the Historiography of Eighteenth-century Britain," *Journal of the History of Ideas* 57 (1996): 297–316.

5. Peiresc's name remains a source of some mystification, often found written in a variety of ways by himself and by others. Its pronunciation, too, is uncertain. Gassendi, in the *Vita*, suggests that the Latin Peireskius conveyed the proper pronunciation. But letters preserved in Leiden offer a different perspective. In the correspondance with Charles de l'Escluse (Clusius) in the Universiteitsbibliothek Leiden (Vulc. 101) the first three letters (1602-4) are signed N.C. fabri de Callas. But in the postscript to that of 25 February 1604, Clusius is asked no longer to address correspondance to S$^r$ de Callas but "Au Sr. de Peirets: chez mons$^r$ le conseiller de Callas" (his father). In a letter written only two days later, however, Peiresc signed "N.C. Fabry." Clusius, upon receipt, nevertheless annotated the overleaf "1604 N.F. de Peiretz." The next two letters from Peiresc (15.ii.1605 and 25.viii.1605) were signed "N.C. de Peirets." In the last preserved letter, of 15 February 1606, the familiar "Peiresc" was scrawled at the bottom. A postscript gives some explanation for this latest change: "Notre petit village de Peiresc s'appelle dans les vieux cadastres Latins *Castrum de Petrisco*." (It was ommitted by Tamizey de Larroque in the printed version of this letter, *Lettres de Peiresc*, ed. Tamizey de Larroque (Paris, 1888–98), 7 vols. 7:956. These letters are printed from the *minutes*, not from the autographs in Leiden, and often lack signatures or postscripts.) For a person with distinct regional, professional, familial, and personal identities the choice of a name was, in fact, a choice. I thank Tom Cerbu for first alerting me to the question of pronunciation.

6. Arnaldo Momigliano, *The Classical Foundations of Modern Historiography*, 54; Laurence Sterne, *Tristram Shandy*, 2.xiv; *Danielis Georgi Morhofi Polyhistor* . . . (Lübeck, 1708), 239; Balzac to Chapelain, quoted in Jean Jehasse, *Guez de Balzac et le Genie Romain 1597–1654*, (Paris, 1972), 422. The best modern treatments of Peiresc are those of Agnès Bresson in the apparatus of her edition of Peiresc's *Lettres à Claude Saumaise et à son entourage* (Florence, 1992); Cecilia Rizza, *Peiresc e l'Italia* (Turin, 1965), Sydney H. Aufrère, *La Momie et la Tempête. N.-C.F. de Peiresc et la "curiosité Egyptienne" en Provence au début du XVIIe siècle* (Avignon, 1990), and the essays of David Jaffé, most recently "Peiresc—Wissenschaftlicher Betrieb in einem Raritäten-Kabinett,"

*Macrocosmos in Microcosmo*, ed. Andreas Grote, (Opladen, 1994), 301–22 and "Peiresc and New Attitudes to Authenticity in the Seventeenth Century," *Why Fakes Matter. Essays on Problems of Authenticity*, ed. Mark Jones (London, 1993), 157–73 and "Aspects of Gem Collection in the Early Seventeenth Century-Peiresc and Pasqualini," *The Burlington Magazine* 135 (1993): 103–20.

7. De Sailly to Peiresc, 23.viii.1632, Paris, Bibliothèque Nationale, Ms. Latin 9340, fol.67v. For discussions of this theme, see Helmut Zedelmaier, *Bibliotheca Universalis und Bibliotheca Selecta. Das Problem der Ordnung des gelehrten Wissens in der frühen Neuzeit*, (Cologne, Weimar, and Vienna, 1992) and Luc Deitz, "Ioannes Wower of Hamburg, Philologist and Polymath. A Preliminary Sketch of his Life and Works," *Journal of the Warburg and Courtauld Institutes* 58 (1995): 132–51.

8. Pierre Gassendi, *Viri Illustris Nicolai Claudii Fabricii de Peiresc Vita* (1st ed. 1641), translated by William Rand as *The Mirrour of True Nobility and Gentility* (London, 1657), 186, 265.

9. For modern accounts of this encounter, see Philippe de Robert, "La Naissance des études samaritaines en Europe aux XVIe et XVIIe siècles," 15–27, and Mathias Delcor, "La Correspondance des savants européens, en quête des manuscrits, avec les Samaritains du XVIe aux XIXe siècle," 27–43, both in *Études samaritaines. Pentateuque et Targum, exègèse et philologie, chroniques*, ed. Jean-Pierre Rothschild and Guy Dominique Sixdenier (Louvain-Paris, 1988); Jean-Pierre Rothschild, "Autour du Pentateuque samaritain. Voyageurs, enthousiastes et savants," *La Bible au Grand Siècle*, ed. Jean-Robert Armogathe, (Paris, 1985) 61–74; J.G. Fraser, "A Checklist of Samaritan Manuscripts Known to Have Entered Europe before A.D. 1700," *Abr-Nahrain* 21 (1982–83): 10–27; and the editor's introduction in Jean-Pierre Rothschild, ed., *Catalogue des manuscrits samaritains de la Bibliothèque Nationale* (Paris, 1985).

10. Peiresc to Scaliger, 15.ii.1606, in *Epistres françaises des personages illustres et doctes à Joseph-Juste de la Scala*, ed. Jacques de Reves (Harderwijk, 1624), 142.

11. Anthony Grafton, *Joseph Scaliger. A Study in the History of Classical Scholarship. II: Historical Chronology* (Oxford, 1993), 181–82, 241.

12. Scaliger, *De emendatione temporum* (1583), 2, quoted in Grafton, *Joseph Scaliger II*, 262.

13. Scaliger himself told students at Leiden that "even if he had been really rich, he would not have built a great library but would have travelled extensively" (quoted in Grafton, *Scaliger II*, 108.

14. Grafton, *Joseph Scaliger II*, 88, 710.

15. Alasdair Hamilton has shown how William Bedwell took up Scaliger's insight by studying Arabic and Turkish together against the common backdrop provided by Islam (*William Bedwell the Arabist 1563–1632* [Leiden, 1985], 85).

16. Arnaldo Momigliano, "A Prelude to Mr. Gibbon," *Sesto contributo alla storia degli studi classici e del mondo antico* (Rome, 1980), 1:249–63.

17. La Popelinière to Scaliger, 4.i.1604, *Epistres françoises*, 303–7.

18. Peiresc to Dupuy, 18.xii.1632, *Lettres de Peiresc*, 3:396. Peiresc's two copies of Scaliger's *De re nummaria* (Leyden, 1616) were also heavily annotated. See H. Omont, "Les manuscrits et les livres annotes de Fabri de Peiresc," *Annales du Midi* 1 (1889): 316–39.

19. Grafton, *Scaliger II*, 242.

20. In 1581 Scaliger wrote to Claude Dupuy asking him to contact their mutual friend in Padua, Gian Vincenzo Pinelli, the later patron of Peiresc, and request him to acquire a Samaritan *Computus* through a Jewish friend in Constantinople. In a casual but revealing aside Scaliger speculated that Coptic texts might also be accessible through the same channel (Scaliger to Dupuy, 4.ix.1581), *Lettres françaises inédites de Joseph Scaliger*, ed. Tamizey de Larroque (Paris-Agen, 1879), 117–18). This casual conjunction of Samaritan and Coptic was to reappear as a research program in Peiresc's correspondance of the 1630s.

21. See Grafton, *Scaliger II*, 413–21.

22. On the back of a broadside published by Marseilles's municipal government and cannibalized by Peiresc for scrap paper, he recorded information concerning the provenance of his Samaritana, including these letters. They fell into the hands first of M. Genebrard, then of M. Pol Hurault the Archbishop of Aix, and finally came to M. Billon, who passed them to Peiresc's friend Gallaup de Chasteuil, from whom they reached Peiresc in August of 1629 (Paris, B.N., Ms. Latin 9340, fol.93r).

23. Originals are in Bibliothèque Nationale, Ms. Sam. 11; Peiresc's Samaritan, Hebrew, and Latin copies are in Paris, B.N., Ms. Latin 9340, fols.95–103.

24. Paris, B.N., Ms. Latin 9340, fols. 102–3.

25. Grafton, *Joseph Scaliger II*, 631 n.48.

26. Grafton, *Joseph Scaliger II* 737 n.30.

27. Leschassier to Peiresc 10.v.1610, Aix-en-Provence, Bibliothèque Méjanes, Ms. 204 (1022), 127–29.

28. For a discussion of this part of the story see my "Les Origines de la Bible polyglotte de Paris: *philologia sacra*, Contre-Réforme et *raison d'état*," XVII$^e$ *Siècle* (forthcoming).

29. Peiresc to Dupuy, 19.v.1628, *Lettres de Peiresc* , 1:617.

30. Peiresc to Aleandro, 27.v.1628, Rome, Vatican, Barberini-Latina Ms. 6504, fol. 216r.

31. Peiresc to Holstenius, 24.ix.1628, *Lettres de Peiresc*, 5:293.

32. Peiresc to Holstenius, 10.xi.1628, *Lettres de Peiresc*, 5:297.

33. Peiresc to Dupuy, 22xi.1628, *Lettres de Peiresc*, 1:751.

34. Peiresc to della Valle, 26.xi.1628, Carpentras, Bibl. Inguimbertine, Ms. 1871, fol. 242r-v; Aix-en-Provence, Bibl. Méjanes, Ms. 213 (1031), 63.

35. Peiresc to della Valle, 26.xi.1628, Carpentras, Bibl. Inguimbertine, Ms. 1871, fol. 242v; Aix-en-Provence, Bibl. Méjanes, Ms. 213 (1031), 64.

36. Peiresc to della Valle, 9.x.1630, Carpentras, Bibl. Inguimbertine, Ms. 1871, fol. 249r; Aix-en-Peovence, Bibl. Méjanes, Ms. 213 (1031), 83.

37. Peiresc to Morin, 8.xi.1632, *Antiquitates Ecclesiae Orientalis*, [ed. Richard Simon] (London, 1682), 190–92.

38. Samuel Petit, *De Epocha annorum Sabbaticorum apud Samaritanos*, Paris, B.N., Ms. Latin 9340, fol. 85.

39. Peiresc to Holstenius, 6.viii.1629, *Lettres de Peiresc*, 5:341. In a letter to della Valle of 7 September 1633, Peiresc addressed precisely this issue of formulating a Phoenician grammar from the well-known passage in Plautus (Rome, Vatican, A.S.V. Archivio della Valle-del Bufalo 52, fol.6r; Carpentras, Bibl. Inguimbertine, Ms. 1871, fol. 251r; Aix-en-Provence, Bibl. Méjanes, Ms. 213, (1031), 93), which Petit himself deciphered by reading the characters as Hebrew and then inserting the missing vowels. Petit sent a copy of this analysis, dated July 1629, to Peiresc (B.N., Ms. Latin 9340, fol. 83). Examples of Petit's Phoenician scholarship preserved in Peiresc's papers also include a brief examination of the relationship between Hebrew, Canaanite, Phoenician, and Punic (fol. 84).

40. Peiresc to Dupuy, 21.vii.1629, *Lettres de Peiresc*, 2:133–34. In January 1630, Peiresc wrote to Gassendi, then in Paris, introducing Petit, whom he described as possessing much that "would link you in friendship with him." Peiresc thought that Petit could make an important contribution towards deciphering the Samaritan texts as he had already, following in the footsteps of Scaliger, "found beautiful things to write about the Computus of the Samaritans" (Peiresc to Gassendi, 18.i.1630, *Lettres de Peiresc* 5:238–39).

41. While his comparison of della Valle's Targum with his own seems to show familiarity with Samaritan, and therefore Hebrew, a letter from François de Gallaup-Chasteuil of 1630 indicates that Peiresc left it to him to comment on a Samaritan computus (Paris, B.N., Ms. Latin 9340, fol. 76).

42. Dupuy to Peiresc, 20.iii.1629, *Lettres de Peiresc*, 2:687. The text is currently in Paris, B.N., Ms. Latin 9340, fols. 51–56.

43. Peiresc to Dupuy 14.iv.1629, *Lettres de Peiresc*, 2:68. Crinesius began, as was common in this genre, with the history of the Samaritans out of Biblical sources and then added the observations of scholars like Scaliger. In doctrinaire Protestant fashion he rejected the newly received view that the Samaritan alphabet was older than the familiar Hebrew letters. It was precisely this style of *parti pris* erudition that Peiresc scorned (*Lashon Shamraita: hoc est Lingua Samaritica* (Altdorf, n.d.) esp. sigs. [A4]-B.

44. Peiresc to della Valle, 2.v.1630, Carpentras, Bibl. Inguimbertine, Ms. 1871, fol. 248v; Aix-en-Provence, Bibl. Méjanes, Ms. 213 (1031), 80–81. It is part of a shipment of letters to the usual

Roman addressees (Cardinal Barberini, Suares, Menestrier, dal Pozzo, Cardinal Bentivoglio, Francesco Gualdo, Christophe Dupuy, and Holstenius) sent on the 15th (see "Les Petits mémoires de Peiresc," ed. Tamizey de Larroque, *Bulletin Rubens* 4 (1896): 99–100; this record of Peiresc's correspondence is preserved in Paris, B.N., Ms. nouv. acq. fr. 5169).

45. Peiresc to della Valle, 9.x.1630, Carpentras, Bibl. Inguimbertine, Ms. 1871, fol.249v; Aix-en-Provence, Bibl. Méjanes, Ms. 213 (1031), 83–84.

46. Denys de Sailly to Peiresc, 3.vii.1632, Paris, B.N., Ms. Latin 9340, fol.81r.

47. De Muis was professor of Hebrew at the Collège Royal and the author of a series of stinging attacks on Morin's "Samaritan thesis" entitled "Assertio veritatig hebraicae" (1631, 1634). He was said to have been encouraged by Richelieu in response to Le Jay's refusal to let him take over patronage of the Polyglot and have it bear his name.

48. Peiresc to de Sailly, 6.vii.1632, Paris, B.N., Ms. Latin 9340, fols. 79–80.

49. In a letter as late 12.i.1603 to his closest friend, J.A. de Thou, Scaliger wrote of Postel, whom he had encountered in Paris in 1562, that "Jamais homme ne m'enseigna tant que celui-là" (*Lettres françaises inédites*, 351).

50. On this see J.G. Fraser, "A Prelude to the Samaritan Pentateuch Texts of the Paris Polyglot Bible," in *A World in Season. Essays in Honor of William MacKane* (Sheffield, 1986), 223–47. Fraser relates the problem of dating to the question of authorship; he suggests a dating between 1613 and 1616, with the third section most likely Peiresc's.

51. Peiresc to Dupuy, 7.xi.1629, *Lettres de Peiresc*, 2:192–93.

52. Peiresc to Dupuy, 18.xi.1629, *Lettres de Peiresc*, 2:203. Much the same information was conveyed to della Valle in a letter of 4 March 1630.

53. Peiresc to Dupuy, 18.xi.1629, *Lettres de Peiresc*, 2:203–4.

54. Carpentras, Bibl. Inguimbertine, Ms. 1821, fol.486r.

55. Carpentras, Bibl. Inguimbertine, Ms. 1821, fol.454r.

56. Paris, B.N., Ms. Latin 9340, fol.49r.

57. Aix-en-Provence, Bibl. Méjanes, Ms.1168, [unfoliated].

58. Carpentras, Bibl. Inguimbertine, Ms. 1864, fol.261r.

# Musical Scholarship in Italy at the End of the Renaissance, 1500–1650
## *From Veritas to Verisimilitude*

### Ann Moyer

Music had been a well-defined and well-established discipline since the Middle Ages. It was among the first fields affected by the extension of humanist methods, including history, into subjects outside the *studia humanitatis*. In fact, the sixteenth-century debates over the nature of music and its study helped define the humanists' studies of philology and history as methods that could be applied to other disciplines. These debates resulted not only in the reclassification of music but also in the development of new ways of classifying subjects more generally, and raised questions about how fields of knowledge related to one another. Thus the field of music and its changes offer an early example of the establishment of new disciplinary definitions and boundaries that occurred during the next two centuries.

The existence of "music" as a discipline meant, of course, that the disciplinary term referred not only to music compositions or performances but also to scholarly writings about music. In this respect music differed from subjects later seen as related to it, such as the visual arts; Europeans produced works of art in great numbers long before formal writings about art were undertaken, let alone identified as a field in their own right. Indeed, studying music could and often did mean the reading of texts rather than the production or analysis of musical compositions. Both the existence of a strong classical and postclassical textual tradition and the ephemeral nature of musical performances themselves worked together to give music a unique relationship to historical analysis. For not only did past writings about music exist in abundance, but they were much more accessible to historical study than were past performances. In addition, the discipline of music was closely identified with a particular philosophical tradition—the Platonic and Pythagorean—which lay behind its claims that it was a master discipline, one both prior and essential to the study of others.

By the late sixteenth century the scholarly methods of late humanism, rather than Pythagorean mathematics, came instead to serve more and more as the ways to define and study music. This art of music was now distinguished from the closely related science of sound.[1] The introduction of humanist analysis meant that for the first time both musical thought and musical style could be studied historically. Yet in the process music—while still seen as an important arena for scholarship and practice—came to be seen less as a master discipline than as one studied with the tools of other fields.

Musical scholarship of the early Renaissance had developed without a break from earlier traditions. Its medieval discourse had been dominated by the universities and the church, as locations for production of texts and the education of the writers. It formed part of the quadrivium, the mathematical portion of the liberal arts, along with arithmetic, geometry, and astronomy. The liberal arts themselves predated the universities as a medieval curriculum and persisted despite the scholastic additions of new fields and new classifications; quadrivial fields were easily labeled as subfields of mathematics. Music, with astronomy, was also often described as an "intermediate science," since it was mathematical but, unlike geometry, required physical objects for its full exercise. Music's core textbook was that of the late Roman scholar Boethius, though modern authors might be used in addition. Its subject was mathematical proportions, especially as seen in the lengths of plucked, sounding strings. In a tradition going back to Plato's *Timaeus* it saw the ratios of the musical consonances (fourth, fifth, and octave: 2:1, 3:2, 4:3) as the basic order underlying the cosmos. These proportions defined beauty as an objective reality that existed both in the object and in human perceptions of its qualities, because these ratios were present both in the beautiful sound or object and in the human soul.

Like other quadrivial fields it had two parts, theory and practice. A whole treatise "De Musica" would be divided into these two parts, though smaller treatises might focus on part of the practical tradition, such as notation, counterpoint, or plainchant. Practical music applied the mathematical proportions that were the heart of the discipline, and was fully post-classical in origin. Except for their rules for composition, these writings included nothing recognizable to modern readers as stylistic criticism; even the mention of good composers by name appeared only very late. Aesthetic response to music was explained in terms of the listener's perception of proportions, mainly in pitch but also of rhythm. There was no sense that these standards might be relative to cultural factors; these were absolutes, linked as much to the order of the universe as gravity is in the modern world.

The changes in the field began in Italy at the end of the fifteenth century with writers such as Franchino Gaffurio and Giorgio Valla, whom later writers identified as the first scholars of their own era.[2] At this time humanists had begun to broaden their editorial interests to include scientific works. Most of the music treatises they translated (often at the request of musical scholars) were—like the writings of Boethius—also Pythagorean in outlook, reinforcing scholarly opinion that the ancients agreed that this was the best philosophical and analytic approach to the subject. During these same years composers were facing rapid changes in style, some of which seemed to contradict musical theory. They looked to ancient sources for answers, hoping to justify their new practices. The importance of such works as Vitruvius's architecture treatise, which emphasized the role of musical proportion in buildings, also heightened general scholarly interest in the field. Humanists themselves became interested in music because they had come to realize that ancient poetry and drama both had been performed with music. They wanted to know about the social settings and other details of ancient music performance, topics not covered by the ancient music treatises. So they turned for assistance to the literature they already knew as humanists. These sources (for example Quintilian) described ancient music as having had dramatic effects on human emotion and even action. Renaissance musicians sought the best ways to duplicate such miraculous effects.

Thus the sixteenth century saw several decades of lively and often acrimonious debate about how to evaluate confusing and conflicting types of evidence and principles of interpretation. It soon became obvious that ancient sources frequently disagreed with one another on terms and definitions and perhaps more, but it was not so easy to find the sources of the disagreements. Part of the solution had to lie in developing better principles for interpreting the sources, principles that would come from humanist textual criticism. But part also came from the use of mathematical analysis and natural philosophy to try and understand the physical behavior of musical instruments. Only after 1550 was there enough accumulated scholarship—and a generation of distinguished scholars—to begin productive efforts at a new synthesis.

This new synthesis developed relatively quickly thanks to such musicians and scholars as Gioseffo Zarlino and Vincenzo Galilei, and such humanists as Girolamo Mei. It developed from both humanist study and natural philosophy in similar ways. These scholars historicized the field's philosophic basis in Boethian and Pythagorean thought, and limited its claims. Zarlino, a noted scholar of ancient musical writings, took the first step. He was able to show from textual evidence alone that not only had ancient styles of composition and performance changed over time, but that the modern practices of polyphony and counterpoint were quite unknown in antiquity, a finding soon confirmed with the first discovery of ancient musical manuscripts.[3] Thus the arguments of his contemporaries comparing ancient and modern music needed to be reconsidered. But Zarlino devoted most of his energy to studying music in terms of mathematics and natural philosophy. To do so with precision he came to distinguish two parts to the discipline: historical and methodical. The latter are the timeless and absolute principles of proportion, along with other axiomatic principles; the former are the particulars of musical practice and custom (*uso*), subject to change over time.[4] This distinction opened up a large part of the field to historical analysis. Musical style could now be studied humanistically and historically, as a cultural phenomenon in its own right.

Despite this distinction Zarlino still understood the field as united by universal harmonic principles, but he was quickly challenged on two fronts. One challenge came from the humanists. Girolamo Mei composed a history of musical modes, the first attempt to compose a historical account of an aspect of musical style.[5] He came to realize that the Pythagoreans had not commanded universal agreement among ancient musical scholars. They had been only one of three schools or sects, though they had composed many treatises and found favor with Boethius; further, one of the rival schools (that of Aristoxenus) had denied any relationship between tuning or musical proportions and the cosmos. Mei's work suggested that modern musicians and scholars should understand the tradition of Boethius in its historical context rather than accepting its absolute claims to truth.

The other challenge came from examining the physical behavior of musical instruments. Vincenzo Galilei found by experiment that he could produce musical intervals such as an octave not just by the traditional means of proportional string length but in other ways that could be measured quantitatively, for example by adding weights to strings.[6] An octave formed this way did not have the ratio prescribed by Boethius or Pythagoras. The traditional ratios were not wrong, he argued, but neither were they universal. These findings supported Mei's arguments for a historicized, rather than an absolute, interpretation of Boethius.

Galilei went on to distinguish a science of sounding bodies from an art of music. This art uses the principles of sound production just as a painter uses colors; the art itself is the product of individual creativity in a given cultural context.[7] By 1600 this distinction between science and art was seen in most Italian writings about music and its relationships with other disciplines. The science of sound accumulated knowledge by experiment and numerical measurement, though now without the earlier claims about the cosmic order. The art of music explained its effects on the listener by means of the text and story, the composer's invention and judgment, and its cultural location in time and place.

Most Italian musical scholars accepted this distinction, either tacitly or explicitly, by about 1600. Giovanni dei Bardi defined music (ca. 1578) as a composition of words, harmony, and rhythm, and mentioned neither science nor mathematics.[8] Lodovico Zacconi distinguished between music-as-science involving number and proportion, used by arithmeticians and architects, and music-as-art, which relates to sounding music and was his chosen subject.[9] Vicenzo Giustiniani, writing about 1628, noted that music ranks high among the arts called liberal, approaching but not reaching the ranks of science.[10]

This transformation had further consequences for the study of music, for the study of related fields, and for the relationships of those fields to one another. Musical scholarship itself showed great continuity in many respects. Professional and scholarly activities remained unchanged; the field's ability to resolve its major questions mitigated against any sense of crisis or quandary over the decline of the Boethian tradition. The replacements for Boethian theories had been long in developing and moved easily into place as the older ones gradually became secondary or faded away.

In fact the Boethian tradition did not disappear from all European thought at the same rate. It departed first from Italian musical thought and more gradually from related disciplines and the scholarship of other geographic regions. Music and poetics retained Platonic connections longer in northern Europe

(Pontus de Tyard and Guy le Fèvre de La Boderie, for example); harmonic proportion remained in analytical use longer in fields such as astronomy (as in the work of Kepler but not of Galileo). Even musical scholars still cited Boethius at times, but with ever fewer references to his cosmological principles. Orazio Tigrini, for example, began his *Compendio della musica* (Venice, 1588) with an invocation of Boethius, the need for reason in the study of music, and music's status as a mathematical field, but did not in fact discuss proportion or Pythagorean ideas; he merely referred interested readers to Zarlino.[11] Giovanni dei Bardi mentioned briefly that according to Plato the world was composed of harmony. Yet he did not even mention the *Timaeus*, preferring to cite passages in the *Republic* or *Laws* that address music as part of society rather than cosmology.[12]

The most important replacement for the Pythagorean tradition was humanist poetics. Poetics, literary criticism, and language study in general had become important and hotly debated subjects in Italian scholarly life. By linking itself to a field with such a high profile, music could shift from Pythagorean to literary scholarship without suffering a noticeable loss in its claims to importance; and in any case musical scholars were no strangers to controversy and debate. The connection between music and poetry was found in a historical (or quasi-historical) narrative of their common origins. According to any number of ancient sources, in the earliest societies poets, musicians, and prophets were identical; their divinely-inspired influence on the earliest mortals was the first impetus to civilization. This image had been used as early as Boccaccio[13] and began to appear regularly in writings on music from about 1500. Despite the appeal of the image of the poet-musician to Renaissance musical scholars, the great distance that separated humanist studies from the quadrivium hampered its development in depth.[14] The humanist Raffaele Brandolini made the first systematic effort to link music and poetry (1513); he did so at the level of theory mainly by favoring musical theory, assimilating metrics into theories of harmonic proportion and buttressing it with a Platonic description of poetic furor.[15] Zarlino's identification of music's historical side gave poetics much more room for development in musical analysis, though he had not intended it to supplant mathematics as it soon did.

The application of poetics had several important effects in the ways writers could discuss and evaluate music. First, the heart of the subject became not sound but musical compositions, a distinction important not only for the development of stylistic criticism but also of music history, as will be discussed below. Second, the use of poetics naturally favored compositions with text— that is, vocal music—over purely instrumental music. Third, within the classification of vocal music the poetic text took precedence over the music itself in both composing and evaluating musical works. A particular piece of music could be judged good or bad on the basis of its ability to communicate the text. This criterion could replace the Boethian emphasis on purity of harmonic intervals as the judge of quality, and certainly offered more interpretive power to the composer. Yet the literary text, not the music, now became the superior evaluative category. Music criticism began to address issues based in literary analysis, such as mimesis and social appropriateness.[16]

An early manifestation in the practical tradition of this focus on the text had been the appearance of rules for matching words with the musical line composed for it, a practice hitherto left to the discretion of the singer. The first systematic presentation of text-setting rules appeared in the 1530s.[17] Some of the principles behind the early rules seem fairly obvious, such as the importance of matching metrical and musical accents. By the century's last decades such rules expanded to describe good mimetic traits as opposed to barbarisms.[18] Such advice assumes that a text was selected before composing the music itself; Tigrini's treatise offers an example. He recommends that composers begin the process of composition by considering the work's subject matter. This subject matter was presumably found in a pre-existing text, since all his examples are settings of poetry of various types: madrigals, French chansons, masses, and so on.[19]

Scholars frequently praised music that expressed the emotions indicated in the text by imitating the intonation patterns used by the spoken voice. This approach could produce some very successful results, as in the development of operative recitative.[20] It was also used in the composition of madrigals, an extremely popular genre, where its excesses found criticism from Vincenzo Galilei and others.[21] Critics tended to focus on results that were too obvious or literal, such as portrayals of sighs and gasps or other exclamations. Also criticized were attempts to dramatize a single word when done at the expense of the sense of the line or passage, such as setting the word "heaven" with high notes or "limping" with an uneven rhythm.[22] While most of this criticism had its basis in the contemporary experience of music composition and performance, critics could also cite ancient authority. Lucretius, for example, had argued that the earliest music had begun with human imitation of sounds in nature, such as the singing of birds, which gradually progressed to real music. Due to this primitive status, the imitation of sounds or physical qualities alone, with no concepts underlying them, was considered inferior to more abstract imitations of meaning.[23] Conversely, Galilei criticized as artificial the practice of building visual imitation into music notation, so that it was seen by the performers but not heard.[24]

Positive suggestions, on the other hand, often seem more vague than helpful. Tigrini advised the composer only that a cheerful text should be set to a mode that is equally cheerful, rather than to one melancholy in nature.[25] Zarlino offered somewhat more specific ways to express some emotions, such as harshness, bitterness, or cruelty (*asprezza, durezza, crudeltà, amaritudine*). Parts should move by whole tone or ditone rather than semitone, a suspension of the fourth or eleventh over the lowest part may be used, and so on.[26]

Music might also support the communication of the text by actively inducing emotion in ways that corresponded with the text's meaning. This possibility required a theory of how music causes emotional effects, and of which emotions it is capable of causing. Such theories, whatever their points of disagreement, shared a common origin. They had developed out of the *laus musicae*, that part of musical writings that had been among the earliest to see humanist influence. Prominent among music's praiseworthy qualities was its ability to cure ailments and affect emotions for better or worse.[27] Musical scholars of the early sixteenth century had explained this ability entirely in terms of harmonic proportions in the listener's soul matched by those of the musical performance, almost as sympathetic vibration. But as composers became increasingly interested in improving their ability to move the listener, these discussions became more specific and more detailed.

Zarlino's explanation of music's emotional effects took both text and pitch into account. He still based his argument on harmonic proportion. Yet he acknowledged that this was only one factor in a larger set of causes, which included the listener's complexion or temperament as well as educational level.[28] For him aesthetic response still had its basis in an objective and external reality, the ratios in the sounding music. He accounted for the variations observed in individual listeners in ways that could be reconciled with the traditional arguments: harmonic proportions are the world's building blocks that underlie such factors as the balances among the humors that make up an individual's complexion.

The case of Zarlino shows that before Galilei theories of sound production were relatively fixed and stable; the space for developing more complex explanations of music's emotional effects therefore lay on the side not of sound production but of its perception. The earliest attempt to discuss perception in any detail seems to have been made about mid-century by the physician and polymath Girolamo Cardano in explaining why the listener's notions of consonant and dissonant intervals differ at times from the mathematical norms. He argues that because the mind's construction is orderly it has a natural delight in order. Thus it recognizes and enjoys the regularity of consonant intervals, though the use of complex rhythms and of pitch intervals larger than an octave make such perception more difficult. Tempered tuning systems work because the mind will categorize slightly imprecise pitch intervals as the nearest available precise one. Thus an interval may be dissonant in its production yet consonant in perception. Further, the mind also naturally delights both in recognition of the familiar and the learning of new things, suggesting that composers should balance and vary these features to maintain the listener's interest. Cardano claimed, however, that music has only general and limited abilities to induce emotions, and would be incomplete without a poetic text.[29]

Vincenzo Galilei's attack on Pythagorean theory had dealt with sound production, splitting it away from music more decisively than had Zarlino; but it also affected the basis in musical scholarship for explaining perception and cognition. A number of humanist literary theorists had meanwhile become interested in such issues, notably in writing about drama (which came increasingly to mean musical drama). They developed notions of catharsis based on Aristotle, in which appeals to the humors replaced the Pythagorean proportions of musical scholarship. By the late sixteenth century a number of writers, such as G. B. Guarini, expanded these theories to claim that drama's ability to purge the listener's emotion included not just pity and fear as Aristotle had described, but also feelings such as melancholy; further, such purgation took place in comedy as well as tragedy. These arguments in turn underlay the development of musical drama or opera at the end of the century. Modern scholars have noted the close ties between these literary theories of emotion and catharsis and those of contemporary medicine.[30]

Even general discussions of music's emotional effects came more and more to use the medical language of humors, complexions, and faculties (with less and less recourse to a purported basis of these notions in harmonic proportion). Mei explained that music moved the "natural passions and affections," which are easily invoked because they are innate.[31] Even Zarlino located "passions of the soul" in the "sensible, corporeal, and organic appetite," varying by individual complexion.[32] Thus music remained linked to fields we would recognize as sciences, though no longer in its old role as a unifying scientific principle; the study of sound production and the nascent field we would call psychology began to move in very different directions.

This transition from mathematics to medicine in the study of perception affected the relative status of music and text in the same way as did the overall rise of poetics in musical analysis: text came to dominate music. When humanists began to write about musical drama in the mid-sixteenth century, they approached the subject from their own initial interest in literature and poetry. It is hardly surprising that their theories about the nature of drama and its effects on the audience began with the meaning imparted in the text rather than with music. Most claimed that human reason is moved by words, whereas music moves only lower faculties, the emotions or passions. Bernardino Tomitano claimed, for example, that "just as the harmony of the poet is known more by the intellect than by the senses, its contrary, music, pleases the ear more than the mind."[33] Sperone Speroni agreed that while poetry and oratory pertain to the intellect, music and painting pertain to the sense.[34] Even the Platonist Francesco Patrizi, who would disagree with these Aristotelians in many other ways, argued that music's moral power lay in its ability to moderate and control the passions and the irrational. Thus it is especially useful in educating the young because its calming effect on these irrational parts of the mind allows learning to take place.[35] The application of humanist poetics to general musical analysis brought these notions more and more into musical discourse.[36] In the early sixteenth century writers on music had agreed that music was the most effective way to move and develop human reason; by 1600 most denied its effect on reason, a faculty they agreed was moved instead by words.

The preference for vocal over instrumental music seen in these critical approaches developed into some actual attacks on the value of instrumental music, produced mostly by the circle of persons in Florence associated with the creation of opera. To Girolamo Mei the rise of instrumental music over vocal music had marked music's overall historical decline.[37] Giovanni dei Bardi attributed the defects of modern vocal music to the influence of the techniques of instrumentalists, particularly string players.[38] Galilei complained that modern musical styles (notably polyphony) so favor instrumental music that modern works are generally better played than sung.[39] Music without words might charm the ear but could not elevate the mind; it merited neither praise nor promotion.

Not all musical scholars went so far, and some answered these charges. Zarlino, for example, responded by suggesting that such critics had been exposed to good letters but deprived of music, and could but condemn something they did not understand.[40] Galilei, despite his arguments about the value of vocal music, continued to write for the lute and to publish polyphonic madrigals; he even suggested elsewhere in the *Dialogo* that purely instrumental music could indeed affect the emotions, without any accompanying text.[41] And rules for the composition of counterpoint continued to form much of the contents of many practical music treatises, rules without reference to any accompanying poetic text, and that used terms such as "imitation" or "subject" in discussing fugue, not mimesis or poetry. Yet as humanist discourse replaced Boethian theory in articulating music's general principles both within the field itself and in its connections to other fields, instrumental music remained at the margins. The shift to poetics made possible the stylistic and historical analysis of musical compositions for the first time; but based as it was in literary studies it had limited abilities to bring this new analysis to music without text, a limitation that reinforced its own privileging of words. The attacks on instrumental music illustrate the irony that music's expansion into the realms of drama and poetry also tended to lower its overall status in some important ways.

Theories of imitation and catharsis, popular as they were, were not the only available explanations for music's ability to move emotions. Some scholars claimed that the association between music and emotion lay entirely in cultural associations that were based in tradition and custom, and therefore varied by time and place. Galilei tended to favor such explanations, though not exclusively. He argued that the usage of musical modes had varied over time among the ancients, so that as customs changed so too did the emotional effects ascribed to a given mode.[42] Ancient modes had different ethnic associations (as seen in their names) because of regional differences in the typical pitch of speaking voice; such regional differences could also be observed by comparing the preferences of Turks or Moors with those of the Italians of Galilei's day.[43] While the tendencies to cultural relativism implied by such a position had limits in practice, this approach granted musical scholars some amount of freedom from absolute judgments of quality in studying the music of other times or regions.

The study of the musical compositions and performances of other times, ancient or recently past, also became a serious undertaking by the end of the sixteenth century. It was noted earlier that much of the discipline of music meant the study of texts about music for several reasons: the long existence of such a discipline, and consequently the abundance of such texts; its Pythagorean philosophical background, which like all Platonic traditions privileged absolutes and ideals over individual manifestations; and the problems of gaining access to music not actually in performance, which made texts much easier to study by comparison. As scholars became more interested in the musical works of the past they had to confront these problems of accessibility, problems that related to the preservation of notated compositions, the interpretation of music notation, and the great gulf separating music notation from actual performance.

Musical composition and performance traditionally belonged not to musical theory but to practice. The practical tradition's links to antiquity were more tenuous than those of theory, and it had a somewhat different sense of its own history. One part of the practical tradition, Gregorian chant, had showed great stability in repertoire and practice over the centuries. Otherwise, however, the music Europeans heard was generally of fairly recent composition. Insofar as writings about practical music addressed historical issues, they had tended to claim that the music of their own day was superior to that of the past.[44]

References to the music of the recent past often overlapped with discussions of contemporary musical style, though modern music historians have long found almost all such passages frustratingly vague and brief. Writers were more likely to name composers, perhaps with a brief list of their praiseworthy qualities, than to discuss specific compositions or musical passages.[45] Most writers seemed unfamiliar with substantial details about earlier music; their praise of a composer from the previous generation might describe him as having purged abuses and restored good music, but seldom defined the abuses or the improvements.[46] It has been plausibly suggested that the continuous and often rapid changes in music notation since the thirteenth century rendered the compositions of earlier generations difficult or impossible to read. Notation conventions changed so quickly at times that as Zacconi remarked, older performers might forget how to read notation they had understood and performed in their youth.[47] Zarlino commented on the ephemeral nature of musical performance in observing that music differs from the plastic arts in that it does not employ things that remain but things that take place.[48] Without modern technologies of audio recording, music notation was the only means of access to music not actually in performance; yet the access it provided to the music of earlier eras was apparently limited at best.

The many disagreements that marked sixteenth-century efforts to discuss earlier music, whether of the recent past or of antiquity, should thus come as little surprise. Scholars differed over facts as well as larger historical narratives, and were not always consistent in their own claims. Consequently the development of a discipline of "music history" proceeded more slowly than in the visual arts or literature, and is usually seen as a product of eighteenth-century scholarship.[49] Scholars at the end of the Renaissance made no attempts to write a full history of music, concentrating instead on the detailed study of smaller topics.

The practical tradition's focus on recent music and the interest of theory in antiquity combined to produce only a restricted interest in the music of the recent past. Many passages on the subject are fairly brief, such as those already mentioned that identify a composer of a generation earlier who had helped raise modern music to its high level.[50] The topic most closely approaching a historiographic debate was the length of time Europeans had composed music using contrapuntal styles. Many scholars argued that the practice was of relatively recent origin. Mei asserted in 1572 that four-voice music was about 150 years old, suggesting though not specifying that he meant polyphony of any sort, not just modern styles of polyphonic composition; Galilei echoed him in his *Dialogo*.[51] Mei was hardly the first to make such a claim. Heinrich Glarean had suggested in 1547 that polyphony had begun no more than some seventy years earlier.[52]

An author's assessment of the age of polyphony was linked in part to the degree of his interest in reforming modern music. Such dating allowed Mei and Galilei to condemn many aspects of contemporary practice as recent corruptions. Yet the matter cannot be reduced to a dispute between ancients and moderns, since not all scholars were willing to treat contemporary composers as a single group. For example, some (though not all) of the composers usually identified in their harmonic style as "chromaticists" saw themselves as applying principles of ancient harmonic genera in their compositions. Both Zarlino and Galilei criticized such efforts, though Zarlino still found more to praise among modern composers.[53]

Zarlino disputed the recent origin of polyphony in the *Sopplimenti*, claiming that in the form of organum at least, it could be dated as early as the ninth century.[54] His argument illustrates how much more comfortably scholars employed texts than musical notation in building historical arguments. Zarlino began by discussing textual evidence of various kinds; first came a reference to organum in Bede, which he compared to references in Platina's *Lives of the Popes* in order to date the reference to A.D. 865. He moved on to Guido of Arezzo's eleventh-century music treatise *Micrologus* with its description of organum, and a decretal of John XXII limiting the use of polyphonic music. Only then did he note that he possessed a parchment manuscript of two-and three-voice songs dated 1397 by someone not the original owner, and another manuscript supposedly still older; the rarity of such manuscripts, he noted, does not diminish their testimony. The brevity of Zarlino's description and his reference to the paucity of medieval music manuscripts suggest that scholars knew relatively little about their existence, let alone how to interpret them for historical studies of musical style.[55]

The possible existence of ancient polyphony also met with dispute. Zarlino argued consistently and in some detail that ancient music was not polyphonic, based both on textual references to music performance and on the logic of ancient pitch systems. Mei and Galilei agreed, but not consistently. Most of their writings took the position that ancient music was monodic song. Yet Mei's criticism of instrumental music offered a different story. Here he dated instrumental music's corrupting influence to an era as early as the age of Plato.[56] A sort of polyphony had begun when instrumentalists ceased to follow the vocal line and began to add variations, a practice Plato condemned. Eventually this corrupted music overwhelmed virtuous music. Both Galilei and Bardi used this narrative in their own critiques of modern instrumental music.[57] In so doing they followed in the rhetorical footsteps of ancient writers from Plato to Quintilian, who suggested that the music of their own age had grown corrupt and called for a revival of the more virtuous practices of earlier antiquity.

Whatever their opinions on such questions about post-classical European music, it was the music of antiquity that held the broader and more general interest. A century of scholarship on ancient writings about music had improved greatly the abilities of musical scholars to work with the textual tradition, and had advanced their understanding of ancient pitch systems. Yet they had only begun to deal directly with the records of ancient compositions. Great as the problems were in interpreting the notation of the recent past, they knew the difficulties were far greater for that of the Greeks and Romans. When they finally found examples of ancient compositions it was clear they could never recover an ancient performance—or copy one—the way that Michelangelo could copy or re-create ancient sculpture. Ancient notation had committed to written record only basic pitches, leaving the rhythm of the music dependent on the meters of the accompanying poetry. This notation was clearly insufficient to describe or reproduce a performance. Scholars by 1600 agreed that modern notation owed its origins to the eleventh-century monk Guido of Arezzo, and that it was (despite its imperfections) far superior to that of antiquity.

Thus if they wished to recover what was knowable about ancient music, they had to employ a wide range of scholarly tools. Further, it was clear that in composing and performing new music they might emulate aspects of ancient music as they understood it, but could not copy or revive it. The resulting integration of some features of ancient performance with an essentially modern practice was hardly unique to music, of course. It was common in the visual arts, notably in religious art and architecture. So too, authors of prose and poetry understood that vernacular works and translations could only adapt, not reproduce, ancient models because of the differences between ancient and modern languages. Yet in all these other fields the actual artifacts of antiquity, the objects of study, were more directly accessible than was ancient music. In both scholarly research and practical application, music required more indirect scholarship in arriving at any sort of historical understanding. Historical investigation and modern application remained closely connected in these efforts.

The first and easiest means to historical understanding (musical or otherwise) came, of course, via the analysis of written sources. Yet texts were not the only evidence for ancient musical practice. Zarlino, Mei, and Galilei all extended their research to include the range of ancient evidence commonly classified as antiquarian: statues, inscriptions, reliefs, painting, and other nondocumentary relics. Zarlino responded to a query from the scholar Giovan Vincenzo Pinelli on the nature of the plectra used with various ancient stringed instruments, based not only on textual evidence but on surviving sculptures and reliefs.[58] Galilei published drawings of hypothetical reconstructions of ancient stringed instruments in the *Dialogo*, reconstructions that may have prompted Pinelli's questions.[59]

Just as historical research and modern application were closely connected, so too did the antiquarian study of artifacts profit from modern experimentation. Galilei's work in particular showed that many questions about ancient music theory could be resolved with the careful study of sounding bodies in controlled ways, and by attempting to apply theoretical principles to actual musical instruments. So too by continuing their investigations of ancient musical instruments scholars could augment considerably their knowledge about ancient music performance. Since no ancient instruments had survived intact and the known fragments were few and inconclusive, scholars were driven to the further study of statues, reliefs, vase paintings, and so on, correlating those visual findings with literary descriptions, and attempting thereby to identify the range of instruments used at various times and places. This research did occasionally result in the building of musical instruments, as in Giovanni Battista Doni's *lyra barberina*, a stringed instrument designed as a hybrid between ancient and modern models in order to perform new music based on similarly hybrid pitch systems.

Doni pursued the study of ancient musical practice with the experience acquired from his long interest in antiquarian studies, in works composed from the 1630s until his death in 1647. His extensive travels and correspondence ensured this scholarship a wide audience both in Italy and across Europe. Doni's research and writings on the history of ancient musical instruments were the first full-scale studies on the subject, and made use of the ever-growing collections of the Church as well as the information collected from his correspondence with other antiquarians.[60] He also wrote extensively on the styles of ancient theater music and ancient pitch systems, and saw his work as continuing along the path set by Mei.

Doni noted that his research methods were of necessity mixed in nature, since the best records produced during antiquity (written and otherwise) had not survived; scholars must do their best to extract information from the imperfect sources that remain.[61] The results rest not upon truth but only probability or verisimilitude, which he acknowledged to be imperfect but not without value if based on sound principles. Doni's methods relied on the scholar's ability to make judgments about ancient habits; such assessments required in turn a combination of historical evidence and comparisons of the little-understood ancient practice with similar, better-known modern ones. For example, in building his case for the attention span of an ancient audience (essential to his larger arguments about the use of music in ancient drama) he combined such evidence as the number of lines in ancient plays, the probable length of the chorus's performance, and the pace of speech, with similar features of modern drama. He concluded that in this respect at least, ancient and modern audiences were similar. This conclusion allowed him both to make further inferences about likely performance practice in ancient drama, and to make recommendations for the improvement of modern theatrical performance.

Doni's research on musical instruments shared with the work of the previous generation the dual goals of antiquarian research and modern application, though he was not a practicing musician. He joined a long line of composers and scholars in wishing to restore to modern practice the broad range of tuning and scale systems used in antiquity, so as to increase music's ability to express texts. This was the purpose for designing the *lyra barberina* and other hybrid instruments. Given his knowledge of ancient instrument designs he was under no illusion that his *lyra* was a copy of one used in antiquity, and while he thought ancient norms might help improve modern music, he knew that modern music differed significantly from that of antiquity. He told several recipients of one of his works, *De praestantia musicae veteris* (1647) that he wished "to incite the moderns to improve this faculty with the example of the ancients" as well as "to bring some light to things formerly left in darkness."[62]

In this regard too Doni's positions developed out of those of the late sixteenth century. Galilei's discussions of music and cultural change had compared musical styles to spoken languages, which were known to have changed over time during antiquity and afterwards. Many of his recommendations for musical reform had involved imitating the methods of good ancient composers, in a modern setting. Thus when Galilei urged composers to be more like the ancients in fitting music to words, he recommended not that they turn to ancient treatises or to fragments of ancient compositions; rather they should study modern speech, just as the ancient musicians and orators had studied their own spoken languages. They should attend modern theatrical productions and pay attention to the manners of speech and the other ways the actors represent different types of characters. In this way the composer could learn how to set any type of text.[63] For these scholars, historical research and modern application were inextricably connected. Imperfect historical knowledge could be tested for its verisimilitude with modern trials and comparison; so too the standards for evaluating modern productions would be improved by the study of antiquity.

The study of music as understood by Doni and his generation in the seventeenth century differed greatly from that of the age of Gaffurio in the late fifteenth century. The center of the field had shifted from proportion and pitch to musical compositions, and the relationships to other disciplines had changed accordingly. Mathematics remained important in studying the science of sound and in the descriptions of pitch and tuning systems, but not in aesthetics; astronomy and cosmology no longer received more than cursory mention. Mathematical explanations were also restricted to analyses of sound production. The study of sound perception now joined other attempts to explain perception, emotion, and reason as part of medical scholarship.

Within the studies that continued to be known as the discipline of music, the most important single transition was the introduction of analytic tools from the humanist tradition. This change made possible for the first time the historical study of musical scholarship and performance. The interest in historical aspects of music also contributed to the expansion of humanistic historical research into broader aspects of culture, notably into the eclectic studies of cultural topics and non-textual sources known as antiquarian. Historical research maintained a close relationship with modern practice and criticism.

Yet these changes in the field were not without irony. Music in the seventeenth century might claim links to a greater number of subjects than ever before, but it had ceased to do so as the key discipline explaining the universal truths that underlay all others. In fact, the study of the art of music came to be spread among any number of fields, from psychology to poetics, but was the major part of none of them. Rather than lending its analytic methods to other fields, it began to borrow them instead. Its reliance on textual analysis limited the ways scholars and critics could discuss instrumental music. Its close connections to literary scholarship also left music vulnerable, by extension, to the distrust of language-based scholarship that became such a notable feature of seventeenth-century thought. The origin of music history cannot be separated from the end of the quadrivium; it marked for the discipline of music serious losses as well as considerable gains.

## Notes

1. For a more detailed discussion of musical scholarship through the era of Vincenzo Galilei, see Ann E. Moyer, *Musica Scientia: Musical Scholarship in the Italian Renaissance* (Ithaca, N.Y.: Cornell University Press, 1992).

2. See for example Zarlino, *Istitutioni harmoniche* (Venice: Senese, 1588; f.p. 1558), 361–62, where Gaffurio is the earliest scholar named. Girolamo Mei names him as a "modern" in his letter to Vincenzo Galilei, 8 May 1572, in *The Florentine Camerata*, ed. and trans. Claude V. Palisca (New Haven: Yale University Press, 1989), 67. Galilei refers to Gaffurio as the first to restore musical studies from the "blindness" in which it had remained since the barbarian invasions: *Dialogo della musica antica, e della moderna* (Florence 1581; facs. New York: Broude, 1967), 1. Orazio Tigrini typically cites authorities chronologically; he moves from Boethius directly to Gaffurio and on to subsequent modern scholars; *Il Compendio della musica* (Venice, 1588; facs. New York: Broude, 1966).

3. The debates over the texture of ancient compositions—whether they were monodic, homophonic, polyphonic, or some combination of these—were far from settled by this discovery. For a discussion of some of the main positions in this debate see D. P. Walker, "Musical Humanism in the 16th and Early 17th Centuries," pt. 4, 306–7, pt. 5, 57–71, reprinted in *Music, Spirit and Language in the Renaissance*, ed. Penelope Gouk (London: Variorum, 1985), 1.

4. Zarlino, *Sopplimenti musicali* (Venice 1589; facs. Ridgewood, N.J.: Gregg, 1966), 1.2.10–12.

5. Girolamo Mei, "De modis musicis antiquorum" (Vatican City: Biblioteca Apostolica Vaticana, Vat. Lat. 5323). On Mei's work as an early manifestation of music history see Frank Ll. Harrison, "The European Tradition," in *Musicology*, ed. Frank Ll. Harrison, Mantle Hood, and Claude V. Palisca (Englewood Cliffs, N.J.: Prentice-Hall, 1963), 14.

6. Galilei, *Dialogo*, 133–34; Galilei, "A Special Discourse Concerning the Diversity of the Ratios of the Diapason," and "A Special Discourse Concerning the Unison," in *Florentine Camerata*, 180–201.

7. Vincenzo Galilei, "Intorno all'uso delle Consonanze," Florence, Biblioteca Nazionale Centrale (hereafter BNC), Galileiani 1, 11r (another copy, 60v): "Nulla dimeno il Musico si come ancora fa l'Aritmetico de numeri, si serve solo nelle sue bisogne d'una determinata et breve quantità de piu noti al senso, non altramente di quello che fa il Pittore de colori." And also: Vincenzo Galilei, "Discorso all'use dell'Enharmonico," Florence, BNC, Gal. 3, fol. 19v: "la qual maniera d'operare molte cose con pochi mezzi l'hanno ancora nell'arte loro usato alcuni eccellenti Pittori, i quali nelle pitture loro non si sono voluti servire piu che di trover' quattro colori. . . . "

8. Giovanni dei Bardi, "Discourse Addressed to Giulio Caccini, Called the Roman, on Ancient Music and Good Singing," in *Florentine Camerata*, 92–95.

9. Lodovico Zacconi, *Prattica di musica . . . [prima parte]* (Venice, 1592; facs. Bologna: Forni, 1967), 2–4; *Prattica di musica, seconda parte . . .* (Venice, 1622; facs. Bologna; Forni, 1967), 8.

10. Vincenzo Giustiniani, "Discorso sopra la musica de' suoi tempi," in *Le origini del melodramma*, ed. Angelo Solerti (Turin: Bocca, 1903), 98–128, esp. 104–15. He also distinguished his subject from mathematical theory as described by Augustine or Boethius. There is an English translation by Carol MacClintock, Musicological Studies and Documents, 9 (American Institute of Musicology, 1962).

11. Orazio Tigrini, *Il Compendio della musica* (Venice, 1588; facs. New York: Broude Brothers, 1966), 1–2.

12. Giovanni dei Bardi, "Discourse Addressed to Giulio Caccini, Called the Roman, on Ancient Music and Good Singing," in *Florentine Camerata*, 92–93.

13. Giovanni Boccaccio, *Genealogie deorum gentilium libri*, 14.8; Boccaccio relies on Varro and Suetonius by way of Isidore. See *Boccaccio on Poetry*, trans. Charles G. Osgood (Princeton: Princeton University Press, 1930), 42–44, 161–62.

14. On the classification systems in use through much of the sixteenth century and the separation of music and poetry in those systems see Bernard Weinberg, *A History of Literary Criticism in the Italian Renaissance* (Chicago: University of Chicago Press, 1961), 1.1–12, 18.

15. Raffaele Brandolini, "De musica et poetica opusculum," (1513) Rome, Bib. Casanatense 805. For a later example of the analysis of poetry with the tools of music see the brief discussion of Bernardino Parthenio's *Della imitatione poetica* (1560) in Bernard Weinberg, *History of Literary Criticism*, 1.280–81.

16. On the application of the terminology of humanist textual criticsm to music see Don Harrán, "Elegance as a Concept in Sixteenth-Century Music," *Renaissance Quarterly* 41 (1988): 413–38; Howard Mayer Brown, "Emulation, Competition, and Homage: Imitation and Theories of Imitation in the Renaissance," *Journal of the American Musicological Society* 35 (1982): 1–48; and, more generally, Barbara R. Hanning, *Of Poetry and Music's Power: Humanism and the Invention of Opera* (Ann Arbor: UMI Research Press, 1980), 23ff.

17. Don Harrán, *Word-Tone Relations in Musical Thought: from Antiquity to the Seventeenth Century* (Neuhausen-Stuttgart: American Institute of Musicology, 1986), 89; they appeared in Giovanni Maria Lanfranco, *Scintille di musica* (Brescia, 1533).

18. On the use of the term "barbarism" in music and a discussion of the sources see Don Harrán, "Elegance as a Concept," 413–38. Before the late sixteenth century most discussions of barbarism in music refer to the text and its pronunciation by the singer, not to the music itself.

19. Orazio Tigrini, *Compendio della musica*, 36–37.

20. For Jacopo Peri's own description of recitative as imitative of speech see his preface to *Euridice* (1600), reprinted in Solerti, *Le origini del melodramma*, 43–49; English trans. in *Source Readings in Music History: the Baroque Era*, ed. Oliver Strunk (New York: Norton, 1965), 13–16.

21. For a summary and bibliography of scholarship on the madrigal in late sixteenth-century Italy, see Anthony Newcomb, "Madrigal" II.7–12, *New Grove Dictionary of Music and Musicians*. On humanistic criticism and the setting of texts see D. P. Walker, "Musical Humanism," pt. 4, 288–308; Harrán, "Elegance as a Concept"; see also Charles Warren, "Word-Painting," *New Grove*.

22. Walker, "Musical Humanism," pt. 4, 290. Walker here refers to Giovanni Battista Doni, *Lyra Barberina* 2.73 (see n.60).
23. Lucretius, *De rerum natura* 5.1348–1416. On the notion of imitation in sixteenth-century music, see Armen Carpetyan, "The Concept of *Imitazione della natura* in the Sixteenth Century," *Journal of Renaissance and Baroque Music* 1 (1946): 47–67; Walker, "Musical Humanism," pt. 4, 289–95. For criticism of imitation of sense without meaning see Girolamo Mei, Letter to Vincenzo Galilei, 17 January 1578, in *Letters on Ancient and Modern Music to Vincenzo Galilei and Giovanni Bardi*, ed. Claude V. Palisca (n.p.: American Institute of Musicology, 1960), 139–40; Galilei, *Dialogo*, 82.
24. Walker, "Musical Humanism," pt. 4, 291; Galilei, *Dialogo*, 88–89. For a discussion of such criticisms in any genre see Harrán, "Elegance as a Concept."
25. Tigrini, *Compendio*, 37.
26. Zarlino, *Istitutioni harmoniche*, 4.32, 419–20; trans. Vered Cohen in *On the Modes*, ed. Claude V. Palisca (New Haven: Yale University Press, 1983), 94–95.
27. Baxter Hathaway, *The Age of Criticism: The Late Renaissance in Italy* (Ithaca, N.Y.: Cornell University Press, 1962), 210.
28. Zarlino, *Istitutioni*, 86–90.
29. Moyer, *Musica Scientia*, 161–68.
30. Hathaway, *Age of Criticism*, 205ff.
31. Mei, "Letter to Galilei, 8 May 1572," in *Florentine Camerata*, 64.
32. Zarlino, *Istitutioni*, 2.8.87.
33. Bernardino Tomitano, *Quattro libri della lingua thoscana* (Padua, 1570), 86; in Hathaway, *Age of Criticism*, 67; the work was written in 1545.
34. Sperone Speroni, *Opere* (Venice, 1740), 1:210–11; in Hathaway, *Age of Criticism*, 231. The work was composed in 1540. Other writers on poetics agreed with this assessment that music moved the passions while poetry affected the intellect; see for example Francesco Benci (1590), in Weinberg, *History of Literary Criticism*, 1:334.
35. Patrizi, *Della poetica* (Ferrara, 1586), 181–83; in Hathaway, *Age of Criticism*, 251.
36. See for example, Giulio Caccini's preface "Ai lettori" to *Le Nuove musiche* (Venice, 1601), in *Le origini del melodramma*, 57.
37. Girolamo Mei, letter to Vincenzo Galilei, 8 May 1572, in *Florentine Camerata*, 72–73.
38. Giovanni dei Bardi, "Discourse Addressed to Giulio Caccini," in *Florentine Camerata*, 113.
39. Galilei, *Dialogo*, 87; Walker, "Musical Humanism," pt. 5, 66.
40. Zarlino, *Sopplimenti musicali* (Venice, 1588), 8.10; Walker, "Musical Humanism," pt. 5, 66.
41. A volume of lute entablature was completed in 1584, the madrigals published in 1587; see Nino Pirrotta, "Temperaments and Tendencies in the Florentine Camerata," in *Music and Culture in Italy from the Middle Ages to the Baroque* (Cambridge: Harvard University Press, 1984), 222. On Galilei's positive statements about instrumental music see Walker, "Musical Humanism," pt. 3, 226–27; Galilei, *Dialogo*, 90.
42. Vincenzo Galilei, "Compendio di Vincentio Galilei: della tehorica della musica" (ca. 1570), Florence, BNC, Gal. 4, fol. 36r–v.
43. Galilei, *Dialogo*, 71; Galilei, "Discorso . . . all'uso dell'Enharmonio," Florence, BNC, Gal. 3, fols. 7r–8v.
44. Jessie Ann Owens, "Music Historiography and the Definition of 'Renaissance,'" *Notes* 47 (1990): 305–30, esp. 307–18; on modern understandings of the length of time compositions remained in performance repertory, see 323–24. See also Georgia Cowart, *The Origins of Modern Musical Criticism* (Ann Arbor: UMI Research Press, 1981), 35–39.
45. James Haar, "A Sixteenth-Century Attempt at Music Criticism," *Journal of the American Musicological Society* 36 (1983): 191–209, esp. 191–93.
46. Owens, "Music Historiography," 309–14.
47. Owens, "Music Historiography," 319, 322; Zacconi, *Prattica di musica* [prima parte], fols. 85r–v, 115v; *Prattica di musica, seconda parte*, 54.
48. Zarlino, *Dimostrationi harmoniche* (Venice, 1571; facs. Ridgewood, N.J.: Gregg, 1966), 22–23.

49. Frank Harrison, "The European Tradition," 15. On the historiography of Renaissance musical style, see Owens, "Music Historiography," 324–30.

50. Owens, "Music Historiography," 309–14.

51. Mei, letter to Galilei, 8 May 1972, in *Florentine Camerata*, 59–60; Galilei, *Dialogo*, 80.

52. Heinrich Glarean, *Dodecachordon* (Basle, 1547), 113; see Owens, "Music Historiography," 317–18.

53. Zarlino, *Istitutioni* 3.80.357–58; Moyer, *Musica Scientia*, 216–17. Galilei alludes to such efforts without naming them specifically.

54. Zarlino, *Sopplimenti*, 16–18.

55. See Owens, "Music Historiography," 318, 323–24.

56. Mei, letter to Galilei, 8 May 1572, in *Florentine Camerata*, 70–73.

57. Galilei, *Dialogo*, 81–83; see Walker, "Musical Humanism," pt. 5, 63–67.

58. Zarlino, letter to G. V. Pinelli, 14 August 1586, Milan, Biblioteca Ambrosiana R 118 sup., fols. 220r–22v.

59. Galilei, *Dialogo*, 124–31; Moyer, 249–52.

60. Giovanni Battista Doni, *Lyra barberina amphicordis*, 2 vols. (Florence, 1763; facs. Bologna: Forni, 1974). Some, though not all of these writings were published during his lifetime; the treatise on the *lyra barberina*, which contained his studies on ancient musical instruments, was published only posthumously, though many of his correspondents were familiar with the work and its contents, as well as with the *lyra barberina* itself. See *Jo. Baptistae Donii Patricii Florentini commercium litterarium.* . . . (Florence: Caesareis, 1754); Francesco Vatielli, *La Lyra barberina di G. B. Doni* (Pesaro: Nobili, 1909); Claude V. Palisca, "Doni, Giovanni Battista," *New Grove*.

61. In his *Lezioni se le Azioni drammatiche si rappresentavano in Musica in tutto, o in parte* (*Lyra Barberina* 2:145–60, esp. 146), he argues that there are two ways of proceeding in studying about theater music: from the authority of written sources whose authors had certain knowledge of the subject; and conjecture based on sound principles ("per via del congetture, ed argomento, cavati dal verisimile").

62. " . . . sì per eccitare i Moderni a perfezionare questa facoltà con l'esempio degli Antichi . . . " G. B. Doni, Letter 156 to Giuseppe Persico, in *Commercium litterarium*, cols. 227–28; "per eccitare i moderni a migliorare questa facoltà con l'esempio degl'Antichi, e dar qualche luce ad una materia tenuta per l'addietro assai oscura. . . . " G.B. Doni, Letter 159 to Mons. Tommasini, Bishop of Città Nuova, ibid., cols. 229–30. Doni was among the scholars who believed that ancient music included polyphony; Walker, "Musical Humanism," pt. 4, 306–7; pt. 5, 57–60.

63. Galilei, *Dialogo*, 89–90; Walker, "Musical Humanism," pt. 4, 291–95.

# NATURAL LAW AND HISTORY
## *Pufendorf's Philosophical Historiography*

### Michael Seidler

Judged by the numerous editions of his main jurisprudential works, Samuel Pufendorf was among the most influential philosophers of the first half of the eighteenth century. Through him, modern natural law became not only a philosophical science but also a discipline with a history. Although he regarded Grotius as the originator of modern natural law and viewed himself (and Hobbes) as working in that tradition, Pufendorf was himself the founder of its historiography, becoming a model for scores of imitators in the following decades. No "mere" historian, however, he approached the history of moral philosophy as an internal participant rather than an outside observer, providing an example of how intellectual traditions or disciplines are defined by those who reflectively participate in them, and revealing tensions between philosophy and its history that persisted into Kant, Hegel, and beyond.

Pufendorf also wrote several kinds of political and national history for which he was equally famous. Yet with one exception, these works saw few subsequent editions and translations. Some were rediscovered by eighteenth- and nineteenth-century German historians, who greeted them with critical admiration, but on the whole they have received little notice, especially in comparison to Pufendorf's other works. Thus, they have been seldom read or studied, either for their own sake or in relation to his philosophical writings. Indeed, they have even been regarded as an independent, second phase of Pufendorf's intellectual development, dictated only by the contingent exigencies of his political career.[1]

Yet this is incorrect. For Pufendorf was as much a philosophical historiographer as he was a historiographer of philosophy. His histories reflect and support particular philosophical and political views, and should thus be seen in relation to his natural law system. They exhibit the same tension between objectivity and subjectivity that characterizes Pufendorf's use of the history of philosophy, and the same effort to reconcile the apparently divergent demands of historical and philosophical truth and moral and political commitment. Setting aside the importance of Pufendorf's understanding of the history of philosophy for the early German enlightenment, I want here to explore this other, philosophical historiographer's role, partly as an entry into the former question.[2]

The two roles are linked through the idea of the historian as agent and as representing agency. Since human agency implies intentionality, purpose, and commitment, all limited and concretely situated, Pufendorf always wrote history "from somewhere." That is, his accounts self-consciously reflected the outlooks of his narratives' protagonists: in the case of his national histories, the rulers who had commissioned them and whose respective political causes he adopted or represented; in the case of the history of philosophy, those whom he regarded as his intellectual predecessors. Yet Pufendorf also rejected mere partisanship and avowed a dedication to factual accuracy and truth. Together, these features challenge the frequent criticism of his historical works as lacking objectivity, and the former, especially, shows such criticism to assume precisely the kind of "view from nowhere" that Pufendorf rejected as incompatible with an understanding of actions from within. The agency perspective of Pufendorf's historiography also allows it to be integrated into his corpus as a whole, giving this an otherwise unnoticed unity, and it helps to explain the modernizing influence he had on the philosophical, political, and historical disciplines to which he contributed.

The following section traces briefly the abiding interest in history that marks Pufendorf's whole life and career. Section 2 examines his main historical works, the types and tasks of history they reveal, and their assessment by posterity. Section 3 shows how the nature of the available evidence helped determine the internal perspective of Pufendorf's histories and resolves the tension between his joint dedication to truth and the cause(s) of his employers. Finally, Section 4 elaborates the agency theme and relates Pufendorf's historiography to his broader natural law theory.

# I

Since Pufendorf offered neither a formal theory of history nor a sustained discussion of the historian's art, his views must be reconstructed from scattered remarks throughout his corpus, including the correspondence and the dedications and prefaces to his works, especially though not solely the histories, and from his actual practices in the latter. Various materials that were until recently unpublished or generally unknown are also relevant, including the historically-based "opinions" commissioned by Pufendorf's employers, two fictional "letters to Rechenberg" responding to French and Dutch reviews of his historical works, and Pufendorf's short treatise on how to educate young noblemen. Together, these sources yield a consistent account.[3]

A survey of Pufendorf's life and career discovers ample evidence of an abiding interest in and cultivation of different kinds of history, from the earliest stages of his education to his final, posthumous works. Thus, Glafey reported in 1739 that already at the Fürstenschule in Grimma, which Pufendorf attended from 1645 to 1650, he kept a methodically organized notebook that excerpted many classical authors.[4] When it was taken by a classmate, Pufendorf reportedly reread all of the authors. This episode, if true, not only attests his love for antiquity but also explains his vast familiarity with ancient sources, as reflected in the more than three thousand citations (from 148 ancient writers) in *De jure naturae et gentium* (1672).[5]

We know little about the next stage of Pufendorf's formal education, his university studies at Leipzig between 1650 and 1656, beside the fact recounted by Pufendorf himself that they included much Aristotle.[6] Still, the image of Leipzig as a stifling, ultratraditional, scholastic backwater rigidly controlled by theological conservatives, which was cultivated by Pufendorf's nineteenth-century biographers and repeated by subsequent expositors, does not do justice to the intellectual stimulation he received there.[7] Indeed, one of his earliest biographers (Peter Dahlmann) reported that

> [d]uring his first academic years [at Leipzig] he very regularly frequented the seminars [*collegia*] and heard many academic teachers, so that he might learn in time to avoid as a poisonous plague the *praeconceptas opiniones* to which one allows onself to be lead too easily if one has had and heard only one professor.[8]

Dahlmann notes also that Pufendorf's favorite subjects at Leipzig were "Jura Divini et Naturalia," and that he linked these with "Studium Historicum, Politicum et Civile."

About the study of history at Leipzig we know that Johannes Strauch, a close friend of Pufendorf's older brother, Esaias, delivered during 1650–52 a set of formal "philological and historical-political lectures" on universal history, and also made informal presentations on Aulus Gellius and Tacitus before the Collegium Gellianum, the academic society to which he and Esaias belonged. When Strauch departed for Jena in 1652, his chair went to Christian Friedrich Franckenstein, who complemented the required lectures on universal history with excursions into Tacitus, Suetonius, Eutropius, and—more significantly in view of Pufendorf's subsequent advice to educators of youth, and his own practice—aspects of modern history.[9] Some influence of Franckenstein is likely, whether or not Pufendorf formally studied under him, since he appears to be the only teacher from the Leipzig period with whom Pufendorf stayed in touch. The study of Tacitus and other ancients (philology), the emphasis on modern history, and the "eclectic" effort to avoid preconceived opinions were all elements that marked Pufendorf's later views and practices, both as historian and as philosopher.

We are somewhat better informed about Pufendorf's unofficial studies at Leipzig, particularly his membership in the Collegium Anthologicum, the extracurricular academic society which he joined in 1655 and before which he gave some fifty lectures prior to his departure for Copenhagen in 1658. Forty of these presentations, which reveal a strong philological interest, dealt with history or cultural history.[10] The thirty on ancient history were externally organized according to the medieval "four monarchies" scheme of universal history, recently reemphasized by Melanchthon and Sleidan, and evinced a predilection for Herodotus. However, they also included a lecture (#6) on the origin of political authority (*Staatsgewalt*) that adumbrates themes in Pufendorf's later jurisprudential works, and one (# 3) that anticipates his explicit rejection of so-called "translation theory" in *De statu imperii Germanici* (1667). Pufendorf's interest in history was apparently unique, since other members of the Collegium tended more exclusively toward theological topics. Moreover, his presentations clearly evince the links between history and natural law mentioned by Dahlmann and characterizing much of his later work.

For both circumstantial and methodological reasons, Pufendorf's first natural law treatise, the *Elementorum jurisprudentiae universalis libri duo* (1660), exhibits more systematic rigor than historical awareness, though Grotius and Hobbes are mentioned in the preface and their positions engaged in the argument.[11] A continued reliance on history is evident during Pufendorf's Heidelberg period, however, in the various academic essays that constituted the transition from the youthful *Elementa* to the mature *De jure naturae*.[12] History was also crucial to the composition of Pufendorf's infamous (though long anonymous) *De statu imperii Germanici*, which arose out of the concrete circumstances of the so-called "Wildfangsstreit" involving the Palatine Elector, Karl Ludwig, and indirectly the Emperor in Vienna, and which based its criticisms of current conceptions of the Empire on a historical analysis of the relations between the Emperor, Pope, and the German estates. Many of Pufendorf's other political opinions and policy justifications at various stages of his career also contained historical accounts, as did the well-known *Les anecdotes de Suède* (mid-1680s), whose acute analysis of Sweden's ongoing transformation from a limited into an absolute monarchy is important for interpreting Pufendorf's views on sovereignty.[13] So did his works on religion, including the pseudonymous (as Basilius Hyperta) attack on the papacy ("Historische und politische Beschreibung der geistlichen Monarchie des Stuhls zu Rom" [1679]), the *De habitu religionis christianae ad vitam civilem* (1687), and the posthumous *Jus feciale divinum* (1695), which undergirded his views on toleration and church-state relations with a particular account of church history. Indeed, practically everything Pufendorf wrote, at every stage of his career, used history in some way.

Pufendorf also lectured on history at both Heidelberg and Lund, usually beginning with Roman history (including Tacitus) and moving on, like Franckenstein at Leipzig, to the modern period, where he examined the conditions and institutions of contemporary states. The Lund lectures were the direct source of his perhaps most famous historical work, the *Einleitung zu der Historie der vornehmsten Reiche und Staaten so itziger Zeit in Europa sich befinden* (1682–1686), which was frequently reissued and "continued" (by others) throughout the first half of the eighteenth century. The overt pedagogical intent characterizing this junction of past and present explains why Charles XI's chancellor, Magnus de la Gardie, planned during the early 1660s to invite Pufendorf to teach rhetoric (*verba*) and history (*realia*) at a new *collegium illustre* in Stockholm, where noble youth would be instructed in practical subjects useful for governmental administration and diplomacy.[14] Finally, Pufendorf's abiding interest in historical studies is convincingly established by the contents of his library, which included over four hundred historical works in many languages—about twenty percent of the total. Ancient history is expectably well represented here, but works on modern history predominate, particularly those on the French civil wars, the Dutch struggle for independence, and the two seventeenth-century English revolutions—all of which helped shape his views on contract, sovereignty, and resistance.[15]

## II

Even in his own time, Pufendorf was regarded as a polyhistor in the broad sense that his historical writings fall into several categories.[16] Thus, *De statu imperii*, the essay on the papacy, and even the posthumous *Jus feciale* may be called theoretical histories because they explicitly use historical details in the service of particular interpretations of, respectively, the Holy Roman Empire, the Catholic Church, and Protestantism. In each case, the immediate context is polemical and the underlying politics affects the historical account.[17]

The *Einleitung* belongs in the category of derivative history because, as Pufendorf explained, he derived "the History of each Kingdom from its own Historians" rather than from independent researches.[18] No history can be totally derivative, of course, if only because it involves the constructive act of using some but not other materials. Moreover, Pufendorf concluded each national history with a concise *raison d'état* analysis identifying the peculiar character, strengths, and current situation of each state. These accounts not only evoked varying, sometimes hostile reactions from their subjects, but they also indicate the multidimensional or, in this sense, narrowly polyhistorical nature of the work. However, Pufendorf's *raison d'état* approach, which is introduced in the preface as a generic analytical tool (replacing the traditional four monarchies scheme which he, like Bodin and Conring, abandoned), also suggests a new kind of secular universal history, particularly in the wider context of his natural law legitimation of the state.[19]

A third category consists of Pufendorf's "original" histories, the monumental works he composed as official state historian of Sweden and Brandenburg. Appointed to the former role in 1677, he wrote first his *Commentariorum de rebus Suecicis libri XXVI ab expeditione Gustavi Adolphi in Germaniam ad abdicationem usque Christinae* (1686) and then its shorter sequel, the *De rebus a Carolo Gustavo Sueciae Rege gestis commentariorum libri VIII*, which, though completed by the time he was "loaned" to Berlin in 1688, remained hostage in Stockholm until he returned to free it in 1694, after which it appeared posthumously in 1696.[20] In Berlin Pufendorf produced, again with remarkable speed, another grand account of European events since mid-century, seen this time through the prism of the Great Elector, the *De rebus gestis Wilhelmi Magni Electoris Brandenburgici commentariorum libri novendecim* (1695). He also began a history of the Great Elector's son, the *De rebus gestis Friderici Tertii, electoris Brandenburgici*,

... *commentariorum libri tres*. This latter work, though unfinished, and unpublished until 1784, is significant because of its detailed account of the English Revolution of 1688, portions of which had already appeared in the previous work.[21]

Finally, in Pufendorf's natural law works we find not only an astounding variety of historical references used to elaborate and illustrate the arguments, but also a hypothetical history of humankind that undergirds his contractarianism. This might seem like a return to theoretical history in a broader sense were it not also the case that Pufendorf sometimes discussed the originary natural state as a historical reality. In such instances, he often employed the Bible as a more reliable historical record than other ancient sources, which supposedly borrowed from it. The many Bibles in his possession indicate more than piety.[22]

Now all these distinctions are ultimately external, since history is always theoretical in the sense of reflecting perspectives shaped by practical intentions. This is so not only for narrowly theoretical history with its more or less overt political purposes, but also, as noted before, for the derivative histories of the *Einleitung* which exemplify Pufendorf's universalistic understanding of political history in terms of real and imaginary, and permanent and temporary states' interests. It holds even for his official state histories, whose annalistic, quasi-mechanical review of affairs of state proceeds beneath the umbrella of more basic political principles; and certainly for the hypothetical history that undergirds Pufendorf's entire natural law system. What ultimately ties these apparently disparate forms of history together is that they are all ways of understanding and facilitating political and moral agency. They bestow an identity on individual and collective agents by orienting them within a context of meaningful past actions, particularly as these are seen to bear upon the present. Accordingly, as the conception of human agency changes in response to altered circumstances—as it did in the early modern era—so do the forms and content of history.[23]

This helps to explain Pufendorf's preference for modern history and his conviction about the importance of historical studies in the formation of the ruling elites who were the primary social agents of his time. According to the preface of the *Einleitung*, "... History is the most pleasant and usefull Study for Persons of Quality, and more particularly for those who design for Employment in the State, ... he who has no Relish for History is very unlikely to make any Advantage of Learning or Books. ... "[24] The ancient historians, Pufendorf grants, are both "usefull and pleasant," but one ought only to begin with them and turn, thereafter, to "the History of later Times," a topic much neglected in the education of youth. For "to understand the Modern History as well of our Native Country, as also [of] its neighboring Nations" bestows a "considerable Advantage," as those "employ'd in States Affairs" well know. Such views recommended Pufendorf to the Swedish nobility studying at Heidelberg, and particularly to Erik Lindeman (Lindschöld), the tutor of Charles X's (illegitimate) son, Gustav Carlssohn, and later of Charles XII. Lindemann was persuaded by

Pufendorf to shift his views on education away from scholastic pedantry toward more practical subjects, and to revise completely his young charge's course of studies.[25] He was also attracted by Pufendorf's integration of divine voluntarism and social contract theory, which offered early modern nation-states new forms of moral and political legitimation.[26] Thus, when de la Gardie planned the new university at Lund he first offered a professorship to Lindeman, and when the latter declined he turned to Pufendorf.

Pufendorf considered a knowledge of (recent) history useful for several reasons. First, it provides a context for present action. Knowledge of a state's past deeds and intentions, and of the genesis of its institutions and policies, is essential for understanding its current condition and relation to other states. Thus, in *De statu imperii*, Pufendorf compared the efforts of anyone trying to explain the irregular structure of the Empire without a knowledge of German history and of the science of politics to "ein Esel beim Seitenspiel."[27] There, as in the essay on the papacy, history offered a strategic revelation, much like the "pickpocket history" [*beutelschneider historie*] Pufendorf mentioned wittily to Thomasius on November 26, 1692, responding to the latter's plan to write ecclesiastical history. For just as pickpocket history teaches one to protect one's purse at book fairs (*auf der meße*), while traveling, and in taverns, both ecclesiastical and civil history contain many "intrigues" which it is useful to bring to light. "In sum, everything is good and useful that can undeceive people and make them see."[28]

Secondly, it is not only others' past that is instructive; one's own history can also provide models of successful action and opportunities for recognizing and avoiding mistakes. Thus Pufendorf wrote to Pregitzer in Berlin, on October 3, 1691, responding to the anticipated revelations in his *Frid. Wilh.* : "it is important for one's fatherland to know its former errors, lest it stumble later against the same stone."[29] Of course, educating a state or even its leaders takes place in the public domain, and the knowledge gained in this way is also available to their opponents. Thus, Pufendorf's *Frid. Wilh.* was criticized for revealing too many of Brandenburg's errors and intrigues, and its publication, circulation, and translation accordingly restricted.

Thirdly, history may provide a motive for proper action. This idea appears clearly in the prefaces and dedications, including, again, that to *Frid. Wilh.*, where Pufendorf wrote shortly before his death that history is "the most splendid among all types of monuments" for preserving "the memory of great men." A published history withstands the ravages of time better than statues, paintings, and mausoleums, no matter how imposing, for it is not confined to one's own country but easily disseminated to anyone anywhere who is capable of understanding it. Also, it sets forth not only "gestorum . . . capita" but also the "indolem" or "habitum animi" of a prince for the admiration and imitation of others.[30] Such statements go beyond flattery or professional self-congratulation, since Pufendorf spoke also of desert in these contexts, implying a historian's obligation to the subjects of his narrative. Rather they reveal his awareness of the historian's power to affect others' actions by exposing them to the anticipated gaze of posterity.[31] Pufendorf's invitation to monarchs and others to attempt to deserve well of history was sometimes matched by the threat that "even those whose rank has exempted them from human punishment can in no way escape the free censure of humankind, no matter how quiet the citizens must be while they are alive, or [no matter how much] a nation of flatterers has learned to cover up even the most shameful crimes. . . . "[32] This notion of history as sanction—one of the few available in the age of absolutism, apart from religion—clearly reveals the historian as a moral agent in his own right.

Pufendorf's concrete historiographical methods, particularly in the long "original" histories, tend at first to obscure these aspects of his work. His significance for German historiography is usually attributed to his extensive use of official documents, which are quoted or paraphrased at length, often with little editorial comment, and with the political realism revealed thereby. Moreover, his formally annalistic accounts deal mainly with external relations among states, that is, with wars, diplomatic negotiations, and alliances.[33] Particular events such as battles are mentioned but not of primary interest; nor are internal affairs of state such as court infighting, political economy, or—despite his comment above—the exploits of prominent personalities as such. In general, therefore, his histories seem formulaic, even dull, as Pufendorf seems to have replaced the official schema of universal history with an equally staid and determined *reportage*.

This is a superficial verdict, however, even though it was expressed by none other than Frederick the Great, who found Pufendorf's *Frid. Wilh.* tedious because it failed to distinguish the essential from the peripheral; and it was not universally shared.[34] Thus, J. P. Ludewig, who had lectured on Pufendorf's *Einleitung* earlier in the century, described *Frid. Wilh.* as "truly pragmatic" and compared its value to that of holy scripture itself. And Frederick's own minister, Count Hertzberg, who first published Pufendorf's *Frid. Tert.* in 1784, praised *Frid. Wilh.* as "unique in its manner, and far superior to all old and new historians in regard to truthfulness."[35] Later assessments have been similarly divided, generally commending Pufendorf's reliance on archival materials but faulting his style, the lack of a broad interpretive stance, and his selectivity or supposed partisanship.[36] Yet Pufendorf recognized most of these criticisms and preempted some of them. The most important for our purposes is the last, for it not only counteracts some of the others but also leads to his most explicit reflections on a historian's proper role.

# III

Pufendorf was a self-conscious historian aware of the problem of evidence. Thus he advised Thomasius in 1688 that the latter should in his planned church history return "to the sources themselves" (i.e., the acts of the councils and the writings of the church fathers) instead of relying only on secondary materials such as older, pre-Reformation church historians.[37] In the case of modern history, however, which dealt with the affairs of states whose (real or imagined) interests were often temporary, he thought that knowledge was "better learned from Experience and the Conversation of Men well vers'd in these Matters, than from any Books whatsoever."[38] This meant, in practice, recourse to original, archival documents at only one remove from direct experience and conversation.

Yet there were difficulties even here, as Pufendorf told the Catholic Count Ernst von Hessen-Rheinfels in 1690, thanking him for the correction of a name, because the first-hand reports of those involved in campaigns usually contained only surnames. Indeed, though he wished that all campaigns included people as "intelligent, curious, and diligent" as the Count himself, who would observe and annotate everything exactly, this was seldom the case; therefore, posterity had to be satisfied with a merely general account of such events. Pufendorf was, of course, primarily interested not in events as such but in their causes, particularly the relations and negotiations between states, for there he found "the instructions and reports of ministers, on which it is finally safe to rely, insofar as there is any certainty in human affairs."[39] Unfortunately, he ran into similar problems here, in that ministers often wrote that they would deliver a full oral report upon their return but then left nothing in the archives, despite a matter's importance. Ironically, therefore, the nature of the archival material on which Pufendorf relied because of his search for certainty not only dictated and facilitated his historical writing, but it also complicated and restricted it.

Leibniz was in this as in other respects a severe but useful critic of Pufendorf, mainly because his criticisms allowed the latter's position to emerge more clearly. Thus he complained about *Frid. Wilh.* that since Pufendorf did not have much personal experience of public affairs, he could only "copy out" (*exscriptorem esse*) the reports of others. This dependence necessarily produced errors, according to Leibniz, since ministers' reports were often based on false rumors and eventually refuted.[40] Yet Pufendorf had far more political experience and savvy than Leibniz allowed, nor was he gullible or careless about his materials. Rather, given his material limitations, he also placed some responsibility upon the reader, commenting at one point that "anyone who understands what history is" could recognize from his treatment the reliability of that particular account.[41]

Leibniz also criticized Pufendorf for attempting to write "secret" as opposed to "public" history, that is, history which revealed not only the deeds (*gesta*) of leaders but also their reasons for acting (*gestorum causae*).[42] He himself considered this too speculative. Pufendorf, in turn, readily admitted that such reasons are difficult to determine, not merely because of the laxity of legates and ambassadors but also because princes, who are not accountable to anyone, consider it superfluous to leave records of their reasoning concerning the administration of things.[43] Moreover, he consciously excluded speculation about others' unknown intentions, thinking it "reckless to probe or interpret others' secrets by conjecturing."[44] Still, he chose to write secret history because he regarded history as an account of actions and not merely of events. The inner, intentional dimension of actions exacerbated the problem of evidence and imposed on him an interpretive or reconstructive task, from the viewpoint of the historical agent. Many elements of his style and method can be explained from this.[45]

Pufendorf's internal histories were necessarily one-sided the way all human actions are limited by experience, perspective, and practical interests. However, this did not necessarily make them parochial or biased. For one, his treatments sought to include the actions and decisions of "the other side" (*adversae partis*) insofar as these came within his protagonists' purview, and he tried sometimes—albeit unsuccessfully—to obtain information from other archives in support of this.[46] Also, though his histories differed from external, public, or universal history in the traditional sense, they tended toward increasingly wider perspectives. Thus he said of *Frid. Wilh.* that because of the breadth and extent of the Great Elector's connections, it offered a "select knowledge of nearly half a century."[47] Ironically, one could describe Pufendorf's approach—in Leibnizian language—as "monadic," referring to a subject's increasingly perspicuous (and thus active) role within a larger whole. This certainly helps to explain the inclusion of certain phenomena that otherwise seem not to fit into the confines of a particular national history, such as the detailed account of the 1688 English Revolution in the history of Frederick III.

The problem of historical evidence may be seen either as a cause or as an effect. That is, a lack of evidence may constrict the historian and lead to a one-sided account, or else the historian may out of partiality or interest constrict the evidence. We have just considered the former possibility. The problem also posed itself in the latter form because of the apparent tension between Pufendorf's self-conception as a historian dedicated, on the one hand, to his sovereign or nation, and on the other hand to the truth.

In an anxious letter to minister Paul von Fuchs written in Greifswald on his way to Berlin, Pufendorf requested an official guarantee of safe passage from the Great Elector. For he had received messages from Leipzig and Wittenberg warning that there might be efforts to abduct and harm him. The basis for these fears were not only certain political opinions which he had rendered while in the service of Sweden, but also the knowledge that he had "...many enemies, at both the Imperial and the Saxon courts, because in my Swedish history I presented the conduct and actions of both sides without concealing anything, just as the records [*acta*] in the royal archive had handed them to me, and as a historian properly should."[48] The official safe-conduct was speedily issued and Pufendorf arrived in Berlin without incident. But he was also concerned about how his Swedish history had played there, and pointed out in the same letter that he had shown the entire manuscript to Falaiseau, the Elector's representative in Stockholm, who had raised no objections. Then he says in an almost matter-of-fact way, "... when a writer's pen expresses the sentiments of the lord he serves he cannot have these attributed to himself, since a piper plays the tune of the one who pays him."[49] While the previous statement presents the historian as dedicated to the truth, the latter seems to characterize him as a quill-for-hire. The two defenses seem incompatible.

One explanation of the latter comment might be to read it not as a principled position but as an admission about the role of the learned in an age of patronage, when the work of scholars was widely used for purposes of political and confessional justification and aggrandizement. A similar statement occurs in a letter to Seilern, where Pufendorf says that a historian "cannot but express the perceptions of his prince or commonwealth, unless—what is most foolish—he wishes to accuse and to condemn himself."[50] The florid and—to us—sometimes abject dedications of literary works were not a mere formality in the seventeenth century but signified essential dependency relationships, as Pufendorf knew from his struggles at Lund after 1672, when his carefully cultivated ties with the Swedish aristocracy (including Lindeman and de la Gardie) protected him. He could also have pointed to the fate of the *Historisches Reichskolleg* planned by Franz Chr. Paullini and Hiob Rudolf, which never secured the requisite support, partly because of its attempt to cross political and confessional lines.[51] The notion of an independent Republic of Letters not beholden to secular powers is a modern anachronism that should not be used to judge seventeenth-century scholars.

Soon after *Gust. Adolph.* was translated into German, in 1688, it was attacked by the son of a certain General Christoph Houwald who complained bitterly to the Elector about the work's portrayal of his father, and demanded that all copies of the book be confiscated and Pufendorf himself arrested until he rewrote the offending passage.[52] The matter was gingerly handled at court by privy counsellor Blumenthal, who eventually succeeded in getting both Houwald and Pufendorf to accept a mutually unsatisfactory solution. The episode is important because it evoked from Pufendorf several letters which lay out explicitly his view of a historian's role. To Blumenthal he wrote that his Swedish history was not a "privatum scriptum" but a public document authorized and approved by the king, who had ordered his historian, as "secretary," to record his predecessors' actions, judgments, and feelings.[53] The letter also avows that Pufendorf has been true to the archives and calls "the whole learned world" to witness that he has not let himself be influenced by his private feelings, but has in fact has practiced the greatest moderation. Pufendorf later reiterated these points to Houwald himself, adding that as a private person he would not have dared on his own to present great potentates' counsels and actions so freely, and to judge them, but that such individuals may surely allow their own and their ancestors' feelings toward those with whom they have interacted to be freely spoken or written about.[54] Then he joined the image of a masterbuilder to that of secretary: a historian is like a builder who expertly constructs a house with given materials according to the owner's specifications and desires.

All this sounds suspiciously like a secular version of the theodicist's invitation to take one's complaint to the top, and in an immediate sense it was. However, Pufendorf's defensive metaphors also work against this reading. To be sure, a historian is commissioned by his prince, whose views and intentions shape and color the resulting work; but just as a secretary's or builder's materials are given, so are the historian's. As the letters themselves maintain, he cannot simply invent reality but must base his accounts on official records that reflect actual deeds and deliberations. Even sovereigns cannot invent history out of whole cloth but are bound by what they and others have done, which the historian is in turn bound to respect: "Since human affairs are not all the same, but history is to represent them as they in fact were, one cannot write about things differently than as one finds them, unless one wishes to pass in the world for a fabulist and flatterer."[55] These are roles that Pufendorf consistently rejected; the historian as secretary or builder is not simply a propagandist. The claims of truth always act as a limitation on and counterweight to other duties and commitments.

*Gust. Adolph.*'s critics included one of its main subjects, the former Queen (and formerly Lutheran) Christina, then living in Rome, who complained that the work contained many things that only Protestants could approve. To this Pufendorf thought it reasonable (*non absurde*) to reply that "a history of the war waged by the Swedes in Germany, which could have been very pleasing to the Roman Court, would have been ridiculous."[56] For princes and commonwealths (and thus their historians) measure their actions not only by the common law (*ius*) of men but also by their own special interests, particularly when religious differences are involved. In case of conflict, each party typically regards its own cause as no less just than that of the other. Accordingly, "... the history of two hostile princes can be written by two persons with the same appearance of truth [*pari specie*], as long as each adjusts himself to the opinions, impressions, and interests of his own prince." Indeed, Pufendorf says, waxing autobiographical, the same skillful individual can write both histories, build the same information into both accounts, and even have one borrow directly from the other, so long as the general perspective is different.[57] This is because a historian's task differs from those of lawyer and judge.

This important comparison introduces the other pole of Pufendorf's self-conception as a historian, namely his commitment to the truth (*amorem veri*).[58] A lawyer is essentially a special pleader, even a propagandist, for his client, whereas a judge renders verdicts from a supposedly acontextual or disinterested metaperspective. A historian, however, though writing from a point of view, describes things as he finds them and leaves judgment to the reader.[59] In the former regard, Pufendorf claimed that although someone writing about wars can seldom use such moderation of language and feeling as to avoid offending many people, he himself had at least not been influenced by private feelings.[60] Indeed, though he allowed that private individuals should share their sovereign's feelings (including hatreds) toward other princes, he admitted to "softening" even these sometimes.[61] This admission need not challenge his veracity as a historian, since it may be seen, at least in part, as a writer's judicious effort to control his readers' possible (over)reaction and, thus, to facilitate their apprehension of the truth. As noted above, Pufendorf rejected the roles of fabulist and flatterer. In fact, he claimed outright that he was a "non-partisan historian," even to the point of exposing his own side's weaknesses and embarrassments.[62]

The preface to *Gust. Adolph.* states that "above all, one demands of a historian that he say nothing false and not fail to say anything that is true."[63] That is, he must say *what* he finds, and he must *say* what he finds. This latter duty to reveal recalls the notion of history as sanction, and of the historian as independent moral agent, in Section 2 above. The preface also asserts—referring to those in Brandenburg who would be displeased by *Gust. Adolph.*'s publication of certain things which they would rather keep silent—that princes are born under the law (*lex*) that both their distinguished (*egregria*) and their evil (*prava*) deeds will come to be known by many, and that history rightfully (*suo jure*) transmits such deeds to the memory of posterity "just as it has found them." Indeed, Pufendorf says, such evenhandedness and dependability, revealing the good along with the bad on the basis of authentic archives, is the main duty of a historian. A similar statement occurs in the dedication to *Frid. Wilh.*, almost ten years later, where Pufendorf proclaimed that it was of little importance to him whether the work met with external success (*extrinsecus cultus*), but that it sufficed "to have clearly exposed to posterity the uncorrupted truth [drawn] from authentic sources."[64]

Pufendorf's independence as a historian is attested by the fact that he was criticized by both sides, and for the same reason. Thus, he confided to Friese that "[w]hen one observes how they have treated the [still unpublished] history of Carl Gustav in Sweden, one can well lose one's appetite to write histories for these people." Indeed, the Swedes were already so upset ("so einen hasz gegen mich gefasset") about earlier characterizations of them in the *Einleitung* (1686) which they did not regard as favorable enough, that Pufendorf debated anxiously about whether to return to Stockholm after his tasks in Berlin had been completed.[65] Yet he was equally apprehensive about the anticipated reaction to *Frid. Wilh.* in Berlin, acknowledging that "it is dangerous to write true things about the powerful," even though he knew that he would also have his defenders and was emboldened by old age.[66] As it turned out, his defenders lost power soon after his death, and the posthumously published *Frid. Wilh.* met with strong criticism in Berlin and other courts for revealing state secrets. Copies were bought up; a planned French translation was halted; and an incomplete, sanitized German translation appeared only in 1710. Pufendorf's unfinished history of Frederick III did not appear until 1784. Historically, both works were made ineffective.[67]

## IV

At the start of *De jure naturae* (I.1.1), Pufendorf contrasts the progress made in the study of the physical world with the lack of attention to "moral entities." This notion was borrowed from his Jena teacher, Erhard Weigel, and referred to the moral categories and distinctions "imposed" on otherwise neutral natural entities by God and, analogously, humans, in accordance with the natural law. As developed by Pufendorf, the theory of moral entities offered a voluntaristic account of value in an increasingly disenchanted world, as it were, and a realistic description of the origin of many legal and social distinctions in an age of absolutism, when individual and collective identities were being mutually redefined. It established a nonmetaphysical, nonteleological moral framework compatible with the emerging natural sciences and allowed for flexible, plural human choice based on particular agents' concrete circumstances. Moreover, many of its categories applied to individual and collective agents alike, including the new, secular nation-states, and provided the terms in which individuals and the wholes they comprised would lodge claims against one another for the next two centuries. Historiography also played an important role in this process, not only as a concrete means for shaping identity in a context bereft of traditional universals, but also as a form of agency modeled on and subject to the natural law's broader demands.

Pufendorf's historiographical principles may be compared with those of Francis Bacon, whom he respected for his defense of innovation against the claims of tradition. Bacon preferred particular over universal history because it rested more solidly on the details of experience; and in the category of just or perfect as opposed to imperfect history, he especially preferred biographies or "lives" (for their "advantage and use"), and "particular relations of actions" recorded by individuals directly involved in them (for their "truth and sincerity").[68] Pufendorf's own, modern histories based on archival research clearly reflect these preferences. Like Bacon, he valued direct experiential reports, including memoirs and travelogues, which were well represented in his library and widely used in *De jure naturae*, mainly as anthropological comparisons.[69] The travelogue is not merely an indicator of Pufendorf's empiricism, however, but also an analogue of his agent-centered conception of historiography as a kind of appropriation of the other, or of the self as other. Through his history, the historian creates or affirms identity or selfhood—his own or of those for whom he writes—by depicting agency defining itself through time. This involves a process of active representation on his part, doubly so in the latter case where he not only makes the nonpresent present but also does so for someone with whom he may or may not, in another sense, wish to be identified. The former kind of representation reveals historiography as an essentially (re)constructive act: not all things can be seen or (re)presented, nor from all perspectives; rather, historiography always involves principled choices. The latter raises philosophical questions about the nature and extent of the representative's or historian's responsibility. In both cases, historiography itself is essentially a social, political, and moral act.

Empirical, agent-centered history is inevitably multifaceted in that there are plural, more or less coincident histories of multiple agents engaged in more or less common causes. Such diversity may suggest the loss of an intellectually or morally normative center, leaving only a neutral methodology to join the various accounts together. Indeed, Pufendorf's use of *raison d'état* analysis seems at first to confirm this conclusion. However, just as self-preservation and need are subsidiary to the overarching sociality law of Pufendorf's natural law theory, so is his states' interest analysis to its legitimation of the state's role in human affairs. The plural, more or less incompatible interests, histories, and identities which a skillful historian may in turn articulate and foster are ultimately assessed in terms of broad normative principles that ensure, above all, the possibility of continued agency. (That is, the sociality law fosters continued social relationships among humans.) Indeed, these principles guide the historian just as they do, or should, the particular subjects of his narratives. The former fulfills his responsibility by recounting the truth as he sees it. Agent-centered historiography is not incompatible with truthfulness, whose claims Pufendorf repeatedly asserted, but only with the dissimulation, intrigue, and flattery he consistently rejected.

There is a parallelism between Pufendorf's commitment to a balance of powers among European states (and his opposition to universal monarchy, whether papal, Austrian, or French), his acceptance of multiple histories from different agents' perspectives, and his openness to philosophical criticism and innovation. While it would go too far to claim that he was committed to pluralism as such, he clearly acknowledged diversity in all these spheres and generally regarded it, along with change and innovation, as more of an advantage than a detriment. This marks him as a modern, perhaps even a liberal. There is, to be sure, a lasting tension between his attempts to establish natural law as a philosophical science with its own disciplinary history, and his openness to perspectival variation and novelty, but this merely solidifies the comparison, as anyone familiar with recent debates can attest. Now as then, resolving the tension seems to require the identification of nonexclusive universals not captured by authority-based traditions (and histories), and respectful of any constant preconditions of rational agency as such. This thought takes us back to Pufendorf's youthful efforts to avoid preconceived or sectarian opinions, and to his lifelong attempt to distinguish what is correct (*das Richtige*) from what is important (*das Wichtige*), that is, to his philosophical eclecticism.[70] His intellectual affinity to that tradition is another story, however. The burden of this paper has been to show that Pufendorf's historiography must be a part of it.[71]

# Notes

1. Leonard Krieger, *The Politics of Discretion. Pufendorf and the Acceptance of Natural Law* (Chicago: University of Chicago Press, 1965), 170–201. Krieger's claim is challenged by Horst Denzer,

*Moralphilosophie und Naturrecht bei Samuel Pufendorf* (Aalen: Scientia Verlag, 1972), 250–51. Abbreviations will be introduced gradually as needed. Unattributed translations are my own.

2. On Pufendorf as historian of philosophy, see Richard Tuck, "The 'Modern' Theory of Natural Law," in Anthony Pagden, ed., *The Languages of Political Theory in Early-Modern Europe* (Cambridge: Cambridge University Press, 1987), 102–7; Fiammetta Palladini, *Samuel Pufendorf. Discepolo di Hobbes* (Bologna: Il Mulino, 1990), 175–88; and Timothy Hochstrasser, "Natural Law Theory: Its Historiography and Development in the French and German Enlightenment, ca.1670–1780," unpublished Ph.D. dissertation (Cambridge University, 1990).

3. For Pufendorf's correspondence we may now consult *Samuel Pufendorf. Briefwechsel*, ed. Detlef Döring (Berlin: Akademie Verlag, 1996), vol. 1 of *Samuel Pufendorf. Gesammelte Werke*, ed. Wilhelm Schmidt-Biggemann, a new, critical edition of Pufendorf's main works. Döring has numbered the letters and provided line enumerations. Accordingly, subsequent citations will be to letter number, page, and lines. E.g., *Briefwechsel*, #16, 24.24–29 = letter #16, p. 24, lines 24–29.

On Pufendorf's political "opinions," see Detlef Döring, "Samuel von Pufendorf als Verfasser politischer Gutachten und Streitschriften," *Zeitschrift für historische Forschung* 13.2 (1992): 189–232. Some of these works are in Detlef Döring, ed., *Samuel von Pufendorf. Kleine Vorträge und Schriften* (Frankfurt am Main: V. Klostermann, 1995), which also contains Pufendorf's "Epistolae Duae Super Censura…ad Adam Rechenbergium," 488–506, and "Unvorgreiffliches Bedenken Wegen Information eines Knaben von Condition," 537–50. The first "letter," responding to a review by Abbé LaRocque in the *Journal des Savants* (1687), is most applicable.

4. Adam Friedrich Glafey, *Vollständige Geschichte des Rechts der Vernunft*, Neudruck der Ausgabe Leipzig 1739 (Aalen: Scientia Verlag, 1965), 121.201–2. On the Fürstenschule at Grimma, see Wolfgang Hunger, *Samuel von Pufendorf. Aus dem Leben und Werk eines deutschen Frühaufklärers* (Flöha: Verlag Druck & Design, 1991), 19–25.

5. See Denzer's citation index in *Moralphilosophie*, 333–57.

6. Pufendorf, *Briefwechsel*, To Boineburg (Jan. 13, 1663), #16, 24.24–29.

7. Cf. Heinrich von Treitschke, "Samuel Pufendorf," in *Historische und politische Aufsätze* (Leibzig: G. Hirzel, 1897), vol. 4, 202–303; and Detlef Döring, *Samuel Pufendorf als Student in Leipzig. Eine Ausstellung von Detlef Döring* (Leipzig: Universitätsbibliothek, 1994).

8. "Vita, Fama, et Fata Literaria Pufendorfiana . . . von Petronius Hartevvigo Adlemansthal [Peter Dahlmann]," in *Samuels Freyhrn. von Puffendorff kurzer doch Gründlicher Bericht von dem Zustande des H.R. Reichs Teutscher Nation* (Leipzig: Gleditsch und Weidmann, 1710), #3, 650.

9. Döring, *Student*, 22–23. Franckenstein's approach was apparently controversial, since he was warned to stick to the four monarchies scheme. Pufendorf's "Bedenken" begins with the claim that the upbringing of youth rests especially "auf diesen 3. Stücken…, daß man sie nemlich (1) zur Gottesfurcht, (2) guten Sitten, [und] (3) *dienlicher Wissenschaft* anführe" [emphasis added]. The third category comprises *verba* (*Eloquence*) and *res* (*historiam veterem et recentem* and *doctrinam moralem et politicam*) and recommends Cornelius Nepos, Phaedrus, Curtius Rufus, Florus, Eutropius, Suetonius, and Tacitus, followed by Herodotus, Diodorus Siculus, Thucydides, Xenophon, Dionysius of Halicarnassus, Polybius, and Livy. See *Kleine Vorträge*, 537, 541, 544–45.

10. See Detlef Döring, *Pufendorf-Studien* (Berlin: Duncker & Humblot, 1992), 165–68, 174; and "Samuel Pufendorf (1632–1694) und die Leipziger Gelehrtengesellschaften in der Mitte des 17. Jahrhunderts," *Lias* 15 (1988): 13–48. Most of Pufendorf's Collegium Anthologicum lectures are included in *Kleine Vorträge*, 21–86.

11. Pufendorf wrote the work in Danish captivity, without access to a library, and since he sought to show that natural law was a deductive science, he employed a quasi-mathematical method consisting of definitions, axioms, and observations. Grotius and Hobbes are cited only in the preface because, "aside from the tedium of frequent citation, we have followed rather their arguments than their authority." *Elementorum jurisprudentiae universalis, libri duo*, vol. 2, trans. by W. A. Oldfather (Oxford: Clarendon Press, 1931), xxx.

12. These essays were collected as *Dissertationes academicae selectiores* (Lund: A. Junghans, 1675), and reissued in 1677 and 1678.

13. Based on an autograph at Wolfenbüttel, Döring, "Verfasser," 206, n.65, now appears to

attribute the anonymous *Les anecdotes* to Pufendorf. So does S. Jägerskiöld, "Samuel von Pufendorf in Schweden, 1668–1688. Einige neue Beiträge," in *Satura Roberto Feenstra*, ed. J.A. Ankum, J.E. Spruit, and F.B.J. Wubbe (Fribourg: University Press, 1985), 570.

14. Jägerskiöld, "Pufendorf in Schweden," 559.

15. On Pufendorf's library, see Detlef Döring, "Die Privatbibliothek Samuel von Pufendorfs (1632–1694)," *Zentralblatt für Bibliothekwesen* (1990): 107–9, and Fiammetta Palladini, "Die Bibliothek Samuel Pufendorfs," in *Samuel Pufendorf und die Europäische Frühaufklärung* (Berlin: Akademie Verlag, 1996), 29–39. Palladini has also completed a critical edition of the catalog of Pufendorf's books which will appear in the Repertorien series of the Herzog August Bibliothek, Wolfenbüttel.

16. Leonard Krieger, "Samuel Pufendorf," in *Deutsche Historiker*, vol. 9, ed. Hans-Ulrich Wehler (Göttingen: Vandenhoeck & Ruprecht, 1982), 7–8; and Hans Peter Reill, *The German Enlightenment and the Rise of Historicism* (Berkeley: University of California Press, 1975), 17ff.

17. On the origin of *De statu imperii* in the politics of the "Wildfangsstreit," see Detlef Döring, "Untersuchungen zur Entstehungsgeschichte der Reichsverfassungsschrift Samuel Pufendorfs (Severinus de Monzambano)," *Der Staat* 33.2 (1994): 185–206. Döring, *Pufendorf-Studien*, 202, n.52, points out the unhistorical nature of Pufendorf's account of *feuda oblata* in this work—which the latter himself admitted to Thomasius (June 19, 1688), *Briefwechsel*, #137, 195.64–67.

18. References to the *Einleitung* are to the English translation by J[ohn] C[rull], *An Introduction to the History of the Principal Kingdoms and States of Europe* (London: J. Gilliflower, 1695), preface.

19. Ulrich Muhlack, *Geschichtswissenschaft im Humanismus und in der Aufklärung. Die Vorgeschichte des Historismus* (Munich: C.H. Beck, 1991), 109–14, 124–26; Gertrude Lübbe-Wolff, "Die Bedeutung der Lehre von den vier Weltreichen für das Staatsrecht des römisch-deutschen Reiches," *Der Staat* 23 (1984): 369–89; and Friedrich Meinecke, *Die Idee der Staatsräson in der neueren Geschichte*, 2nd ed. (München: R. Oldenbourg, 1925), 299.

20. Moritz Ritter, *Die Entwicklung der Geschichtswissenschaft an den führenden Werken betrachtet* (Munich and Berlin: R. Oldenbourg, 1919), 196–202. *Gust. Adolph.* made use of an existing (unpublished) work by Bogislaw Chemnitz (pseud. Hippolithus à Lapide), Swedish state historian before Pufendorf. The latter respected Chemnitz's work but thought it was marred by hatred of the Hapsburgs.

21. Pufendorf, *Frid. Tert.*, bk. 1, #s 49–68, correspond to *Frid. Wilh.*, bk. 19, #s 84–99. Writing to Seilern (March 5, 1690), *Briefwechsel*, #175, 262.34–35, Pufendorf mentioned his "molestissimus . . . labor" in the archives; and in mid-November 1690 (*Briefwechsel*, #194, 293–94.12–14) he excused his late reply to Landgraf Ernst von Hessen Rheinfels with: " . . . wann ich den gantzen tag mit extrahiren und schreiben mich müde gemacht, so will mir gar nicht vonstatten gehen, auff den Abend brieffe von importantz zuschreiben, zumahl mein kopff sich das geringste von dem Acht à 9 stündigen schlaff [nicht] abbrechen laßen will...." Pufendorf had an assistant (*Briefwechsel*, To Thomasius, March 14, 1688, #130, 184.23–28), but his correspondence during the 1690s reveals a growing concern about time and health. This may explain some of his longer, undigested excerpts from the archives; he was in a great hurry.

22. See *Samuel Pufendorf's "On the Natural State of Men,"* trans. Michael Seidler (Lewiston, NY: Edwin Mellen, 1990), including the "Introductory Essay," 25–36. On the supposed dependence of ancient authors on the Bible as a historical document, see Pufendorf's Collegium Anthologicum lecture no.13, in *Kleine Vorträge*, 50–53.

23. Though Pufendorf himself did not use the term, many commentators have referred to such history as "pragmatic." See Krieger, *Discretion*, 172–73, and Reill, *German Enlightenment*, 41–42.

24. This preface, like most others, is unpaginated. Denzer, *Moralphilosophie*, 252, notes that Bodin, whose works Pufendorf possessed (Döring, "Privatbibliothek," 107 and 133, n.42), had also emphasized the usefulness of history for "gute Staatsführung" and "richtige Politik." Hans Rödding, *Pufendorf als Historiker und Politiker in den "Commentarii de rebus gestis Friderici Tertii"* (Halle a.S.: Max Niemeyer, 1912), 16, describes Pufendorf's demand of history as: "[k]ein antiquarisches, sondern politisches Wissen." Cf. Horst Denzer, "Leben, Werk und Wirkung Samuel Pufendorfs. Zum Gedenken seines 350. Geburtstages," *Zeitschrift für Politik* (1983): 174.

25. Jägerskiöld, "Pufendorf in Schweden," 559–62.
26. Cf. Lübbe-Wolff, 387: "Seit dem 12. Jahrhundert hatten deutsche Kaiser die Aufnahme des römischen Rechts begünstigt, weil der justinianische Absolutismus ihre Zentralisierungs- und Rationalisierungsbemühungen zu fördern geeignet schien." Thus, any challenge to such justifications was welcomed by those who claimed political independence from the Empire, and Pufendorf's natural law doctrine played well in both Sweden and Brandenburg for this reason.
27. Samuel Pufendorf, *Die Verfassung des deutschen Reiches* [Severinus de Monzambano, *De statu imperii Germanici*, 1667], trans. & annot. Horst Denzer (Stuttgart: Reclam, 1976), dedication, 6.
28. Pufendorf, *Briefwechsel*, #223, 349.29–32.
29. Ibid., #208, 323.28–30. Cf. To Adam Rechenberg (Dec. 6, 1690), #196, 298.32–34.
30. *De rebus gestis Friderici Wilhelmi Magni, Electoris Brandenburgici commentariorum libri novendecim* (Berlin: J. Schrey & H.J. Meyer, 1695), dedication, 2–3.
31. Pufendorf, *Briefwechsel*, To Houwald (March 26, 1689), #160, 240–41.57–61.
32. Pufendorf, *Frid. Wilh.*, dedication. Cf. the preface to *Gust. Adolph.*, where Pufendorf speaks of "Historia, cujus liberrimam censuram nemo Principum nisi recte agendo effugerit."
33. Ritter, *Die Entwicklung*, 197.
34. Cf. Rödding, *Pufendorf als Historiker*, 75, n.2.
35. See J.P. Ludewig's *Germania princeps postcarolingica sub Conrado I. orientalium Francorum rege* (Halle and Magdeburg: Zeitler, 1710); Franz X. von Wegele, *Geschichte der deutschen Historiographie seit dem Auftreten des Humanismus* (München: R. Oldenbourg, 1885; reprint New York: Johnson Reprint Co., n.d.), 511, 514–15; and Rödding, *Pufendorf als Historiker*, 2, n.3.
36. Joh. Gust. Droysen, "Zur Kritik Pufendorfs," in *Abhandlungen von Joh. Gust. Droysen zur neueren Geschichte* (Leipzig: Beit & Comp., 1876), 307–86; Ludwig Geiger, *Berlin 1688–1840. Geschichte des geistigen Lebens der Preussischen Hauptstadt*, 1. Band, 1. Hälfte (Berlin: Gebrüder Paetel, 1892), 122–29; and Ernst Salzer, *Der Übertritt des grossen Kurfürsten von der schwedischen auf die polnische Seite während des ersten nordischen Krieges in Pufendorfs "Carl Gustav" und "Friedrich Wilhelm"* (Heidelberg: Carl Winter, 1904). Droysen, Salzer, Wegele, and Rödding are the main older sources for Pufendorf's historiography. They and other secondary literature are briefly reviewed by Döring in *Pufendorf-Studien*, pp. 144–51.
37. Pufendorf, *Briefwechsel*, (Dec. 30, 1688), #158, 235.26–28.
38. Pufendorf, *Einleitung*, preface.
39. Pufendorf, *Briefwechsel*, (Nov. 18/28, 1690), #194, 294.20–21, 28–29, 32–36. Cf. "Epistolae Duae," 465: "Am aufschlußreichsten und sichersten sind freilich die Aussagen der an den Geschehnissen unmittelbar Beteiligten."
40. Letter of Leibniz on Oct. 18, 1728 [sic], quoted by Wegele, *Geschichte*, 514, n.1. Droysen, "Zur Kritik Pufendorfs," 375–76, rejects Leibniz's comments and describes him as "in Sachen der grossen Politik doctrinär, voller Voreingenommenheiten, ohne Verständniss der realen Macht." More generally on the troubled Pufendorf-Leibniz relationship, see Döring, *Pufendorf-Studien*, 134–42.
41. Pufendorf, *Briefwechsel*, (March 16, 1689), #160, 240.35–36.
42. See Gottfried Wilhelm Leibniz, *Codex Iuris Gentium (Praefatio)* (1693), in *The Political Writings of Leibniz*, ed. & trans. Patrick Riley (Cambridge: Cambridge University Press, 1972), 167–68.
43. Pufendorf, *Briefwechsel*, (Mar. 5, 1690), #175, 262.33–39.
44. Pufendorf, *Gust. Adolph.*, preface.
45. Pufendorf's expositors often note the active way in which he molded his material. Thus Krieger, *Discretion*, 190, speaks of Pufendorf's "figurative rather than literal correspondence to the sources"; and Droysen, "Zur Kritik Pufendorfs," 349–50 (cf. 368), describes Pufendorf's method as "discussive." Wegele, *Geschichte*, 519, comments: "Man kann nicht sagen, dass Pufendorf die Ereignisse erzählt…es kommt ihm vielmehr darauf an, die jeweilige Situation klar zu stellen und die Verhandlungen offen zu legen, durch welche sich die dargestellten Thatsachen vollziehen. Er will nicht zeigen, wie die Ereignisse sich für sich und durch sich selbst gestalten und entwickeln, sondern wie sie denjenigen, die auf einer Seite die Fäden in der Hand halten, erscheinen, oder doch wie sie von ihnen erfasst, verstanden und verknüpft werden."

46. Pufendorf, *Gust. Adolph.*, preface. According to Salzer, *Der Übertritt*, 6, n.15, Pufendorf tried through Christina to get materials from the Vatican, and he made a special trip to Cassel, in 1684, on his way to Holland to arrange for the publication of *Gust. Adolph*. He was unsuccessful in both attempts and later complained about the the "ohnzeitige sorgfalt" and "lächerliche behutsamkeit" of courts, which attempted "solche dinge zu secretiren, die in augen der gantzen welt passiret." Also see Pufendorf, *Briefwechsel*, To Thomasius (Sept. 18, 1690), #144, 204.44–48, and To Landgraf Ernst (March 29, 1690), #176, 264–65.32–38.

47. From Pufendorf, *Frid.Wilh.*, bk. 1 (quoted by Wegele, *Geschichte*, 516, n.3).

48. Pufendorf, *Briefwechsel*, (January 19, 1688), #123, 172.48–50.

49. Ibid., #123, 171.40–42.

50. Ibid., #175, 262.50–51.

51. See Franz X. von Wegele, "Das historische Reichscolleg," *Im neuen Reich. Wochenschrift für das Leben des deutschen Volkes* 11.1 (1881): 941–60; and Döring, *Pufendorf-Studien*, 145.

52. See Pufendorf, *Briefwechsel*, #s 154, 160, 162, 171, and for background, Joh. Schultze, "Der Geschichtschreiber Samuel v. Pufendorf," in *Forschungen zur Brandenburgischen und Preussischen Geschichte, Neue Folge* 55 (1944): 169–80.

53. Pufendorf, *Briefwechsel*, (Dec. 7, 1688), #154, 226–27.12–23. Cf. To Friese (Oct. 26, 1692), #222, 347.10–11.

54. Ibid., (Mar. 16, 1689), #160, 239–40.11-25.

55. Ibid., #160, 241.61–63. Cf. #119, 172.50–52 and #222, 347.7–9.

56. Ibid., To Seilern (March 5, 1690), #175, 262.53–54, 59–65.

57. Pufendorf transferred significant portions of material from *Carl Gustav* to *Frid. Wilh.*, and from the latter to *Frid. Tert*. Salzer, *Der Übertritt*, 86–87.

58. Pufendorf, *Briefwechsel*, To Pregitzer (Oct. 3, 1691), #208, 324.30–32. Döring's text has "amorem recti," with "amorem veri" as a variant. Cf. *Frid. Wilh.*, dedication: "Mihi sufficit, incorruptam veritatem, e genuinis fontibus, candide posteritati exposuisse, quo nomine nemo quid offensae ostendere potest, . . . "; and "Epistolae Duae," 493, where Pufendorf describes himself as "simplicem veritatem profess[us]."

59. Pufendorf, *Briefwechsel*, To Blumenthal (Dec. 17, 1688), #154, 227.19–21. Also, To Friese (Oct. 26, 1692), #222, 347.13–16; and *Gust. Adolph.*, preface.

60. Pufendorf, *Briefwechsel*, To Seilern (Mar. 5, 1690), #175, 262.28–29; To Blumenthal (Dec. 17, 1688), #154, 227.21–23; "Epistolae Duae," 493; *Gust. Adolph.*, preface; and the posthumous *De rebus a Carolo Gustavo Sueciae rege gestis commentariorum libri septem* (Nüremberg: Christopher Riegel, 1696), bk. 1, #1, 5: " . . . apud cordatos ex hoc uno commendationem scriptioni meae expecto, quod eadem sincera fide ex indubiis documentis deprompta est, nullo affectu aut praejudicio interpolata."

61. Pufendorf, *Briefwechsel*, To Seilern (March 5, 1690), #175, 262.47–48. Also, To Friese (Oct. 26, 1692), #222, 347.9–16; and To Dahlberg (May 25, 1692), #219, 343.69-72. On Pufendorf's political sympathies, see Lars Nilehn, "On the Use of Natural Law. Samuel Pufendorf as Royal Swedish State Historian," in *Samuel von Pufendorf, 1632–1982*, ed. Kjell A. Modéer (Lund: Bloms Boktryckeri AB, 1986), 62ff.

62. Pufendorf, *Briefwechsel*, To Dahlberg (May 25, 1692), #219, 343.77, 83–85; and To Fuchs (Jan. 19, 1688), #123, 172.48–52.

63. Pufendorf, *Gust. Adolph.*, preface; and Pufendorf, *Briefwechsel*, To Rechenberg (Dec. 6, 1690), #196, 298.33–34.

64. Pufendorf, *Frid. Wilh.*, dedication. Also Pufendorf, *Briefwechsel*, To Pregitzer (Oct. 3, 1691), #208, 323.24; and Pufendorf, "Epistolae Duae," 489 and 494, where he speaks of "historia, quae testis est temporum."

65. Pufendorf, *Briefwechsel*, To Friese (Oct. 26, 1692), #222, 347.3–7. See Pufendorf, *Einleitung*, ch. 13, #18, 535; and Pufendorf, *Briefwechsel*, To Pregitzer (Oct. 3, 1691), #208, 324.32–37.

66. Pufendorf, *Briefwechsel*, To Pregitzer (Oct. 3, 1691), #208, 323.27–28. Also, Pufendorf, *Frid. Wilh.*, dedication.

67. On the fate of *Frid. Wilh.*, see "Vita . . . Adlemansthal [ . . . Dahlmann]," #42, 786–87, and

#43, 795. Wegele, *Geschichte*, 510–11, 520–21, reports that Leopold's historian in Vienna, S.G. Wagner, disapproved that Pufendorf cared more about the demands of historiography than about the reputation of his prince.

68. Francis Bacon, *The Advancement of Learning* [1605], ed. J. E. Creighton, World's Great Classics, vol. 28 (New York: Colonial Press, 1900), bk. 2, chs. 7–8, 54–57. Wegele, *Geschichte*, pp. 476–77.

69. Döring, *Pufendorf-Studien*, 169. At "Bedencken," 550, Pufendorf recommends "Voyagen zu lesen, und solche Bücher, die fremde Völcker, Landschafften, Natur und Beschaffenheit beschrieben [sic]" as "das anmuthigste, nützlichste und keuscheste Divertissement" of youth.

70. For this distinction, see Wilhelm Schmidt-Biggemann, *Theodizee und Tatsachen. Das philosophische Profil der deutschen Aufklärung* (Frankfurt am Main: Suhrkamp, 1988), 203–5. Also see Pufendorf, "Bedencken," 549, on thinking for oneself.

71. Thanks to Fiammetta Palladini for helpful remarks on an earlier version of this essay.

# Eighteenth-Century Anthropology and the "History of Mankind"

## Anthony Pagden

> En découvant la loi du temps comme limite exterieure des sciences humaines, l'Histoire montre que tout ce qui est pensé le sera encore par une pensée qui n'a pas encore vu le jour.
>
> —Michel Foucault,
> *Les mots et les choses. Une archéologie de sciences humaines*

## I

The history of the modern human sciences has been shaped, in ways which the history of the natural sciences have not, by what those disciplines have since become. As Loren Graham, Wolf Lepenies, and Peter Weingart have observed, "the humanities could almost be defined as those disciplines in which the reconstruction of a disciplinary past inextricably belongs to the core of the discipline."[1] This is inescapably the case with disciplines such as economics and psychology and anthropology which no one very much before the mid nineteenth century would have recognized as a discipline at all, or would have been taken to be some other kind of inquiry altogether. Each of these terms has a history in which each described something which either covers no part of the modern project or once referred to something now quite marginal to it. Psychology was closer to what we would describe as the philosophy of mind, economics was the study of household management. The term "Anthropology," as Kant used it when he spoke of an *Anthropology from a pragmatic point of new* was what would today be described as critical psychology. In most histories of the modern disciplines which bear these names, however, these early activities are either treated as formative stages in what would, at some later stage, emerge as the "true" discipline, or as false starts and consequently as no part of the history at all. Both the teleological view and the more openly, and currently more acceptable, historicist view have had the effect of closing off the possibilities which these earlier, more inchoate but also less bounded, projects frequently offered. In this essay I wish to look at one of the possibilities modern anthropology has left behind—the possibility for projection—not because I believe that it can, or should, be recovered but because it might tell us something which we do not fully understand about the, on the face of it, bewildering, but persistent, concern with what European cultures have for long (and in most places still do) looked upon as the "primitive."

Since the formal division of the social sciences into their modern faculties takes place outside the historical period with which I am concerned I shall frequently use the more comfortable term "human sciences" to describe those areas of inquiry whose subject is human social behaviour. I want to examine two features of these sciences. The first, which I shall deal with last, is the close association, amounting to interdependence, of the human sciences upon a particular kind of history, what in the eighteenth-century was frequently called the "history of mankind" or the "history of civilization." The second is the distinction, appropriated from Dilthey by Radcliffe-Brown in 1952 to characterize the modern discipline of social anthropology, between the "ideographic" and the "nomothetic": between, that is, the brute data which is the human or social scientist's subject matter, and the theories she devises to make sense of them.[2]

## II

Modern anthropologists, modern sociologists, modern political scientists tend to assume that some direct experience of their subject matter is a necessary condition of understanding it. It is the field-work which grants the theoretician privileged access to his data. What Vincent Crapanzano calls the "constitution of the ethnographer's authority," derives almost entirely from "his presence at the events described, his perceptual ability, his "disinterested" perspective, his objectivity and his sincerity."[3] It is, to use Clifford Geertz's pun, the authority of the "I" witness which has the unique power of guaranteeing the literal truth of what he or she says.[4] It could also be argued that it is not only the ethnographer's identity which is constituted in this way: it is also the ethnographer's subject matter. Cockfights may indeed take place on Bali and the Azande may perhaps believe that twins are birds, but it is only because of what Geertz and Evans Pritchard have written that these items of someone's everyday experience have come to form part of a recognized canon of ethnographic information. Recent social science has collapsed the distinction between the ideographic and the nomothetic much as it has collapsed the distinction between the observer and the observed, between hypothesis and verification, between subject and object.

Things, however, were not always so. Anthropology, properly understood, has been given many genealogies. But whether we trace it back to Montesquieu and Vico, Ferguson and Kames, Buffon or De Gérando, Blumenbach, or only so far as Fustel de Coulanges and Durkheim, there has always existed an assumption that there must exist an initial separation between raw ethnographic data and some kind of theory of society. There were those who collected and those who thought about the significance of what others had collected. The terms "ideographic" and "nomothetic," as Radcliffe-Brown described them, are Neo-Kantian (although they are not used in this way by either Dilthey—Radcliffe-Brown's immediate source—much less by Kant). The conception, however, is far older even than Kant himself. Here, for instance in 1625 is Samuel Purchas, the compiler of one of the earliest collection of traveller's narratives, *Purchas His Pilgrimes*, addressing his potential reader about the nature of his project.

> What a World of Travelers have by their own eyes observed in this kind, is here ... delivered, not by one preferring Methodically to deliver the History of Nature according to rules of Art, nor Philosophically to discuss and dispute; but as in way of Discourse, by each Traveler relating what in that kind he has seen. And as *David* prepared materials for *Solomon's* Temple or (if this be too arrogant) as *Alexander* furnished *Aristotle* with Huntsmen and Observers of Creatures, to acquaint him with their diversified kinds and natures ... so here Purchas and his Pilgrims minister individual and sensible materials (as it were with Stones, Birches and Mortars) to those universal Speculators for their Theoretical structures.[5]

On the one side of Purchas's dichotomy stands the observer, the mere field worker, the "ideograph," impartial because uniformed; on the other the Baconian natural historian, the "universal Speculator," the "nomothete."

The objectifying habit (if I may call it that) which made the ideographic/nomethetic distinction possible in the first place, has clearly been embedded in European culture for far longer than any of the available genealogies of the human sciences would seem to suggest. Objectification, the ability to observe from the standpoint of the listening, rather than the speaking, subject,[6] is an integral part of Western culture's passion for taxonomy. From Herodotus, through Giovanni da Pian del Carpini's description of the Mongols in the thirteenth century, through Giovanni Ramusio and Bernardino de Sahagun in the sixteenth, Richard Hakluyt and Samuel Purchas in the seventeenth, down to the *Recueil des voyages* in the eighteenth, the collection of data—largely for immediate political or religious ends—provided the European reader with some kind of access to cultures remote from his own.[7] At least they offered a reminder that such cultures did exist.

But unlike the modern anthropologist or ethnologist, the premodern observer's direct engagement with his subject, was a cause for suspicion. The eighteenth-century orientalist, Abraham-Hyacinthe Anquetil-Duperon, was, said the Neapolitan *philosophe* Fernando Galiani, "all that a traveller should be, exact, precise, incapable of creating any system, incapable of seeing what is useful and what not. Thus one should amass detail. To order such material, however, is another matter."[8] Contact with the data could amount to contagion. Travellers—because they travelled—had a natural predisposition to the bizarre, something which, from the stance of the universal speculator made their testimony deeply unreliable. "For in this race of authors," complained Shaftesbury, "he is ever completed and of the first rank, who is able to speak of things the most unnatural and monstrous."[9] Direct contact with the world could distort one's image of it. The traveller was also more inclined to notice, and describe difference than he was similarity, and Shaftesbury's concern was to demonstrate a degree of common agreement among men, a "common" and a universal "moral" sense. Francis Hutcheson, in many respects a disciple of Shaftesbury, was equally repelled by what he called the "Absurdity of the monstrous Taste, which has possess'd both Readers and Writers of Travels," who are indifferent to all that is inherently virtuous in mankind on the grounds that "These are but common Storys.—No need to travel to the Indies for what we see in Europe every day."[10] But, of course, it was precisely the confirmation that "what we see in Europe every day" could be seen equally in other, and indeed all, parts of the globe that both Shaftesbury and Hutchenson were searching for. Men were far more interested in human sacrifices than they were in the humdrum business of family love. And because of this concern with the unusual they were rarely, if ever, able to distinguish between what was truly universal, and thus worthy of study, and what simply amounted to mere detail. A fascination with "exciting Horror and making Men Stare" led inevitably to a more general failure to distinguish between what was intrinsic to any particular society and what was not. Travel writers dealt, ultimately, in minutiae. "The contemplative man," wrote Diderot, who professed a personal horror of travel, "is a sedentary being; the traveller is either ignorant or a liar. He who has been granted genius as his lot, is contemptuous of the minute details of experience, and he who experiences is almost always without genius."[11]

The limiting case of this is offered by Joseph Francois Lafitau's *Moeurs de sauvages américains comparées aux moeurs des premiers temps* (1724). This is perhaps the first text to attempt to combine a substantial collection of data which the author himself had collected "in the field" with significant measure of "universal speculation" as to its meaning. Lafitau was widely read during the eighteenth century—he is cited by Vico, Voltaire, Ferguson, Smith, Turgot, and many others—but except for Vico (who cites him only as evidence for the superiority of his own method) and Voltaire (who made fun of his claim that "Red-Indians" were red because their mothers wore red face-paint), all of these writers read him exclusively as a source. Precisely because the arguments he made for the superiority of his method rested on his claims to ethnographic integrity, that method went almost unnoticed, until the early twentieth century, and then it was hailed only on the grounds that its author had anticipated certain modern anthropological conceptions: rites of passage, for instance, the place which relative age plays in descent systems and the existence of distinctive kinship vocabularies.[12]

The nomothetic aspect of the human sciences—at least as it emerges in the eighteenth century—has a more complex history. This begins in the late seventeenth century and it supposes, in the first instance, that the only possible way of transcending the challenge posed by scepticism on the part of those who wished to preserve some claim to certainty in human affairs was an appeal to a minimal, and seemingly self-evident, ethical claim. This took one of three forms: the self-preservation of Hobbes and Grotius, the primitive sociability of

Pufendorf, and—what from a distance came out as much the same thing—Shaftesbury's common "moral sense."

All such claims were, of course, arrived at by way of a thought experiment which involved the construction of a pre-social stage in man's history. This effectively shifted the understanding of what a natural universal law was away from that of supposedly innate—and of course theocentric—cognitive order to one based upon a concept of human needs. Humans, on all three of these accounts, needed to be social beings in order to survive both psychically and psychologically. But since needs could clearly only be judged in the particular instances in which they arose, they could, equally clearly, only be known through specific inquiries into specific cases. All three arguments for some minimal, and thus irrefutable, basis for sociability—and in particular the last two—were therefore, heavily dependent upon a body of ideographic data. Ancient history and modern travel accounts were ransacked for the evidence that all men everywhere and at all times displayed some instinct for sociability. "Every authentic and well-written book of voyages and travels," wrote Cook's companion George Forster, in his translation of Andrew Sparrman's *Voyage to the Cape of Good Hope* (1785),

> is, in fact, a treatise of experimental philosophy. . . . It is the modern philosophers chiefly, and the living instructors of our own times, who have mostly had recourse to these treasures as containing the best materials for the purpose of building their systems, or at least, as being best adapted to the support and confirmation of their doctrines.[13]

The theory of natural sociability based upon needs, if treated historically, could offer an explanation as to why societies were so very different from one another over both space and time. It could also—and this was to become crucial—be incorporated into a general theory of progress. For needs grow exponentially. Changing needs make for changed persons. Humankind is, therefore, a category of being we can only hope to understand in terms of a process over time. We are, as Kant put it, not merely rational, but also temporalizing beings, and such beings can understand reasoning only under the form of time.[14] Authority in both the cognitive and the social order had been constructed by human agents. They had not, as in the infamous Christian and absolutist systems, been given from the outside by some superior agent. Man, as Kant said, "was not meant to be guided by instinct and instructed by innate knowledge, he was meant to produce everything out of himself."[15] To understand him, therefore, one had to unravel the process of construction. History thus became the point at which the ideographic and nomothetic converged, only a narrative of things in time could provide the basis for a science which sought to impose order upon the scattered and allusive data of the human condition. "History" therefore became, in Foucault's words "the mother of all the sciences of man," which "constitutes a favored environment which is both privileged and dangerous. To each of the sciences of man it offers a background, which establishes it and provides it with a fixed ground and, as it were, a homeland (patrie)."[16] "The progress of the human spirit," wrote Condorcet in the *Esquisse d'un tableau historique des progrès de l'esprit humain*,

> Is subject to the same general laws which we can observe in the individual development of our faculties, for it is the consequence of that development, considered as it manifests itself in a great number of individuals gathered together in society. But the result which each instant presents which depends upon that presented by previous incidents, and at the same time influences those times which will follow. This account (tableau) is therefore historical. . . .[17]

And in all such accounts, not merely Condorcet's, the difference between peoples in the past—in the European past that is—and peoples in remote cultures is collapsed into a single notion of "savagery"—a term, it should be noted, whose origins are botanical and strictly nonpejorative. The distant "savage" thus becomes the laboratory in the field, a mode of access to modern European man's remote ancestor. As the reviewer of Lafitau's *Moeurs des sauvages américains* in the *Mémoires de Trevoux* of 1724 observed, Lafitau's method had made "distances in time analogous to distances in space."[18] The metaphysician and the moralist, argued Condorcet, had been concerned with the behaviour of individuals. The historian of mankind, by contrast aimed at a study not only of "the diverse societies which have existed at the same time," but also of those which existed "in the succession of time." This was, it should be said, unashamedly a history of "civilization."[19] For that is, as Kant understood it, simply what *Anthropologie* or the human sciences, at this stage in their own internal history, were.

This, however, raised, the problem of the necessarily relationship between the ideographic and the nomothetic in a particularly acute manner. All of the "universal speculators" I have been discussing took a strongly ethnocentric or present-centric view of the "savage." This is not banal retrospective moralizing. It is the claim, we find made by Lafitau and others with direct experience of cultures widely different from their own—so different in some cases that, as Lafitau recognized, our conceptual vocabularies might not be sufficient to describe them—that if the kind of material such cultures provide is to be of any use for any kind of scientific project, then they had, first, as Lafitau put it, to be judged by their own "habits and customs."[20] Lafitau's own project had been to demonstrate the truth of Christianity, and thus the falsity of Bayle's scepticism, by showing that all peoples share comparable religious beliefs and practices. But his objective was, on the basis of a meticulous sociology, and a detailed reading of classical history, to demonstrate the truth of Christianity. It did not, as previous works had done, begin with the assumption that Christianity, and the Christian Thomist tradition of natural law, was true, and then elaborate explanations which could account for why so many different peoples seemed so flagrantly to disregard it. The tendency, which in Lafitau's view all previous writers on *les sauvages* had displayed, to treat ideographic data simply as data, detached, that is, from the historical specificities of each individual culture, could only in the end result in the proliferation of arid abstractions. As another comparativist, Nicole Antoine Boulanger complained in 1766, "The Philosophers, Metaphysicians and Jurists in default of any history have sought to create [their natural men] through reason alone." What was really needed was a "real human being in a real state."[21] It should be said that the states of nature used by the theorists of the seventeenth century, like those used by Rousseau and others in the eighteenth, have generally been taken to be entirely, and recognizably, counterfactual. The fact that Locke and Hobbes identified them with America, Rousseau with the Vaud, and so on have been assumed to be mere decoration.[22] They could be anywhere or any time. Their function was not demonstrative but heuristic. But that is not how contemporaries read them. Adam Smith's attacks—and those of his Neapolitan contemporary Francesco Antonio Grimaldi—on Rousseau, for instance, takes his claims about the state of nature literally. So, too, did Hume and Ferguson. And these people were, it does not need saying, neither willfully obdurate nor, themselves, unfamiliar with such conceptual strategies. You did not, of course, need to go "into the field" to avoid the charge of mere abstraction, but you did—as Boulanger and several others recognized—need to collapse the distinction between brute data and "universal speculation." If the ideographic data in this kind of human science proved to be recalcitrant, you did not simply need, as Kant claimed, more theory.[23] You also needed a theory about what kind of data it was that you had. Giambattista Vico, of course, famously made much the same kind of observation. All those who would use the evidence of (in this case) past societies to sustain their theories, he argued, had to be sensitive to what the causes were which governed past agents, and not merely impute to them an anachronistic causation of their own, as in Vico's view, the natural law theorists had done. The scholars, he complained "want whatever they themselves know to be as old as the world."[24] Yet "we can scarcely understand and are absolutely unable to imagine how the man of Grotius, Puffendorf and Hobbes must have thought."[25] And if we could not do that we could not easily assume that this man's inference from the discomfort of his primitive condition would be that he ought to abandon the state of nature and form a civil society. The point Vico is making—or at least one of the points he is making—is the claim now familiar to those who worry about counterfactuals that possible worlds had to be properly true to some kind of antecedent factual reality.

## III

There is a further dimension to this account. Scepticism—in particular Cartesianism—together with Lockean sensationalism, had, as Jean d'Alembert among others, recognized, performed the invaluable service of dumping the whole scholastic edifice which had been built up from Aquinas through to

Suárez on the notion of innate ideas. But Descartes himself had also attempted to replace it by an overly austere claim that the only usable knowledge had to possess the certainty of mathematics.[26] Everything else belonged to the realm of opinion. And while opinion might be interesting it could never count as knowledge. This made any kind of human science unworkable, for in the end "opinion" was all that any such science had to rely upon. The task—as Donald Kelley explains in this volume[27]—was to invest opinion with a new dignity. Probability was, after all, the closest you could get to truth in human affairs, and human reason, as Locke had said, was only a potent enough instrument so long as man refrained from using it to "plumb the Ocean of Being." While the model of the natural sciences might be a useful one, no account of human behavior could hope to reproduce the mathematical certainty or range of those sciences. For the human scientist, unlike the natural, is crucially what he is studying. As Michel Leris once observed of his own attempts to be a "true" ethnographer, "it is always our own kind (*nos semblabes*) which we are observing, and we cannot adopt towards them the same indifference of, for instance, the etymologist."[28]

In our post-Khunian world, I suspect that we have largely lost sight of the full significance of this crucial distinction between the possible objectives of the natural and human sciences. But when Charles Taylor published his now famous paper "Interpretation and the Sciences of Man" in 1971, he was conscious that his conclusion that "Human science looks backward. It is inescapably historical" was, although certainly not new—indeed Aristotle had said something similar—was nevertheless, "radically shocking" to the predominantly American Social Science community, to which it was addressed, because it broke with a tradition which aspired to some kind of indivisibility in both scientific method, and what would count as scientific conclusions.[29] When Locke, Condillac, Condorcet, and Buffon made similar distinctions, that tradition was still unstable. But I think that the insistence of their claims indicates the force with which they could all see the kind of reductionism to which "scientism," in particular the highly seductive claims of Grotius and Hobbes, could ultimately lead. To want to create a science of man which would attempt to establish first causes was part of what Vico called the arrogance (*boria*) of scholars, and it could only result in metaphysics.

The *sciences de l'homme* or the "history of mankind" belonged, then, not with the new astronomy and physics which, in their different ways had made such an impact on the seventeenth-century natural-law theorists. It belonged, instead—as Purchas (and even more clearly perhaps Giovanni Ramusio) had understood—with those new kinds of Baconian science (cartography, hydrography, and geology) which had emerged during the sixteenth and seventeenth centuries and which, as Antonio Pérez Ramos has argued, by aiming not at epistemic certainty, but at a measure of descriptive accuracy, had generated "a whole cluster of now attainable 'objects of knowledge.'"[30]

The human sciences were to provide an account of effects or—to use an earlier language—"secondary causes" by means of an examination of their origin, as far as that could be known, and of their development. Such sciences had two properties. They were in the first instance progressive. All, that is, were concerned with establishing the state of the present, and in particular a present condition which, during the second half of the eighteenth century came increasingly to be described as a "civilization." The progress they described was assumed to be, as much for Montesquieu as it had been for Aristotle, from the simple to the complex. Dugald Stewart, for instance, observed of Smith's "Dissertation on the Origin of Languages" (1793), that this was "a specimen of a particular sort of inquiry, which so far as I know, is entirely of modern origin" and which addressed the question: "by what gradual steps that transition has been made from the first simple efforts of uncultivated nature to a state of things so wonderfully artificial and complicated."[31]

All such histories existed—as Lafitau's work had done—at a junction between an account of an historical process and the description of data collected from the field, an enterprise which became, archetypically, Vico's "jurisprudence of mankind" based upon an understanding of the "common customs of peoples." But here Vico's project in the *Scienza nuova*, Condorcet's in the *Esquisse d'un tableau historique des progrès de l'esprit humain* and Kant's in the *Idea for a Universal History with a Cosmopolitan Purpose*, begin to look quite unlike any part of any history of modern human science. The purpose behind this was to reveal, in Condorcet's formulation, to the *philosophe politique*, the "true nature of all our own prejudices."[32]

And this brings me to the second property of such histories: their concern not with individuals but with collectivities. Cartesian skepticism may, as D'Alembert in the *Discours préliminaire* pointed out, have "lifted the yoke of scholasticism," but it had replaced it by nothing but the most extreme form of solipsistic reasoning.[33] The antifoundationalism of much of the thinking of the late Enlightenment, that of Kant himself, of Diderot, Herder, and of course of Kant's own "Newton of the moral world," Rousseau, derives from a rejection of the sheer and humanly impossible self-centeredness of the Cartesian project, a self-centeredness which would have made any kind of *human* science impossible. Knowledge of the human world required not merely a high degree of epistemological modesty and was inescapably the examination of collective acts, it had itself to be a collective act. The conjectural historian could no more be detached from his own place in the scheme he was attempting to describe, than he could be distinguished, as a rational being, from his object of study. Or, as Taylor puts it, "a study of the science of man is inseparable from the an examination of the options between which men must chose."[34] This is, as Onora O'Neil has shown, most powerfully true of Kant; but Kant's own position owes much, as he himself made clear, to Bacon, whose "in commune consultant" provides the epigraph for the *Critique of Pure Reason*.[35] It is this emphasis not only on the collectivity as the true object of study, but on the place of the historian, which explains the centrality in many such conjectural histories of an account of the origin and development of language, for the invention of language was, of course, the supreme collective act. "The invention of the arch," wrote Condorcet, "was the work of a man of genius; the creation of a language was the work of an entire society." The Third *Epoque* of the *Esquisse* runs from the invention of agriculture, which by "attaching man to the soil he cultivates"[36] makes for the fixity which was, for Condorcet as much as it had been for Aristotle, a necessary condition of civility, to the creation of the alphabet which demands what he calls a "double doctrine."[37] Similarly the invention of printing at the end of the Seventh *Epoque*, is identified as the final stage before the arrival of the true age of scientific understanding, because printing, of course, makes collective cognitive action, finally accessible to all.[38] Human history has, of course, its ups and down, its false starts, and its catastrophic reversals. But Condorcet, in common with most, if never quite all, of the "Historians of mankind," was always insistant that ultimately it was a narrative of irreversible progress towards increasing scientific detachment—the final unmasking of all our prejudices.

Few modern human sciences would, I suppose, take the unmasking of prejudice to be their goal. But for the Enlightenment, the "history of mankind" was a project with a clear and declared political purpose. It was, ultimately to teach man how to live a better collective life. And it was to achieve this by allowing him to see the story of his existence for what it was. In this it shares something with the human sciences as they have been conceived by a number of modern practitioners from Foucault to Clifford Geertz as "efforts to render various matters on their face strange and puzzling."[39]

There is another—and for any disciplinary history still more unsettling—aspect to these projects. The modern human sciences are conceived as, to use Taylor's phrase, "ex post facto understanding." For their eighteenth-century creators, however, and in particular for Condorcet and Kant, they were also believed to possess important predictive possibilities. Both the *Esquisse* and the *Idea for a Universal History with a Cosmopolitan Purpose*, end with general outlines for a future state of mankind, and both were written with that purpose in mind. (Condorcet's longer text, which was never written would have been overwhelmingly concerned with such prospects.) For if once, it was argued, we are empowered to see the world as it is, by properly understanding how it came to be, then there is no reason why the progression which had begun with one thought-experiment—the state of nature—should not end with another. Furthermore both Condorcet's utopian Tenth *Epoque* and Kant's far more modest Ninth Proposition are not merely predictive. They are also prescriptive. "These observations," declared Condorcet at the beginning of the *Esquisse*, "on what man has been, on what he is today lead directly to the means of ensuring and accelerating the new developments which he may still hope to make."[40]

Underpinning these claims was another: that man must progress in a way which is discoverable from past experience, because, as Kant remarked, "It must be assumed that nature does not work without a plan and purposeful end, even amidst the arbitrary play of human freedom." The philosophical mind must, therefore, be able to produce the ultimate "universal history" following, as he puts it an "a priori rule." This is not, Kant insisted, intended to supersede the "task of history proper, that of empirical composition," but rather to come at the data such history might offer "from a different angle."[41] The outcome of this encounter would prove, as he said in *The Conflict of the Faculties*, to be a "narrative of things imminent in future time, consequently as a possible representation a priori of events that are supposed to happen there." What form these events will have is not clear. But it is clear that for Kant the two components of any "practical" anthropology, with which I began this paper—what Kant himself describes as "brute data" and philosophical reflection—the ideographic and the nomothetic—could only be fully integrated not in any vision of the modern, disengaged discipline, but in some projection for a future state— "designated and divinatory and yet natural."[42]

I end here, in the future with the observation that, from the point of view of a history of a discipline it is important not merely to identify the shape of the development, the genealogies, the origins and so on of recognizable intellectual enterprises. It may also be important to look more attentively than we perhaps do at the objectives that, in the course of time, those projects have been compelled to shed. For the "history of mankind" in the eighteenth century, was founded upon the assumption that even if there was no evidence for the existence of a purposeful God, nature clearly had a purpose, and that mankind was, because of his participation in nature compelled to develop in certain ways and not in others. For Condorcet or Kant to have acknowledged that their own intuitions about the historicity of the sciences of man implied, in Taylor's words, that "the very terms in which the future will have to be characterized, if we are to understand it properly, are not at all available at present" would have constituted a denial of the possibility of any kind of "anthropology."[43] For us it seems (almost) self-evident.

## Notes

1. Loren Graham, Wolf Lepenies, and Peter Weingart, eds., *Functions and Uses of Disciplinary Histories* (Dordrecht, 1983), xv.
2. Radcliffe-Brown, *Structure and Function in Primitive Society* (London, 1952), 1.
3. Vincent Crapanzano, "'Hermes' Dilemma: The Masking of Subversion in Ethnographic Description,'" in James Clifford and George E. Marcus, eds., *Writing Culture. The Poetics and Politics of Ethnography* (Berkeley, Los Angeles, London, 1986), 51–76.
4. Clifford Geertz, *Works and Lives. The Anthropologist as Author* (Stanford, 1988), 73–101.
5. Samuel Purchas, *Hakluytus posthumus, or Purchas His Pilgrimes* (London, 1625), vol. 1, To the Reader.
6. See C. Bally, *Le Langage et la vie* (Geneva, 1965), 58, 72, 102.
7. I have attempted to give some account of the classificatory methods used in these texts in *European Encounters with the New World. From Renaissance to Romanticism* (New Haven and London, 1993), 83–86.

8. Quoted by Girolamo Imbruglia, "Tra Anquetil-Duperon e L'Histoire de deux Indes. Liberta, dispotismo e feudalismo," *Rivista Storica Italiana* 106 (1994): 141.

9. Anthony Ashley Cooper, Third Earl of Shaftesbury, *Characteristics of Men, Manners, Opinions, Times*, ed. John M. Roberts (Gloucester, MA, 1963), 1: 222–23.

10. Francis Hutchenson, *An Inquiry into the Original of our Ideas of Beauty and Virtue*, 3rd. ed. (London, 1729), 205–7.

11. Quoted in Imbruglia, "Tra Anquetil-Duperon." For Diderot's views on travel, see Pagden, *European Encounters*, 156–62.

12. A. van Gennep, "Contributions à l'histoire de la méthode ethnographique," *Revue de religion* 67 (1913): 320–38; A. Metraux, "Les Précurseurs de l'ethnologie en France du XVIe. au XVIIIe. siècle," *Cahiers d'histoire mondiale* 67 (1963): 721–38; R. Needham, "Age, Category and Descent," *Bijdragen tot de taal-land-ne volkenkunde*, Deel 122, Anthropologica, 8: 1–35.

13. Andrew Sparrman, *A Voyage to the Cape of Good Hope, towards the Arctic Polar Circle and around the World* (London, 1785), iii.

14. Onora O'Neil, *Constructions of Reason. Explorations of Kant's Practical Philosophy* (Cambridge, 1989), 22.

15. *Idea for a Universal History with a Cosmopolitan Purpose*, in *Kant's Political Writings*, ed. Hans Reiss (Cambridge, 1970),43.

16. Michel Foucault, *Les Mots et les choses. Une archéologie de sciences humaines* (Paris, 1966), 382–83.

17. Marie-Jean-Antoine-Nicolas Condorcet, *Esquisse d'un tableau historique des progrès de l'esprit humain*, ed. Alain Pons (Paris, 1988), 80.

18. *Mémoires de Trevoux*, (Trevoux, 1724), 1572.

19. Condorcet, *Esquisse d'un tableau historique*, 80.

20. On this, see Anthony Pagden, *The Fall of Natural Man. The American Indian and the Origins of Comparative Ethnology* (Cambridge, 1986), 198–209.

21. Nicole Antoine Boulanger, *L'Antiquité dévoilée par ses usages* (Amsterdam, 1772), 1:8.

22. For the claim that Locke's state of nature was intended to describe a real place which was, in most respects, identical with contemporary America, see James Tully, *An Approach to Political Philosophy: Locke in Contexts* (Cambridge, 1993), 137–76.

23. *On the Common Saying: "This May be True in Theory, but It Does Not Apply in Practice,"* in Kant's *Political Writings*, 61.

24. The best known formulation of this claim is in Giambattista Vico, *Scienza nuova*, paragraph 127.

25. Vico, *Scienza nuova prima* (First New Science), paragraph 251.

26. On this, see Judith Shklar, "Jean d'Alembert and the Rehabilitation of History," *Journal of the History of Ideas* 42 (1981), 643–64.

27. See above, chapter 1, note 22.

28. Michel Leris, "L'ethnographie devant le colonialisme, *Cinq études d'ethnographie* (Paris, 1969), 85.

29. Charles Taylor, *Philosophy and the Human Sciences. Philosophical Papers 2* (Cambridge, 1985), 15–57.

30. Antonio Pérez Ramos, *Francis Bacon's Idea of Science and the Maker's Knowledge of Tradition* (Oxford, 1988), 35.

31. Adam Smith, *Essays on Philosophical Subjects*, eds. W.P.D. Wightman and J.C. Bryce, in *Glasgow Edition of the Works and Correspondence of Adam Smith*, vol. 3 (Oxford, 1980), 292.

32. Condorcet, *Esquisse d'un tableau historique*, 80.

33. Jean le Rond d'Alembert, *Discours préliminaire de l'Encyclopédie*, in *Oeuvres complètes de D'Alembert* (Geneva, 1967), I:36.

34. Taylor, "Interpretation and the Sciences of Man," 54.

35. O'Neil, *Constructions of Reason*, 6–9.

36. Condorcet, *Esquisse d'un tableau historique*, 106.

37. Ibid., 118.
38. Ibid., 178.
39. Clifford Geertz, "The Strange Estrangement: Taylor and the Natural Sciences," in James Tully, ed. *Philosophy in an Age of Pluralism. The Philosophy of Charles Taylor in Question* (Cambridge, 1994), 83.
40. Condorcet, *Esquisse d'un tableau historique*, 80.
41. Immanuel Kant, *Idea for a Universal History with a Cosmopolitan Purpose*, 53.
42. Kant, The *Conflict of the Faculties (Der Streit der Fakultaten)*, trans. Mary J. Gregor (Lincoln and London, 1990), 141–42.
43. Taylor, "Interpretation and the Sciences of Man," 56.

# Part IV
# Natural Sciences

# FRANCIS BACON AND THE REFORM OF NATURAL HISTORY IN THE SEVENTEENTH CENTURY

## Paula Findlen

> "Do you really think it is easy
> to provide the favourable
> conditions for the legitimate
> passing on of knowledge?"[1]

In the final decade of the sixteenth century, Francis Bacon (1561–1626)—younger son of Elizabeth I's Lord Keeper of the Seal, and an English lawyer of middling years with some modest success and even larger aspirations—set out to redraw the map of knowledge. All this is well known, since Bacon belongs to the canon of individuals whose innovations comprise our image of the scientific revolution. In this capacity, Bacon has been hailed variously as the father of induction, the first serious proponent of experimentation, and the blueprint for modern man. Bacon, the disgraced Lord Chancellor who thought that dying of pneumonia while attempting to ice a chicken was equivalent to Pliny's demise in sight of Vesuvius (both giving their lives for curiosity, as he wrote in his deathbed missive to the Earl of Arundel[2]), is an especially compelling figure in the history of Renaissance thought. He left behind reams of instructions on how to investigate nature and an equally lengthy list of directives on the proper demeanor of the natural philosophers entrusted with his new method. As Bacon argued repeatedly in his many works, his goal was to make natural philosophy a profession, a business and a matter of state; it was to have primary rather than secondary status among the variety of human endeavors. This contrasted dramatically with its current state, as Bacon described it: "natural philosophy was never any profession, nor never possessed any whole man, except some monk in a cloister, or some gentleman in the country, and that very rarely; but became a science of passage to season a little young and unripe wits, and to serve for an introduction to other arts . . . ," he wrote in his *Filum Labyrinti*.[3] Instead Bacon proposed to make natural philosophy important and permanent. Science was to become something that civil society would take seriously.

Central to Bacon's reform of natural philosophy was his image of natural

---

Thanks to Deborah Harkness, Mordechai Feingold, Alix Cooper, and audiences at Harvard, the Folger Shakespeare Library, and the Renaissance Society of America for their bibliographic suggestions and comments.

history as a foundational discipline. As many scholars have noted, natural history became increasingly important to Bacon as the years passed.[4] In his forties, it was simply one part of a vast program of reform, a stepping-stone on the royal road to knowledge. Natural history was not an especially elevated component of his original project but a convenient rubric under which to accumulate particular experiences that would yield better and more lasting philosophy. As Bacon entered his sixties, less sanguine about his ability to involve others in the work of his project and more conscious of the difficulties of completing his encyclopedia within his own lifetime, natural history occupied more of his time and energy and, accordingly, assumed greater importance in his writings. Natural history is "the foundation of all," he affirmed in the *New Organon* (1620).[5] Elsewhere he suggested that while natural history could exist without natural philosophy, the latter would essentially vanish without the former.[6] This was a message that Bacon's heirs—the early members of the Royal Society and the communities of naturalists in Oxford and Cambridge during the second half of the seventeenth century—would take to heart.

Despite the importance of natural history to Bacon's program of intellectual reform, it remains a relatively undervalued aspect of his work. In the scores of books and articles produced on Bacon in the last few decades, I have yet to find a single one that explains why Bacon chose natural history as his main point of departure.[7] In many respects, it was not an obvious choice for him to make. Natural history was not yet a strong tradition in England in the late sixteenth century relative to its popularity on the continent; its heyday in that country lay in the 1660s through 1690s and beyond rather than in this earlier period. And Bacon himself evinced little direct interest in late Renaissance natural history. Certainly he was familiar with Aristotle, Pliny and Dioscorides but, of modern naturalists, he mentioned only Conrad Gesner, Georg Agricola and José Acosta by name. Not once did the name of Ulisse Aldrovandi—whose three-volume *Ornithology* (1599–1603) appeared precisely during the period in which Bacon accelerated his project—grace his works. Instead Bacon displayed greater knowledge of the writings of encyclopedists such as Francesco Patrizi, Bernardo Telesio, and Girolamo Cardano, the natural magic tradition as practiced by Giovan Battista della Porta and Giordano Bruno, and the chemical philosophy of the Swiss physician Paracelsus. Yet arguably these writers contributed equally to the tradition of natural history since it was not yet a field with strictly defined boundaries.[8]

Of course Bacon often chose not to cite authors whom he read, putting into practice his advice that knowledge be born from experience rather than authority, so we cannot simply take the absence of names as a literal indicator of what he did or did not know. And yet for anyone versed in the natural history tradition, Bacon's own natural history seems striking devoid of the common markers that we associate with this genre: no lengthy descriptions of flora and fauna; no discussions of etymology, number of toes and fronds, or other classificatory devices; no humanist embellishments nor even pictures grace such works as the *Sylva sylvarum* (1627). Natural history had come to mean something quite different to Bacon than it had traditionally connoted; like natural philosophy it was in need of reinvention. This transformation absorbed other disciplines into natural history, ostensibly making them more useful to society, public sciences rather than private indulgences. Ever conscious of genre, Bacon acknowledged that his own natural history did not conform to prior definitions of this subject. As he confessed in his final work, "For this writing of our *Sylva sylvarum* is (to speak properly) not natural history, but a high kind of natural magic. For it is not a description only of nature, but a breaking of nature into great and strange works."[9] Natural history, in other words, provided a rubric under which to pursue forms of inquiry that were incapable of full redemption in their current state. It provided the *via media* between the academically exalted field of natural philosophy and both the occult sciences and the crafts traditions (each in their own way problematic for the creation of the socially elevated, morally unambiguous and publicly useful form of science that Bacon championed). Drawing upon these various fields of knowledge and sanitizing them through their relabelling as "natural history," Bacon strove to create an approach to nature that would make its study and its practitioners more deserving of public recognition.

This essay explores Bacon's attitude towards natural history, as it developed from the 1590s through the 1620s. It considers Bacon's reform of natural history in light of interest in nature at the Elizabethan and Jacobean courts and more generally among the English aristocracy. Bacon's immediate context provides important background to his ideas. As we shall see, Bacon's distaste for certain aristocratic values proved to be an essential part of his critique of natural history and, more generally, natural philosophy. The courtly pursuit of science—a leisurely pastime with no discernible contribution to the public good—was precisely what Bacon disliked most about the current state of natural knowledge. Science was a pleasurable though inconsequential pursuit when instead it ought to be a matter of great affair.[10]

But how was this transformation to be effected? Central to Bacon's program of reform was the idea of *discipline*. To create a new natural history, the natural philosopher had to cultivate a different demeanor. Bacon's naturalist was to become the model knower. Imagining Renaissance alchemists, naturalists and natural magicians as cultivators of "trivial" arts, Bacon proposed to purge all scientific inquiry of suspect philosophies and humanist embellishments. In his *Thoughts and Conclusions* (1607), he recommended that his readers cultivate "a disciplined firmness of mind." This was also the site of his famous phrase, "Nature cannot be conquered but by obeying her," suggesting that nature itself provided the preparatory discipline for its observers. By the time his *New Organon* (1620) appeared, Bacon advised natural philosophers "to restrain themselves, till the proper season, from generalization."[11] Only through such acts of self-denial could science truly realize its potential, allowing the faculty of reason to overcome that of imagination, making memory its tool.[12] If natural history could not be reformed than science would remain as it had always been, according to Bacon: a subject fit to exercise the minds of children and a leisurely pursuit for gentlemen who delighted in esoteric wisdom and exotic phenomena.

Reforming natural history became an essential step in giving natural philosophy a greater place in the realm of knowledge. To do this, Bacon demanded that natural history give up some of the humanist traits that had made it so popular with the upper classes in the preceding century. Lack of method, as Bacon defined it, had led to the undisciplined state of current knowledge. "What they lack is any art or precepts to guide them in putting their knowledge before the public," he wrote in *The Masculine Birth of Time* (1603).[13] Discipline was a means of narrowing the enterprise, making natural history a proper "history" of nature rather than a repository of the encyclopedic imagination; without it, the natural philosopher could not recognize the true laws of nature. It also dignified natural history by making it a nobler enterprise.

## Natural History at the Elizabethan Court

Before examining Bacon's views on natural history, we should consider the state of natural history in Elizabethan England. In his *Thoughts on Human Knowledge* (1604), Bacon commented on the "small part" played by natural history in the university curriculum. Elsewhere he wrote that scholastic philosophers knew "little history."[14] In contrast to continental Europe, where professorships in *materia medica*, museums of natural history and botanical gardens had flourished since the 1530s, no university in England taught any aspect of natural history as a separate field of study until well after Bacon's death. Despite John Gerard's attempt to found a botanical garden at Cambridge in 1588, there was no botanical garden at any English university until the establishment of the Oxford Botanical Garden in 1620. The chair in botany to accompany this garden did not appear until 1669.[15] By 1605, when Bacon wrote *The Advancement of Learning*, he noted the beginnings of such a material culture: "We see likewise that some places instituted for physic have gardens for the examination and knowledge of simples of all sorts, and are not without the use of dead bodies for anatomical observations." Nonetheless Bacon concluded that this was not enough to satisfy his idea of experience. "But these respect but a few things."[16]

In the 1570s, when Bacon studied at Cambridge, the only means by which natural history entered the curriculum was via the teaching of Aristotle and Pliny in the third year course in philosophy or in informal conversations with professors such as the physician John Caius who published his *History of Rare Animals and Plants* in 1570.[17] For all these reasons, Bacon was right to point out that natural history had been undervalued by the academic establishment.[18] Despite the paucity of formal instruction in this area, however, it is worth noting that some of the foremost naturalists in Tudor England—William Turner,

John Caius, William Gilbert, Thomas Moffet and Thomas Penny—studied at Cambridge like Bacon. So we cannot entirely discount the role of the universities in enhancing the status of natural history. If nothing else, the medical and more generally the arts curriculum provided a setting in which professors could inform students of this ancient discipline and introduce them to new research in this area.

While natural history made few inroads in the English universities, it became increasingly popular at court and among the upper classes.[19] On the continent the disciplinary status of natural history had been established by its attractions for patricians and its subsequent institutionalization; in England this was accomplished primarily by its value as a cultural commodity as well as its growing importance in the imperial ambitions of Elizabethans. As Mordechai Feingold observes in his study of the place of science in Elizabethan and Jacobean England, "Like heraldry, botany and horticulture were frequently pursued in a leisurely fashion by those of a certain class and were considered as basic as good manners."[20] By the late sixteenth century natural history was one of several sciences that had achieved noble status. Unlike alchemy and mathematics, practiced mostly notably in England by the Elizabethan magus John Dee (1526–1608), it had the virtue of accessibility. Cultivated by nobles, physicians, herbalists, gardeners and court poets, it was a field of study that did not demand special skills from its practitioners or a privileged state of being to be initiated into its mysteries. While mathematics was practiced by university scholars and the merchant elite, and alchemy by a handful of adepts, natural history appealed to a broad sector of the educated populace.

Natural history was made accessible to Elizabethans not only through the introduction of its ancient works into the university curriculum but also through the increased circulation of texts and the growing presence of naturalists in the city, in the homes of the upper classes, and at court. During the second half of the sixteenth century natural histories increasingly found their way into the private libraries of Elizabethans.[21] By Bacon's day one could find many nobles who owned copies of Pliny as well as indigenous natural histories such as William Turner's *A New Herbal* (1551–68), John Gerard's *Herball, or Generall Historie of Plantes* (1597), William Gilbert's *On the Magnet* (1600), and Edward Topsell's *Historie of Foure Footed Beastes* (1607). Such works were placed alongside the lengthy natural histories written by such continental naturalists as Conrad Gesner, Ulisse Aldrovandi and Charles de l'Ecluse, and travel narratives such as the 1577 translation of Nicolás Monardes's *Joyful News Out of the New Found World* and Thomas Harriot's *Brief and True Report of the New Found Land of Virginia* (1590) that highlighted New World flora and fauna. Certainly Bacon's knowledge of Gesner and Acosta was a product of the new place of natural history in Tudor libraries, and its intersection between humanist encyclopedism and dreams of global conquest.

Bacon's reading of Acosta's natural history of the Indies, in which Aristotle became an object of derision when his theories failed to live up to brute experience, was in keeping with his image as a patron of the English voyages of discovery at the turn of the century. Like many prominent Elizabethans he found the image of an English empire to be particularly appealing; economic expansion, political might and mastery of nature all came together in the possession of the New World.[22] Bacon's continued interest in Harriot's projects suggests that he particularly followed the scientific work done by those involved in the establishment of an English colony in North America. Harriot (1560–1621) and the artist John White had popularized the natural history of North America upon their return to England in 1587, laden with specimens and illustrations.[23] The intellectual riches of Walter Ralegh's expedition to Virginia, which quickly found their way into the work of leading naturalists such as John Gerard and Thomas Moffett, must have fueled Bacon's fantasies about the knowledge that lay just beyond his grasp. Yet it also frustrated him precisely because it lay in someone else's hands—men such as the court physician Moffett or the court astrologer Dee whom Bacon did not trust to transform the wonders of the Americas into the hard "facts" of natural history. Natural history became a much discussed subject among the English nobility and merchants as books and objects entered their homes as a result of the increased importance of travel among the upper classes. Yet mere possession was not enough; *how* one read Acosta or Harriot or what one did with natural objects mattered. From this perspective, Bacon's program of natural history was as much an attempted reformation of the reading and collecting habits of Elizabethans as a call for a new philosophy of nature.

In his writings Bacon assumed that his audience had some familiarity with the literary tradition of natural history. This knowledge both aided and hindered his project. While almost never citing specific works, save for an occasional mention of the ancients who founded this discipline, Bacon nonetheless worried that his readers would misunderstand his project as the one with which they were already familiar. For this reason, he constantly underscored its difference: "Lest men for want of warning set to work the wrong way, and guide themselves by the example of the natural histories now in use, and so go far astray from my design."[24] Such fears reflected Bacon's view that while natural philosophy was relatively unknown to the general reading public, natural history instead suffered from being too popular. Readers of Pliny and Gesner might mistake these texts as the inspiration for his work when instead Bacon hoped to redirect their attention away from the ancient traditions of studying nature and towards the new natural history that took travel and exploration as its point of departure.

Bacon could measure the success of natural history among his contemporaries not only by the growth in published volumes but also by the number of works dedicated to Tudor monarchs and their close associates. Several decades before Bacon proclaimed natural history a "royal work," the community of naturalists implicitly had given it that label by dedicating their histories to various monarchs.[25] Edward Wooton addressed his *On the Differences of Animals* (1552) to Edward VI. William Turner (ca.1508/10–1568), one of the first Englishmen to study natural history on the continent, directed the three installments of his herbal to no less a patron than Elizabeth I. Two years later, Mathias de L'Obel (1538–1616) followed suit with his *New Plant Controversies* (1570). Court counselors and favorites also were perceived as appropriate patrons. John Gerard (1545–1612) dedicated his popular *Herball* to Elizabeth's most trusted counsellor William Cecil, Lord Burghley, who had also briefly patronized Turner; from such episodes we know that Bacon's uncle evidenced some interest in natural history in addition to his well-known pursuit of alchemy.[26] Thomas Moffet chose Mary Herbert, Countess of Pembroke, as his preferred patron while also enjoying the support of Lord Willughby and the Earl of Essex (Bacon's important patron). Thus by the time Bacon conceived of natural history as the key to his new philosophy, it had already received widespread support and approval among the very people from whom he sought support.[27] Living in Essex's principal residence—York House on the Strand—during the Earl's stay in Ireland, Bacon surely came into contact with the naturalists who flocked to this court favorite, while observing at close hand the activities of his relatives the Cecils in the same part of London.[28] Bacon's predecessors and contemporaries may not have dignified natural history in the way that he intended, but they had certainly ennobled it and perhaps planted the seeds of the idea that would lead Bacon to highlight this particular discipline in his reform of natural philosophy.

By the 1570s, the decade when Bacon first began his studies, interest in natural history accelerated due to increased contact between England and the continent as well as the expeditions to the Americas. Bacon himself was to benefit from this situation since his stay at the French court (1576–79) in the service of the English ambassador probably brought him into contact with French collectors of natural curiosities such as Bernard Palissy and the royal surgeon Ambroise Paré. A few decades later, the same diplomatic channels allowed John Tradescant the Elder to collect plants for Bacon's cousin Robert Cecil to place in his garden at Hatfield.[29] At the same time that Bacon enjoyed his first taste of travel, foreign naturalists, particularly from Italy and the Low Countries, were making their way to England. The Flemish naturalist Mathias de L'Obel first arrived in England in 1567, bringing with him the manuscripts of his mentor Guillaume Rondelet. His presence in London until his death in 1616 facilitated communications between English and continental naturalists, contributing to the sort of naturalist's republic of letters that Bacon later advocated in his own writings.[30]

While L'Obel connected Protestant naturalists across the channel, intense curiosity about Italy made English nobles more aware of the centrality of natural history to Italian elite culture. They rediscovered nature in their travels and studies in that region, where they visited the famous "theater of nature" of Ulisse Aldrovandi in Bologna, the botanical gardens in Pisa and Padua and the grand ducal menagerie in Florence. They participated in this new research when they attended Fabricius's anatomical lectures in Padua, where the Aristotelian program of natural history was being revived, and botanized in such notable locations as Monte Baldo near Verona. Of all these activities, the public and private gardens cultivated by Italian patricians most captivated their imagination. As early as the 1570s Elizabethan gardens were imitating their Italian counterparts. In subsequent decades botanical enthusiasts such as Burghley, whose garden in Theobalds on the Strand was the delight of his contemporaries, employed Italian gardeners and English gardeners familiar with the Italian style, and cultivated the services of the Italian physicians whom Elizabeth employed.[31] Natural history was so clearly a preferred leisure activity in Italy that the educated Elizabethans could not help but be drawn to it as they modeled themselves after Italian courtiers; it coincided well with the pastoral lifestyle that they celebrated. By the 1590s, when Bacon began to mention natural history in his writings, the increased circulation of printed texts, the growing presence of naturalists in London and the desire for cultural emulation all made natural history attractive to the English nobility.

The tangible importance of natural history in Elizabethan England was found not only in the pages of printed texts, replete with emblems, adages and morals, but also in the uses of nature at court. Given Bacon's position—the son of Elizabeth's Keeper of the Seal, a Member of Parliament and an aspiring barrister who participated in several court masques in the 1590s—we must think of the court as an important setting in which Bacon observed the growing popularity of natural history among his contemporaries. There natural history was a lavish enterprise that was essentially indistinguishable from many other cultured pursuits. Nature was the subject of lengthy poems such as Philip Sidney's *Arcadia* and Edmund Spencer's *The Faerie Queene*, and the backdrop for popular images such as Nicholas Hilliard's *Young Man Among the Roses* (ca.1587–88). Natural history also became the source of many of the emblems that decorated the closets and banquet houses of Elizabethans. All this was done in imitation of the Queen, whose dresses increasingly looked like a mobile natural history. We need only recall the "Hardwick" portrait (ca.1599), in which Elizabeth I appeared wearing a dress embroidered with the marvels of nature, or the "Rainbow" portrait (ca.1600–03) in which Robert Cecil and John Davies probably played an important role, to realize how literally natural history was woven into the fabric of elite society.[32]

Despite his alleged aversion for this sort of display Bacon himself participated in the decorative arts of natural history. At the end of 1599 he offered the Queen a New Year's gift of a white satin petticoat, embroidered with snakes and fruits—images that appeared in the Hardwick portrait that same year and which, in addition to being favored by the Queen, allowed Bacon to obliquely suggest that she embodied the dubious virtues of Eve.[33] Bacon also did a great deal of his writing in a house decorated with emblematic images of nature; in anticipation of Elizabeth's visit to Gorhambury in 1577, his father Nicholas had added a wing to the family estate decorated with enamelled glass windows portraying various allegories of nature. Moreover Nicholas had placed a statue of Orpheus at the entrance to the estate garden—an image that would reappear prominently in Bacon's own allegory of philosophy—and constructed a banquet house filled with portraits of the famous philosophers, rhetoricians and astrologers of the ages, from Aristotle to Copernicus.[34] Thus Bacon's reaction to certain strands of natural history, which privileged the exotic and allegorized each image, emerged from full knowledge of the courtly approach to nature as a participant in its games.

The prominence of natural history in Elizabethan England and its relation to other disciplines surely influenced Bacon's choice of subject. Natural history offered many advantages from his perspective. First, it was a relatively new and untried field of study, open to alteration and rapidly expanding as a result of the voyages of exploration and discovery. While Aristotle and Pliny had offered many interesting theories and observations about the natural world and encouraged their readers to collect information, they had never really told naturalists how they should observe or what their observations would add up to; moreover their viewpoint seemed increasingly provincial as a changing geography made the physical limits of their knowledge more discernible. In short, it was an encyclopedia without much form or shape, bulging with new information and in need of some intellectual guidance.

Second, natural history was an unspecialized domain that, at its core, produced a simple, expandable and relatively unprivileged form of knowledge. The accessibility of natural history appealed to Bacon who argued that the "leveling of wit" was one of his primary tasks. Rejecting alchemy and natural magic as they were then practiced, Bacon observed that while both might yield insight into the nature of things neither the alchemist nor the magician would be a good model for his ideal observer. Since they specialized in secretive arts, they could not successfully manage the creation of public knowledge, Bacon's primary goal. Likewise the mathematician was an inappropriate figure because his knowledge, though potentially utilitarian, was also not easily conveyed. By contrast, the naturalist worked within the public domain; his was a communicative and collaborative enterprise that drew upon many different sectors of society.

Finally, Bacon approved of natural history because it was a noble enterprise. Natural history had occupied some of the greatest minds of antiquity and been patronized by two of its greatest monarchs, Alexander the Great and King Solomon. It was a noble rubric for fairly ignoble work—the gathering and recording of data—and its management and patronage could be presented as an exalted undertaking. While natural history was not as intellectually elevated as natural philosophy, it may have been socially preferable. Bacon's attention had first been caught by the increased visibility of natural history among the nobles with whom he consorted—his uncle Lord Burghley, his cousin Robert Cecil, and patrons such as Essex and Elizabeth I. He had probably observed its presence on the margins of the university curriculum during his student days and continued to note its popularity as he made his way through Gray's Inn and ultimately the court. For all these reasons, Bacon imagined the reformation of natural history as a fairly obvious conduit to lead his audience along the path he intended. He would persuade them of the merits of his fact-gathering enterprise, convince them that natural history could be something other than courtly entertainment, and reclaim science as worthy endeavor. With this in mind, he set down some guidelines for his readers to follow.

## The Flight from the Trivial

When Bacon began to compose his new philosophy, he advised his readers that his natural history would be unlike any seen before. Describing natural history as a formative discipline for the development of a philosophy based on a more severe induction, he wrote: "For neither Aristotle, nor Theophrastus, nor Dioscorides, nor Caius Plinius, ever set this before them as the end of natural history."[35] What did he mean by this? Scholars have often focused on the methodological aspects of Bacon's concept of induction at the expense of other intellectual issues. To understand Bacon's inductive science, we need to examine more carefully his reaction to late Renaissance natural history and the culture that produced it. We also need to consider the emergence of Bacon's ideas within their immediate political context: the death of Elizabeth I in 1603 and the accession of James I, a male monarch after some fifty years of female rule. As we shall see, Bacon's attack on the trivial linked a historical critique of knowledge with an aesthetic commentary on Elizabethan court culture that had, in Bacon's analysis, feminized the leading minds of England, making them more receptive to aimless and unserious pursuits rather than solid and sober knowledge.[36]

Despite Bacon's lack of specific references to contemporary natural histories, he offered a fairly detailed portrait of the state of Renaissance natural history which indicates that he understood the genre well. One of Bacon's most persistent criticisms of natural history was that it was indiscriminate:

> So in natural history, we see there hath not been that choice and judgment used as ought to have been; as may appear in the writings of Plinius, Cardanus, Albertus, and diverse of the Arabians; being fraught with much fabulous matter, a great part not only untried but notoriously untrue, to the great derogation of the credit of natural philosophy with the grave and sober kind of wits.[37]

Natural history, as Bacon perceived it, had yet to fulfill its potential. In its current state, it was entertaining but untrustworthy knowledge. While the occult sciences impugned the status of natural philosophy through the grandiose and unrealized claims of their practitioners—here Bacon surely had contemporaries like Dee and Bruno in mind as well as Paracelsus—natural history trivialized science by giving itself over wholly to pleasure. Pleasure was in itself a female quality. "And therefore knowledge that tendeth but to satisfaction is but as a courtesan, which is for pleasure and not for fruit or generation."[38]

How had natural history, one of the most ancient scientific disciplines, gone so awry? In pondering this question, Bacon invoked history as a means of understanding how knowledge, when not set on its proper course, might become debased. History became a powerful tool in assessing the evolution of this form of knowledge, invoking an Edenic image of knowledge corrupted by humanity over time. Despite Bacon's stated aversion for mixing religion and philosophy, he argued that the Fall was the original source of the debilitated state of natural history because it was in Adam's time that man had lost dominion over nature.[39] The antiquity of this field made its errors even more glaring since time only magnified these imperfections. Like many humanists before him, Bacon imagined his history of natural history as a dialogue with the past in which he interrogated the originators of the discipline on their intent and highlighted their shortcomings. As Bacon told Isaac Casaubon in 1609, "So I seem to have my conversation among the ancients more than among these with whom I live."[40] This was undoubtedly a reflection of Bacon's difficulties in communicating with his contemporaries as well as a sign of his political astuteness—the dead were safe competitors who listened attentively and had no power to respond.

To varying degrees, Bacon condemned both Aristotle and Pliny for neglecting to provide proper guidelines for this field. True, Aristotle offered a variety of skills that Bacon drew upon. Yet in each instance Bacon qualified his praise by condemning the result. Aristotle had offered a method for philosophizing but his emphasis on universals did not give particulars their due. He had written, Bacon acknowledged, "so accurate a history of animals" yet neglected to provide a criteria for evidence that measured up to Bacon's legalistic standards.[41] He had even given Bacon the inspiration for his own approach to past authors by generally refusing to mention them save to criticize them.[42] Yet in doing so Aristotle had set himself up as a tyrannical authority, becoming one of the idols that Bacon hoped to topple. Despite his many disagreements with his mentor Plato, Aristotle had not moved far enough away from his point of origin. Like Plato he created "light" philosophy that did not rigorously seek truth as its final goal: "scraps of borrowed information polished and strung together."[43] While admiring Aristotle's political skills that had won him the patronage of Alexander the Great, a feat Bacon hoped to emulate, the English natural philosopher discarded the vast majority of his natural history, tainted by its Platonic origins and its predilection for syllogisms.

Pliny offered Bacon a different set of problems. On the one hand, the Roman naturalist had recognized the importance and the scope of his discipline. In *The Advancement of Learning*, Bacon uncharacteristically halted his constant criticism of the ancients to praise Pliny as "the only person who ever undertook a Natural History according to the dignity of it; though he was far from carrying out his undertaking in a manner worthy of the conception."[44] Bacon's kinship with Pliny lay in more than just the curious circumstances of their deaths. Both shared an encyclopedic and operative view of natural history.

Like Pliny, Bacon imagined natural history to cover all fields of knowledge. Pliny's decision to include such subjects as agriculture, the arts and inventions in his *Natural History* provided an important precondition to the third part of Bacon's natural history: things artificial. His emphasis on the little things in nature also appealed to Bacon's preference for the neglected and unglamorous aspects of natural history. Yet Pliny even more than Aristotle erred in his approach to natural history. Systematic in his collection of information, he was indiscriminate in his inclusion of every improbable tale of nature's wonders.

Bacon's meditations upon the ancient creators of natural history led him to speculate about the historical evolution of this discipline. History was particularly important to him because he, like many younger Elizabethans, perceived his age to be crucially different from the past, the dawn of a new historical consciousness.[45] As Bacon observed, Aristotle's disdain for particulars and Pliny's preference for curiosities had had a long-lasting impact on natural history. Notwithstanding Pliny's diligent accumulation of information, his "twenty thousand noteworthy facts" were the beginnings of a culture of natural history that had no consistent criteria of evidence. What few cautions Aristotle had offered about prioritizing information disappeared in Pliny's enthusiasm for quantity rather than quality. Anything and everything entered Pliny's *Natural History* and very rarely did Pliny display skepticism regarding the veracity of these reports. The impact of Pliny's approach was made fully apparent in Bacon's assessment of Renaissance natural history. "Nothing duly investigated, nothing verified, nothing counted, weighed, or measured, is to be found in natural history: and what in observation is loose and vague, is in information deceptive and treacherous."[46] Centuries of developing natural history as more of a literary than a scientific pursuit had rendered the desire to make it a repository of truth marginal to its construction. This was precisely why Bacon so despised the cultivation of curiosities.

Bacon's history of natural history moved quickly from an exploration of ancient writings to a broad-based indictment of contemporary society. As he argued, the perilous moral state of society at the end of the century had brought natural history to a critical point in its uneasy development; it had become so tainted by current intellectual practices that it was on the verge of collapse. The culprits in the disintegration of knowledge were not the ancients per se nor even the excessive desire to imitate them, but the undue emphasis placed on wit in Renaissance culture. Wit became Bacon's primary nemesis, the source of all the trivial arts that he hoped to expunge from the pursuit of knowledge. Its importance in late Renaissance society and its historical validation by the ancients had given full reign to the powers of the imagination, rewarding philosophers who invented the most fantastic and elaborate systems of knowledge rather than those who approached the truth. In Bacon's opinion, this accounted for the popularity of many naturalists, philosophers and magicians. Creators of pleasing genres, they had entertained the upper classes for centuries with their speculations. To signal how different his own work would be, Bacon announced the dawn of a new era, "a birth of Time rather than a birth of Wit."[47] The historical consciousness of this new age not only made it critical of previous efforts to study nature but self-conscious of the limits of human ingenuity. Time provided a certain perspective which wit served to obscure. That perceived clarity of vision provided the basis from which Bacon launched a full-scale critique of Renaissance natural history. Optimistically he imagined Jacobean society as one not wholly given over to pleasure and excess and therefore an appropriate setting in which to reclaim nature from culture.

Bacon's dislike of the current state of natural history stemmed particularly from the evolution of what William Ashworth has aptly dubbed the "emblematic world view."[48] The humanist embellishments that expanded the scope of natural history in the late sixteenth and early seventeenth centuries certainly heightened the literary qualities of Pliny's discipline; they were central to the work of such well-known naturalists as Gesner, Aldrovandi and their English imitators. Natural history had become an elaborate enterprise whose popularity derived not so much from its subject as its manner of presentation; in the process nature had been *denatured*. Bacon's goal entailed an absolute rejection of the highly rhetorical style of his contemporaries; the disciplining of language was one of the initial steps in the reclamation of nature. In a famous passage in his *Preparative Towards a Natural and Experimental History*, he declared:

> First then, away with antiquities, and citations of testimonies of authors; also with disputes and controversies and differing opinions; everything in short which is philological. Never cite an author except in a matter of doubtful credit: never introduce a controversy unless in a matter of great moment. And for all that concerns ornaments of speech, similitudes, treasure of eloquence, and such like emptinesses, let it be utterly dismissed.[49]

Such sentiments were the logical culmination of Bacon's desire to reform the university curriculum so that rhetoric and logic would be taught *after* a sound introduction to natural philosophy rather than before as was traditionally done. Here and elsewhere Bacon attacked the trivium (grammar, rhetoric and logic), arguing that the raw ingredients of nature rather than the language in which they were presented and the format in which they were debated formed the core of natural knowledge.[50] They were the core of the trivializing arts that Bacon hoped to eliminate from science precisely because they were the forum in which free license had been given to the imagination.

Bacon's critique of the style of Renaissance natural history prefigured his dismissal of its content. As he was quick to observe, natural history done for its own sake could be pleasing but it served no higher purpose. Once natural history became the foundation of natural philosophy, however, it no longer could exist in an undisciplined state. Purged of its vices, the ancient encyclopedia of nature seemed on the verge of collapse, supplanted by a new encyclopedia based on a new, more evidentiary definition of history. "For I well know that a natural history is extant, large in its bulk, pleasing in its variety, curious often in its diligence; but yet weed it of fables, antiquities, quotations, idle controversies, philology and ornaments . . . and it will shrink into a small compass. Certainly it is very different from that kind of history which I have in view."[51] In opposition to the humanist encyclopedists who saw the expansion of the literary content of natural history as their goal, Bacon instead proposed to separate the fictive aspects of this field of knowledge from its factual components. The remnants of the latter would form the core of his own natural history.

In his discussion of the theoretical framework that informed Renaissance natural history, Bacon singled out two practices as particularly harmful to the search for truth: the desire to highlight nature's curiosities and the predilection for occult explanations. While Bacon considered a collection of monsters and other errors of nature—the second division in his tripartite natural history— essential to a proper natural history, he did not believe that this should include the phenomena that contemporaries, following Pliny, had called the "jokes of nature" (*lusus naturae*). Renaissance naturalists' predilection for *lusus* as an explanatory category for a wide range of difficult phenomena, from fossils to loadstones to snowflakes, struck Bacon as one of the most egregious examples of making the trivial important. "For small varieties of this kind are only a kind of sports and wanton freaks of nature; and come near to the nature of individuals," he observed. "They afford a pleasant recreation in wandering among them and looking at them as objects in themselves; but the information they yield to the sciences is slight and almost superfluous." Elsewhere he observed that they were of no "serious use towards science."[52] The lack of differentiation between play and science, in Bacon's view, led to the valorization of fantastic creatures such as the Scythian lamb—a woolly zoophyte that allegedly survived by eating the grass around it. Instead Bacon called the vegetable lamb a "fabulous notion," testimony to human ingenuity rather than the powers of observation.[53]

Bacon's attack on the jokes of nature signaled the beginnings of a wholesale assault on any explanations that made the external appearance of nature a hieroglyph of some deeper and hidden meaning. He linked his criticisms to a moral indictment of intellectual pleasure, labeling the jokes of nature "wanton" and "lascivious."[54] The moral turpitude of the encyclopedists became a general point of departure for his attack on any philosopher who sought to make nature obscure. In his *Thoughts and Conclusions*, for example, Bacon chastised those who "made a cult of the incomprehensibility of nature."[55] The Paracelsian doctrine of signatures, which Bacon singled out for particular scorn, embodied all the problems of this approach to nature. Its elaborate system of occult correspondences that made all resemblance meaningful not only confused the natural and the divine, two categories Bacon hoped to separate, but also conflated real relationships among the different parts of nature with false connections. Summing up his disgust with Paracelsian natural philosophy, Bacon wrote: "It is we who deserve pity for spending our time in the midst of such distasteful trivialities."[56]

For Bacon, the ideas of Paracelsus and their elaboration in the work of disciples such as Oswald Croll signaled the beginning of an expanded discourse about nature that exemplified the full degeneration of learning over time. Bacon's ambivalence about the ancients—who, however misguided they were, were nonetheless the originators of human knowledge—stands in marked contrast to his wholesale attack on foreign natural philosophers who had contributed to symbolic approaches to nature in the preceding century. With the exception of a few criticisms of Gilbert, who had died by the time Bacon began to publish his work, he was noticeably and prudently silent about the English. Rather than mentioning figures such as Dee and Moffett—men with whom he surely had some contact in and around London—Bacon used his detailed critiques of foreign philosophers to air his concerns about the direction of intellectual life in England.

While Plato had at least refined the art of conversation, Bacon felt that his Renaissance counterparts offered no such skills. Agrippa was "a trivial buffoon" whose initial praise of the occult sciences and his subsequent condemnation of them only highlighted how little he knew. Telesio, Fracastoro, Bruno and Gilbert were all writers of comedies, a term Bacon frequently used to signify the inconsequentiality of a given philosophy. Cardano was a spider weaving his webs in which to capture innocent readers, unaware of how little the ideas in his *On Subtlety* (1550) and *On the Variety of Things* (1557) owed to nature.[57] The very images of nature summoned forth in Cardano's titles became watchwords in Bacon's critique of learning. In *The Advancement of Learning* Bacon characterized contemporary knowledge as "subtle, idle, unwholesome," linking Cardano's philosophy with leisurely pursuits and moral corruption. In the *Novum Organum* he criticized naturalists for their "great and indeed overcurious diligence in observing the variety of things. . . ."[58] The addition of yet another subtle phenomenon or one more account of variety in the encyclopedia of nature moved natural history further away from its proper but heretofore unrecognized goal, the building of natural philosophy. Instead Bacon proposed a natural history that would describe the fundamental properties of nature, which is surely why he picked such subjects as hot and cold, winds and sound in his list of proposed natural histories.

Bacon's distaste for contemporary natural history stemmed from a broader dislike of what Patricia Fumerton has called "the trivial selfhood of the aristocracy in the English Renaissance." From this vantage point, nature had become one of the many devices through which elites expressed their identity, weaving elaborate natural imagery into their clothes, banquets and masques, and privileging exotica—the curiosities that Bacon so despised—as a form of cultural capital. "Trivial arts *saturate* this world with the pretty clutter of the fragmentary, peripheral, and ornamental," writes Fumerton.[59] Thus Bacon's critique of variety had an additional layer of meaning. In an age in which the court physician Thomas Moffett could devote an entire poem to the silkworm and Ovid's *Metamorphoses* could be served up as dessert, natural history was much more an extension of culture rather than nature; as Bacon stated in his *Description of the Intellectual Globe*, it was indeed "exquisite knowledge."[60]

The last decades of the sixteenth century made this frivolous taste for nature materially apparent as some of the early financiers of English forays abroad began to accrue "queer foreign objects" like the ones Thomas Platter found in Sir Walter Cope's home in London in 1599; Cope not coincidentally was a good friend of Bacon's cousin Cecil.[61] The cabinets of curiosities created by English aristocrats were, in Bacon's view, the ultimate trivialization of natural history, turning natural objects into ornaments that enhanced one's social standing but did not further knowledge (what Bacon described as "the most splendid costly"). Bacon imagined this approach to nature as the product of prolonged female rule; lacking a firm leadership it was no wonder that natural history, like English society, "indulges to excess in matters superfluous. . . . " The feminization of knowledge also accounted for the current distaste for the ordinary and ignoble parts of nature: "such fastidiousness being merely childish and effeminate."[62]

Much to Bacon's dismay the abuse of nature's material wealth only accelerated under the reign of James I. In 1625 we find John Tradescant the Elder writing to the secretary of the Navy to request, on behalf of the Duke of Buckingham, "Any thing that Is strang."[63] As Bacon would have argued, so great was the power of female rule and so seductive was the delight in the trivial that even the arrival of a male monarch could not quickly dissipate its influence. In a famous *Letter of Advice* written to Buckingham in 1616, Bacon pronounced: "Let the vanity of the times be restrained. . . . " While masques and revels—the very medium in which Bacon had initially presented his ideas to the Elizabethan court in the 1590s—were appropriate for a queen's court, he argued kings needed recreation of a "more manly and useful deportment."[64] As we know, Bacon was unsuccessful in his plea to make the Jacobean court less enamored of the trivial pursuits that had flourished under Elizabeth; the Earl of Arundel to whom Bacon directed his last letter was surely the greatest collector of paintings and curiosities in his day. Herein lay the failure of Bacon's carefully laid plans, conceived in the final decade of Elizabeth's rule when many of the younger generation of courtiers were hopeful about what the arrival of a new monarch might bring. Despite James I's initial revulsion at the public extravagances of Elizabethan ceremonial, the number of masques increased, virtuoso culture thrived and, as a result, nature became even more curious.[65]

## Sober Pursuits

Natural history is "not meant to be pleasant," declared Bacon, in what was probably his most succint and humorous statement of his philosophy.[66] Bacon's attempts to tame natural history were indeed a direct response to the place of science in Elizabethan England. On the one hand, the utility of natural knowledge had become apparent by the late sixteenth century as the English ventured further and further away from their own shores, fulfilling Bacon's vision of the naturalist-explorer, a new Columbus who circled the globe in search of knowledge. On the other hand, neither the Elizabethan nor the Jacobean aristocrats to whom Bacon addressed his work saw the increased importance of natural history in their mercantile adventures as a reason to discard the courtly uses of nature that enhanced the humanist tradition of natural history. As a result, few of Bacon's contemporaries were sympathetic to the radical reformation of knowledge that he proposed. Even the universities seem to have been reluctant to accept his offer to revise their curriculum.

Despite the lack of encouragement Bacon persisted in his work to the end of his life; neither political disgrace nor intellectual isolation convinced him that the English monarch and his favorites would be unreceptive to his ideas. In concluding this examination of Bacon's attitude towards natural history, we should ask ourselves what the origins were of his commitment to the sober and disciplined pursuit of knowledge. First, we need to consider Bacon's social position. A member of the House of Commons who appeared on Elizabeth's lists of New Year's gift givers as a "gentleman," Bacon did not belong to the highest stratum of society but to the level below it. Not until after Elizabeth's death was he knighted, in 1603; the series of preferments that James offered him from then until his impeachment for accepting bribes in 1621 were never enough to compensate for his perceived sense of inferiority by the rules of the court. Certainly this alone and the burdens of being the younger and unexpectedly impoverished son of the Lord Keeper of the Seal might have led Bacon to exhibit what his biographer Rawley called a "gravity and maturity above his years."[67]

Bacon's religious upbringing surely played an important role in his attitude towards nature. While rejecting his mother's Puritanism, Bacon nonetheless approached nature as a second Scripture. Mishandling this book of God was akin to blasphemy if not heresy; the philosophers who spun fanciful tales had been possessed by the devil and therefore needed an exorcist. Describing himself as a "more cautious purveyor" of nature's facts, Bacon wrote: "I interpose everywhere admonitions and scruples and cautions, with a religious care to eject, repress, and as it were exorcise every kind of phantasm."[68] This attitude mirrored well Bacon's distaste for enthusiasm of any kind which was, in its origin, a religious principle. Thus the "severe and original inquisition of knowledge" arose not simply from Bacon's legalistic standards of evidence but more deeply from his conviction that nature like scripture could yield one truth. "[W]hereas to find the real truth requireth another manner of severity," he wrote in *The Advancement of Learning*.[69]

The extent to which Bacon described the gathering of disciples as a conversion experience and the disciplining of their minds as a rite of initiation suggests that the model for his scientific community lay somewhere between a reformed religious order and an ideal polity, an apt image for an Elizabethan discontent with the yoke of female rule. "You know I am no courtier," wrote Bacon in his 1616 letter to Buckingham and, in a sense, this was true since he seems to have been profoundly uncomfortable with many aspects of court life, most notably its delight in the fragmentary.[70] His philosophy, he wrote, "teaches the people to assemble and unite and take upon them the yoke of laws and submit to authority, and forget their ungoverned appetites."[71] The restoration of order lay at the heart of Bacon's program and this was precisely why knowledge and the knowledge-making community needed to be disciplined. How else could one create "fit assistants"?[72] Throughout his works Bacon mixed political and religious metaphors, portraying himself both as a priest of nature and as one of the "faithful secretaries" in the cabinet of knowledge whose civic task was to record the laws of nature.[73] What has not been fully understood is the extent to which these images were directed at James I who also enjoyed the image of himself as Solomon well before Bacon publicized it in his writings. With James lay the hope that natural history—and therefore natural philosophy—might amount to something. James alone did not provide the impetus for Bacon's ideas, which were formed at least a decade before he ascended to the English throne, but he was certainly the reason that Bacon could announced with greater confidence after 1603: "It is time...."[74]

As is well known, Bacon's disciplining of natural history had little immediate effect. Not until the 1640s did natural philosophers take up his call and not until the 1660s, with the inauguration of the Royal Society and the crowning of Bacon as the society's intellectual father, did Baconianism find an institutional niche in early modern scientific culture.[75] From this perspective, we might justly view Bacon's original program to be of short-lived and somewhat isolated duration since the Royal Society cared not a whit for the intricacies of Elizabethan and Jacobean politics. Yet I nonetheless think that we need to take Bacon's program of reform seriously because it offered such a powerful model for the creation of disciplines in the seventeenth and eighteenth centuries. The process by which Bacon reinvented natural history became a template for the history of such disciplines as philosophy, physics and chemistry, for as natural philosophy ceased to be the rubric under which all these forms of inquiry were subsumed, it was the story of the discipline itself that remained. We may never fully understand why Bacon chose natural history as the first subject upon which to enact this process; as I have suggested throughout this essay, most likely its unusual position between the universities and the court, between natural philosophy and natural magic, between the old forms of knowledge and the new, made it a particularly attractive choice. In the course of the seventeenth century natural history was to become the dominant discipline in the pursuit of scientific knowledge; every major academy, from the Accademia dei Lincei to the Royal Society to the Paris Academy of Sciences, and many minor ones made natural history their common project, implicitly confirming Bacon's choice of subject.[76] In doing so, they also revised the structure of natural history, inspired by the power of this discipline as a framework for new and empirical knowledge.

## Notes

1. Francis Bacon, *The Masculine Birth of Time* (1603), in Benjamin Farrington, *The Philosophy of Francis Bacon* (Chicago: The University of Chicago Press, 1964), 62 [hereafter cited as Farrington].
2. *The Letters and the Life of Francis Bacon*, ed. James Spedding (London: Longman, 1861–74), 7:550.
3. Bacon, *Filum Labyrinti, sive Formula Inquisitionis*, in *The Works of Francis Bacon*, ed. James Spedding, Robert Leslie Ellis, and Douglas Denon Heath (London: Longman, 1857-58), 3:499 [hereafter cited as *Works*]. Bacon's politicized and public view of science is briefly outlined in Julian Martin, "Natural Philosophy and Its Public Concerns," in *Science, Culture and Popular Belief in Renaissance Europe*, ed. Stephen Pumfrey, Paolo L. Rossi, and Maurice Slawinski (Manchester, Eng.: Manchester University Press, 1991), 100–18.
4. Paolo Rossi, *Francis Bacon: From Magic to Science*, trans. Sasha Rabinowitz (Chicago: University of Chicago Press, 1968), 215–16; Graham Rees, "Francis Bacon's Semi-Paracelsian Cosmology and the *Great Instauration*," *Ambix* 22 (1975): 168.
5. Bacon, *Novum Organum*, in *Works*, 1:152.
6. Bacon, *Parasceve*, in *Works*, 4:252.
7. Among the books on Bacon, the most useful for my research have been: John E. Leary, Jr., *Francis Bacon and the Politics of Science* (Ames, IA: Iowa State University Press, 1994); Julian Martin, *Francis Bacon, the State, and the Reform of Natural Philosophy* (Cambridge, Eng.: Cambridge University Press, 1992); Antonio Pérez-Ramos, *Francis Bacon's Ideal of Science and the Maker's Knowledge Tradition* (Oxford: Clarendon Press, 1988); Rossi, *Francis Bacon*; James Stephens, *Francis Bacon and the Style of Science* (Chicago: University of Chicago Press, 1975); Brian Vickers, ed., *Essential Articles for the Study of Francis Bacon* (Hamden, CT: Archon Books, 1968); and B.H.G. Wormald, *Francis Bacon: History, Politics and Science, 1561–1626* (Cambridge, Eng.: Cambridge University Press, 1993).
8. There is a lively debate about whether or not natural history actually was a "discipline" because of its amorphous intellectual character. Certainly it was never a strong discipline in the sense that it delimited a carefully circumscribed field of knowledge; for this reason, it declined markedly in the late eighteenth and nineteenth centuries, as such scientific fields as botany, biology, comparative anatomy, geology and paleontology emerged. Yet I have preferred to designate natural history as an early modern discipline because naturalists from this period rarely confined themselves to the study of only *one* domain of nature (e.g. botany, zoology, mineralogy) but tended to work across these fields, seeing the historical description of nature as a unified endeavor. This is also apparent in the evolution of professorships in natural history from the mid-sixteenth century onwards, which also demanded the teaching of several areas of natural history. For an interesting discussion of the emergence of disciplines, see Rudolf Stichweh, "The Sociology of Scientific Disciplines: On the Genesis and Stability of the Disciplinary Structure of Modern Science," *Science in Context* 5 (1992): 3–15. Thanks to Alix Cooper for this reference.
9. Bacon *Sylva sylvarum*, in *Works*, 2:378. For a discussion of Bacon's view of natural magic, see William Eamon, *Science and the Secrets of Nature: Books of Secrets in Medieval and Early Modern Culture* (Princeton: Princeton University Press, 1994), 285–91.
10. The connection between Bacon's politics and science has been critiqued in a very interesting article: Markku Peltonen, "Politics and Science: Francis Bacon and the True Greatness of States," *The Historical Journal* 35 (1992): 279-305. Peltonen sees an essential contradiction between the necessity of war in achieving political greatness, a national goal, and the necessity of peace in advancing science, an international goal. I would argue that this practical distinction did not dilute the metaphorical equivalence Bacon saw between success in the realms of politics and knowledge. His successful version of natural philosophy had all the masculine and warlike characteristics that Peltonen attributes solely to Bacon's politics. In other words, the fact that Bacon used similar language for both subjects is worth noting, and potentially suggests even deeper affinities.

Also Bacon's insistence on making science a royal project indicates that he imagined a role for the state in the reformation of knowledge.

11. Bacon, *Thoughts and Conclusions*, in Farrington, 99, 93; idem, *Novum Organum*, as discussed in Lorraine Daston, "The Moralized Objectivities of Nineteenth-Century Science" (unpublished paper). John Leary has also noted Bacon's emphasis on discipline in his *Francis Bacon*, 147, 166, 195; see also Farrington, 18.

12. This approach to knowledge resonates well with some of the features of seventeenth-century English science noted in the work of Robert Boyle, who was influenced by Bacon's work; see Steven Shapin, *A Social History of Truth: Civility and Science in Seventeenth-century England* (Chicago: University of Chicago Press, 1994).

13. Bacon, *The Masculine Birth of Time*, in Farrington, 61.

14. Bacon, *Cogitationes de Scientia Humana*, in *Works*, 3:187; *Valerius Terminus*, in *Works*, 3:285.

15. Mordechai Feingold, *The Mathematician's Apprenticeship: Science, Universities and Society in England 1560–1640* (Cambridge, Eng.: Cambridge University Press, 1984), 33; Charles Webster, *The Great Instauration: Science, Medicine and Reform 1626–1660* (London: Ducksworth, 1975), 125; Phyllis Allen, "Scientific Studies in the English Universities of the Seventeenth Century," *Journal of the History of Ideas* 10 (1949): 239.

16. Bacon, *The Advancement of Learning*, in *Works*, 4:287.

17. Allen, "Scientific Studies," 220; F.D. and J.F.M. Hoeniger, *The Development of Natural History in Tudor England* (Washington, D.C.: Folger Books, 1969), 43.

18. Nonetheless when Bacon himself contemplated endowing chairs at Oxford and Cambridge—a plan he was unable to realize because of his financial debts—he considered natural philosophy the more appropriate subject; see Barbara Shapiro, "The Universities and Science in Seventeenth Century England," *Journal of British Studies* 10 (1971): 51.

19. For an overview of this subject, see Paula Findlen, "Courting Nature," in *Cultures of Natural History*, ed. Nicholas Jardine, James Secord, and Emma Spary (Cambridge, Eng.: Cambridge University Press, 1995), 57–75 (also other essays in this volume). Probably the strongest version of the natural history tradition evident in England before the seventeenth century regarded the role of herbals and other vernacular medical books that often contained lore about plants, animals, and minerals.

20. Feingold, *The Mathematician's Apprenticeship*, 18. For a comparative view of continental natural history, see Findlen, *Possessing Nature: Museums, Collecting and Scientific Culture in Early Modern Italy* (Berkeley: University of California Press, 1994).

21. Antonia Maclean, *Humanism and the Rise of Science in Tudor England* (London: Heinemann, 1972), 220.

22. For a discussion of the literary impact of the Americas on English writing, see Jeffrey Knapp, *An Empire Nowhere: England, America and Literature from Utopia to The Tempest* (Berkeley: University of California Press, 1992).

23. In his *Comentarius solutus sive pandecta, sive ancilla memoria* (1608), Bacon noted that he wished to talk with Ralegh, Harriot and Northumberland, "themselves being already inclined to experim[en]ts"; James Spedding, *The Letters and the Life of Francis Bacon*, 4:63. On Harriot's and White's natural history of Virginia, see Paul Hultan, *America 1585: The Complete Drawings of John White* (Chapel Hill, NC, 1984).

24. Bacon, *Parasceve*, in *Works*, 4:252.

25. Bacon, *Works*, 3:322; 4:101, 251.

26. Maclean, *Humanism and the Rise of Science*, 213–14, 221. On Cecil as patron, see Feingold, *The Mathematician's Apprenticeship*, 204.

27. Space does not permit a discussion of the ways in which the legal culture of Renaissance England may have also contributed to Bacon's choice of natural history as the most evidentiary form of science. This subject has been treated well elsewhere; see Shapiro, "The Concept 'Fact': Legal Origins and Cultural Diffusion," *Albion* 26 (1994): 1–26; and also his *Beyond "Reasonable Doubt" and "Probable Cause": The Anglo-American Law of Evidence* (Berkeley: University of California Press, 1991).

28. John Nichols, *The Progresses and Processions of Queen Elizabeth* (London: John Nichols and Son, 1823), 3:441.

29. Martin, *Francis Bacon*, 28. Mea Allen, *The Tradescants: Their Plants, Gardens and Museum 1570–1662* (London: Michael Joseph, 1964), 36.

30. Hoeniger, *Natural History in Tudor England*, 55–56.

31. John Dixon Hunt, *Garden and Grove: The Italian Renaissance Garden in the English Imagination: 1600–1750* (London: J. M. Dent and Sons, 1986), esp. 104–5.

32. For a discussion of the natural symbolism in Elizabeth's costumes, see Roy Strong, *Gloriana: The Portraits of Queen Elizabeth I* (London: Thames and Hudson, 1987), esp. 151, 157–60.

33. Nicholls, *Progresses and Processions*, 3:457.

34. Neville Williams, *All the Queen's Men: Elizabeth I and Her Courtiers* (New York: Macmillan, 1972), 75; Nichols, *Progresses and Processions*, 55–59. The splendors of Gorhambury are also described in Anthony Esler, *The Aspiring Mind of the Elizabethan Younger Generation* (Durham, NC: Duke University Press, 1966), 185.

35. Bacon, *Parasceve*, in *Works*, 4:254.

36. This aspect of my discussion of Bacon is especially indebted to Evelyn Fox Keller, "Baconian Science: The Arts of Mastery and Obedience," in her *Reflections on Gender and Science* (New Haven: Yale University Press, 1985), 33–42. Fox Keller argues that Bacon's view of knowledge is hermaphroditic, and that Bacon exercised a strong desire to suppress the feminine aspects of knowledge in his call for a masculine science.

37. idem, *Valerius Terminus*, in *Works*, 3:288.

38. Ibid., 290.

39. Bacon, *Novum Organon*, in *Works*, 4:247–48.

40. Spedding, *Letters and the Life of Francis Bacon*, 4:147.

41. Bacon, *Novum Organum*, in *Works*, 4:94. For a discussion of Bacon's approach to evidence, see especially Shapiro, *Probability and Certainty in Seventeenth Century England* (Princeton: Princeton University Press, 1983); and Daston, "Baconian Facts, Academic Civility, and the Prehistory of Objectivity," *Annals of Scholarship* 8 (1991): 337–63.

42. Bacon, *The Refutation of Philosophies* (1608), in Farrington, 113.

43. Bacon, *The Masculine Birth of Time*, in Farrington, 64.

44. Bacon, *The Advancement of Learning*, in *Works*, 4:287.

45. Stephen L. Collins, *From Divine Cosmos to Sovereign State: An Intellectual History of Consciousness and the Idea of Order in Renaissance England* (Oxford: Oxford University Press, 1989), 34. Certainly I would not wish to argue that Bacon's interest in history was purely generational. It also had a strong intellectual component, emanating from his close readings of such authors as Machiavelli, and was a tool through which to proscribe political as well as intellectual behavior.

46. Pliny, *Natural History*, pref.; Bacon, *Novum Organum*, in *Works*, 4:94. This is another instance in which Bacon's legal assessment of natural history is apparent.

47. Bacon, *Novum Organum*, in *Works*, 4:77.

48. William B. Ashworth, Jr., "Natural History and the Emblematic World View," in *Reappraisals of the Scientific Revolution*, ed. David C. Lindberg and Robert S. Westman (Cambridge, Eng.: Cambridge University Press, 1990), 303–32.

49. Bacon, *Parasceve*, in *Works*, 4:254.

50. Bacon was one of the first authors to use "trivial" in its modern sense (inconsequential, unimportant). *The Compact Edition of the Oxford English Dictionary*, 2:3412 cites Thomas Nashe (1589) and William Shakespeare (1593) as the earliest sources for this usage.

51. Bacon, *The Advancement of Learning*, in *Works*, 4:299 (see also 4:508).

52. Bacon, *Parasceve*, in *Works*, 4:255; *Novum Organum*, in *Works*, 4:166. For a more in-depth discussion of the evolution of the category *lusus naturae*, see Paula Findlen, "Jokes of Nature and Jokes of Knowledge: The Playfulness of Scientific Discourse in Early Modern Europe," *Renaissance Quarterly* 43 (1990): 292–331.

53. Bacon, *Sylva sylvarum*, in *Works*, 2:531.

54. Bacon, *Phenomena universi*, in *Works*, 3:688; *Parasceve* and *Descriptio globi intellectualis*, in *Works*, 4:255, 508.

55. Bacon, *Thoughts and Conclusions*, in Farrington, 127.

56. Bacon, *The Masculine Birth of Time*, in Farrington, 66.

57. Ibid., pp.63, 70, 85.

58. Bacon, *Advancement of Learning*, in *Works*, 3:285; *Novum Organum*, in *Works*, 4:166.

59. Patricia Fumerton, *Cultural Aesthetics: Renaissance Literature and the Practice of Social Ornament* (Chicago: University of Chicago Press, 1991), 1, 171.

60. Thomas Moffet, *The Silkwormes and Their Flies* (1599), ed. Victor Houliston (Binghamton: Medieval and Renaissance Texts and Studies, 1989); on Ovid, see Nichols, *Progresses and Processions*, 2:123, n.3. Bacon, *Descriptio globi intellectualis*, in *Works*, 4:508.

61. Arthur MacGregor, "The Tradescants as Collectors of Rarities," in idem, ed., *Tradescant's Rarities: Essays on the Foundation of the Ashmolean Museum 1683 with a Catalogue of the Surviving Early Collections* (Oxford: Clarendon Press, 1983), 17.

62. Bacon, *Novum Organum* and *Descriptio globi intellectualis*, in *Works*, 4:106, 508.

63. MacGregor, "The Tradescants as Collectors of Rarities," in idem, *Tradescant's Rarities*, 20.

64. Spedding, *Letters and Life of Francis Bacon*, 6:23, 25. Bacon also attacked masques as "but Toyes" in his *Essays*; Fumerton, *Cultural Aesthetics*, 237.

65. For an overview of the Jacobean court, see R. Malcolm Smuts, *Courtly Culture and the Origins of a Royalist Tradition in Early Stuart England* (Philadelphia: University of Pennsylvania Press, 1987), esp. 104–6, 153; on virtuoso culture, the best study remains Walter E. Houghton, Jr., "The English Virtuoso in the Seventeenth Century," *Journal of the History of Ideas* 3 (1942): 51–73, 190–219.

66. Bacon, *Parasceve*, in *Works*, 4:255.

67. William Rawley, *The Life of the Right Honourable Francis Bacon* (published 1670), in *Works*, 1:4.

68. Bacon, *Great Instauration*, in *Works*, 4:30. On nature as scripture, see his *Parasceve*, in *Works*, 4:261.

69. Bacon, *The Advancement of Learning*, in *Works*, 3:365.

70. Spedding, *Letters and the Life of Francis Bacon*, 6:27.

71. Bacon, "Orpheus; or Philosophy," in *Works*, 1:722, as quoted in Collins, *From Divine Cosmos to Sovereign State*, 146.

72. Spedding, *Letters and the Life of Francis Bacon*, 7:376.

73. Bacon, *Parasceve*, in *Works*, 4:262.

74. Bacon, *Thoughts and Conclusions*, in Farrington, 91.

75. On this subject, see Theodore M. Brown, "The Rise of Baconianism in Seventeenth Century England," in *Science and History: Studies in Honor of Edward Rosen*, ed. Erna Hilfstein, Pawel Czartoryski, and Frank D. Grande, Studia Copernicana, 16 (Wroclaw: The Polish Academy of Sciences, 1978), 501–22; Shapiro, *Probability and Certainty*, esp. 66–73; and Webster, *The Great Instauration*.

76. For more on this subject, see Shapiro, "History and Natural History in Sixteenth- and Seventeenth-century England," in Shapiro and Robert G. Frank, *English Scientific Virtuosi in the Sixteenth and Seventeenth Centuries* (Los Angeles: William Andrews Clark Memorial Library, 1979), 1–55; Joseph M. Levine, "Natural History and the Scientific Revolution," *Clio* 13 (1983): 519–42; Alice Stroup, *A Company of Scientists: Botany, Patronage, and Community at the Seventeenth-Century Parisian Royal Academy of Sciences* (Berkeley: University of California Press, 1990); and Jardine, Secord, and Spary, *Cultures of Natural History* (Harold Cook's article lists a number of his thoughtful essays on this subject for England and the Netherlands).

# From Apotheosis to Analysis
## Some Late Renaissance Histories of Classical Astronomy

Anthony Grafton

At the end of March 1609 Johannes Kepler finished writing his *Astronomia nova*. On the title page he inscribed a short quotation from the *Scholae mathematicae* of Petrus Ramus. There Ramus had proposed a reform of astronomy—a reform based on a polemical history of the exact sciences. He argued that the wise priests of ancient Egypt and Babylon had created an infallible astronomy by devoting themselves to direct observation of the stars and precise recording of the results. They devised no hypotheses and constructed no geometrical models of the movement of the planets. By doing so, they created an astronomy which enabled them to predict the positions of stars and planets at any given point, precisely and effortlessly.

Unfortunately, Ramus explained, the Greeks and Romans had conquered the East and appropriated the results of eastern science. Worse still, they had transformed these, making the lucid, simple and exact empirical astronomy of the priests on their pyramids into a chaotic, complicated and imperfect science. Ramus identified Hipparchus and Ptolemy as the chief sinners. They had devised a geometrical method, according to which a set of imaginary crystalline spheres, revolving at set rates, generated the motion of each planet. The models they created were sophisticated. But their search for a geometrical world system that consisted entirely of spheres and would enable them to predict planetary motions necessarily ended in failure. Unlike the near eastern priests, they did not collect the vast amounts of precise data needed to give their science a foundation. Their effort to develop models distracted them from their proper job.

These problems, Ramus argued, had long been evident. As early as the fifth century AD, Proclus revealed significant gaps and contradictions in classical astronomy. Nonetheless, astronomers in Islam and the west continued to insist, over the centuries, on developing their science in the geometrical manner first imposed on it by the Greeks. Copernicus went so far as to speculate as to whether the sun stood still and the earth moved with the other planets—a hypothesis that Ramus found completely bizarre. Ramus promised the scholar who succeeded in reconstructing the ancient "astronomy without hypotheses" a chair at the Collège du Roi in Paris, the innovative institution, dedicated to mathematics and the humanities, where he taught.[1]

Kepler commented on this passage scornfully and in some detail. He himself, he claimed, deserved the promised chair, on the basis of the discoveries he had just made about the motion of the planets. But there was no danger that he would make this claim good, for Ramus had died as one of the victims of the massacre of St Bartholomew's in 1572. Kepler, however, also took the trouble to set out a counter-history of astronomy—which suggests that he took a serious view of Ramus's influence, if not of his ideas. Copernicus, Kepler argued, devised no hypotheses: he simply described the world, in the conviction that he had discovered the truth. But the Nuremberg theologian Andreas Osiander, who saw the *De revolutionibus* through the press, disagreed. He added an anonymous forward in which he described the Copernican system as a hypothesis, hoping in this way to remove any suspicion of heresy from the book. In a copy of *De revolutionibus* bearing manuscript annotations by Hieronymus Schreiber, Kepler found the proof for his assertion that Osiander wrote the anonymous preface.[2] Then he went on to argue for the truth of Copernicanism. In any case, he pointed out, a reformation of astronomy must rest not only on the collection of all relevant data, but also on the continued application of the methods of Greek mathematics. Kepler revealed that Ramus's history of science, which traced the decline and fall of astronomy, had no solid basis. He himself, rather than the Egyptian priests, was the first astronomer who had the right to claim that he had developed an "astronomy without hypotheses." Astronomy, more generally, did not decline, but developed slowly over time, and even Ramus could not prevent this progress from continuing in the future.

More than ten years ago Nicholas Jardine rightly pointed out the originality and elegance of Kepler's historical work, showing that he combined considerable detective gifts as a researcher with massive intelligence as an interpreter of texts. Kepler, he argued, founded modern history of science as well as modern astronomy. Earlier scholars had, on the whole, traced this history in celebratory rather than analytical contexts: for example, in inaugural lectures at the outset of university courses on astronomy. Following well-established tradition, they had generally confined themselves to telling stories about Thales and his well— or his monopoly on olive presses—or to sending up a cloud of fine adjectives and metaphors. The point of such texts, after all, lay not in reconstructing the history of astronomy as it really was, but in offering legitimation for its study in the present, by providing it with the splendid genealogy that conferred social and cultural status in the Renaissance.[3] Ramus's theory mattered chiefly because it helped provoke Kepler to develop his counter-arguments.[4]

Jardine was certainly right to argue that Kepler studied the history of astronomy in a profound and original way. In fact, however, his predecessors and opponents reached suggestive results as well. Even Ramus' wildest speculations rested in part on an original reading of ancient sources, which he owed in part to a gifted astronomer.[5] And even Kepler's most critical analyses rested in part on researches carried out by an astrologer and natural philosopher widely regarded as eccentric or even mad. Remarkably enough, the astronomer and the natural philosopher knew each other, corresponded, met, became friends—and then became enemies. The present essay traces the history of their friendship, the development of their two histories of astronomy and astrology, and the impact of both on Kepler.

## Rheticus and Cardano

Kepler named both of his protagonists in the *Astronomia nova*. The astronomer, Georg Joachim Rheticus (1514–1576), he characterised as an erudite madman.[6] True, Rheticus had brought Copernicus's thesis to the attention of the public for the first time with a text that Kepler respected and even reprinted, the *Narratio prima*. Later, however, he tried to determine the movements of Mars. Kepler, proud of his own success in this notoriously difficult enterprise, mercilessly ridiculed his predecessor's effort to solve this technical problem with magical help:

> Once, when Rheticus found himself stuck fast in his observation of Mars, and could see no way out, he took refuge in consulting the oracle of his guiding spirit: perhaps in the intention of sounding its wisdom (in God's name), perhaps out of helpless longing for the truth. The irritable protector seized the troublesome researcher by the hair, smashed him against the ceiling above and then dropped him, so that his body fell to the floor. He then gave his oracular verdict: "that is the motion of Mars."[7]

Perhaps Kepler exaggerated or invented this account. Nonetheless Rheticus certainly did take an interest in various forms of magic as well as astronomy and mathematics.

The eccentric natural philosopher, by contrast, Kepler described as a gifted mathematician. In fact, he consistently cited the mathematical work of the Italian physician, astrologer and philosopher Girolamo Cardano (1501–1576) as profound and authoritative. Given Kepler's intolerance for Rheticus's eccentricities, this tolerant attitude occasions mild surprise. Cardano was and remains widely known for his efforts to preserve and continue the traditions of astrology and natural magic developed in fifteenth-century Italy. He dedicated himself to close study of such absorbing subjects as the magical properties of human fat. An Italian who lived through the years in which the Counter-Reformation gradually took hold and the intellectual climate of the peninsula became chillier and chillier, he nonetheless drew up a horoscope for Jesus—a venture that brought the Inquisition down upon him and cost him a long period of house arrest in Rome at the end of his life. Cardano's analysis of his own horoscope revealed the weaknesses of his character so vividly and immodestly that his seventeenth-century reader Gabriel Naudé, himself no model of Attic brevity or prudence, condemned him for bringing himself into universal discredit.[8]

The most serious attack on Cardano's status as a natural philosopher, moreover, took place long before Kepler wrote, and he knew the results very well. In 1557 Julius Caesar Scaliger devoted more than 900 pages to revealing some of the mistakes in Cardano's popular book *De subtilitate*. These *Exercitationes*, in turn, became a set book on natural philosophy in German universities, and belonged to the young Kepler's favorite reading matter.[9] In the 1620s he admitted that he had been carried away, as a young man, by his enthusiasm for Scaliger's speculations.[10] Despite Kepler's admiration for Scaliger's sharp style, rigorous method and bold philosophy, he also recognized Cardano's superiority as a mathematician—an area in which Scaliger lacked solid competence. Eventually, as we will see, Kepler also came to take Cardano seriously as a guide to the historical development of ancient science.

Cardano and Rheticus came into contact between 1538 and 1543. Cardano was preparing a second, improved edition of his astrological works, the *Libelli duo*. Already in the first edition of 1538 he had drawn up and commented on ten horoscopes in order to ground his astrological principles on examples. He now needed further materials. The Nuremberg publisher Johannes Petreius proposed to publish Cardano's Latin works, with additional matter if possible. Rheticus collaborated with Petreius—on the publication of Copernicus, for example. At some point he sent the Italian astrologer four horoscopes for the new edition.[11]

The two men's shared interest in astrological history may well have brought them together. In his *Supplementum Almanach* of 1538, Cardano argued that the movements of the sphere of the fixed stars determined the fates of cities and kingdoms. The head of Algol, for example, had spent 400 years above Greece and Asia, and caused the destruction of these lands by the Muslims. Now it hovered over Apulia and Naples, threatening Italy. By contrast, the star of the second magnitude at the end of the tail of Ursa Major had been over Rome at the time of the city's founding, and accounted for the martial temperament of the Roman people.[12] Rheticus, for his part, loved such schemata. He believed in the 1540s that the Copernican system contained a solution to the riddles of world history as well as those of astronomy. Fifteen years later, he would propose an astrological system quite like Cardano's in a dedicatory letter to the Emperor Ferdinand II.[13]

In any event, Cardano himself publicized his collaboration with the Protestant astronomer. In commenting on the horoscope of Savonarola that figured, with 66 others, in Petreius's 1543 edition of his *Libelli*, he wrote enthusiastically: "Who could describe more precisely what actually took place? One might be tempted to believe that this was made up, if my friend George Joachim had not sent me these last four horoscopes after I finished my little book. I saw to it that they were added."[14] The four horoscopes in question stated the birth times, characters and fates of Savonarola, Pico della Mirandola, Georg Peurbach and Albrecht Dürer. Clearly Cardano assumed that Rheticus had reliable information about Florence as well as Nuremberg; in fact, the intellectual networks that connected the two cities buzzed with messages.

In the mid-1540s the correspondents met. Rheticus made the obligatory trip to Italy and visited Cardano, no later than March 1547, in Milan. In the third edition of his *Supplementum* and in his *Aphorismi astronomici* Cardano reported in detail on their discussions. They tried to place the rules of astrology on a firmer basis. According to Cardano, their conversations rather resembled those of Sherlock Holmes and Doctor Watson. Twice in the *Aphorismi* dialogues between Cardano and Rheticus appear. According to Cardano, the German astrologer offered his Italian friend horoscopes to interpret. Cardano realized immediately that the first subject, a man, suffered from melancholy and worked as a counterfeiter; he foresaw an execution in the near future. The second subject, a woman, he diagnosed with equal rapidity as a witch and confidence woman, though he could not determine her ultimate fate precisely. In both cases Cardano portrays Rheticus as suffused with astonishment and admiration: "Unde hoc?," "How could you know that?," he asked, jaw dropping almost audibly—and so gave the cue for a detailed account of Cardano's methods.[15] Rheticus also gave his friend another group of celebrity horoscopes, including those of Regiomontanus, Vesalius, Agrippa, Poliziano and Osiander. Cardano printed these and discussed them in the third edition of the *Supplementum*. The birth dates of any number of famous men, as found in modern reference works, rest on no older and more reliable source than this horoscope collection. Cardano's astrological collaboration with Rheticus paradoxically contributed more than a mite to the development of modern history.[16]

Cardano's name does not appear in Rheticus's letters until the middle of the 1550s. In his 1554 commentary on Ptolemy's *Tetrabiblos*, the one surviving ancient textbook of astrology, Cardano made fun of the superfluous complications that the German astronomer Regiomontanus had introduced into his tables, in order—Cardano claimed—to enslave subsequent astronomers.[17] In doing so he continued a polemic that he had begun as early as 1547, when he charged in his commentary on Regiomontanus's horoscope that the German had plagiarized Bianchini and others.[18] Rheticus now allowed himself one dry remark: "Cardani Cardano sunt simillima."[19] This single comment on an intellectually ambitious text of more than 300 pages suffices to show that the relation between the two men had cooled off. The preserved documents do not explain why their friendship came to an end. Perhaps Rheticus was wounded by the role Cardano assigned him in his dialogues: perhaps the comments on Regiomontanus which provoked his one explicit remark made him genuinely angry. At all events, they were never reconciled. In the 1560s Rheticus encouraged another astrologer and natural philosopher, Taddäus Hayck, to compete with Cardano in the development of a system of physiognomics. The Italian, he now declared bitterly, claimed that he had mastered all the mysteries of facial interpretation, but his boasts were empty.[20] The erstwhile collaborators and friends were now open enemies. It will become evident that they not only came to dislike one another, but also took radically different positions on issues that mattered very much to both of them.

## Two Genealogies

Originally, Rheticus and Cardano began from the same historical premises. In his treatise *De temporum et motuum erraticarum restitutione* Cardano argued that the astronomy of Thales, the earliest Greek philosopher, came from Egyptian sources.[21] In the *Aphorismi astronomici* he suggested that Ptolemy had continued an ancient tradition that derived from the Near East: "The difference between Ptolemy's precepts and those of all later writers is greater than that between emeralds and mud. Probably he possessed the wonderful texts of the Chaldeans and Babylonians, from many centuries before, and those of the Egyptians."[22] An anonymous reader of Cardano's book underlined the last sentence forcefully in a copy of the book preserved in the British Library (53 b 7).

Rheticus's thoughts moved in the same direction, but with greater speed and force. In April 1551 he fell under suspicion of having raped a student. Instead of defending himself before the university court in Leipzig, he fled to Prague, and then to Cracow. Here he became a doctor, taking a special interest in the Paracelsian philosophy of nature. His fascination with the development of astronomy never left him, but it now took a surprising turn. For fifteen years Rheticus had ruminated over how to rebuild classical astronomy. He believed firmly that the positions of the fixed stars in Ptolemy's list were often wrong, and that these errors in turn explained other ones in Greek astronomy.[23] Suddenly he realized that the ancient Egyptians must have possessed exactly the knowledge modern Europeans needed and lacked, and that he himself was in an ideal position to reconstruct it. In his *Natural History* the elder Pliny described the making, transportation and purpose of the obelisks of ancient Egypt. He held that the Egyptians had dedicated these shafts to the sun god: the word obelisk, indeed, also meant "sun."[24] A few inferences more and Rheticus was on the move:

> Pliny bears witness that the obelisks were dedicated to the sun god. The sun is also queen and ruler of the heavenly state; all other stars move in accordance with her rhythm. She is the eye of the world, whose light illuminates everything. Obelisks alone make it possible to observe and record all the laws of the heavenly state. Obelisks alone will open experts' eyes and give them the light they need to record the observations required for the history of the planetary motions and to seek out the proofs needed to determine the motions, enabling them to make useful observations of the planetary motions at all times.[25]

Rheticus ridiuled the pathetic astronomical instruments of the Greeks: "armillary spheres, rules, astrolabes and quadrants are human inventions, which require great effort from their users and offer no defense against error."[26] In sharp contrast with these comic pieces of apparatus stood the obelisk, the simplicity of which mirrored its divine origins and absolute infalliblity:

> The obelisk was erected by divine instruction and easily provides all of this. . . . We have shown in detail that the obelisk is a product of philosophy, which

reveals a method for establishing the positions of the sun, the moon, the planets and other stars. From these the regular motions of the stars and all the laws of the heavenly state can be determined. Thus astronomy, geography, and the part of physics which deals with the effects of the stars, a discipline most important for human life, can be constructed, reinforced, and propagated.[27]

Rheticus's interpretation of Pliny swarms with problems. Pliny believed that the Egyptians dedicated their obelisks to the sun god. Moreover, he said that the inscriptions on the obelisks in the Roman Campus Martius and Circus Maximus contained an interpretation of nature "according to the Egyptian philosophy."[28] But he nowhere stated, explicitly or implicitly, that the pillars themselves functioned as astronomical instruments. Perhaps Rheticus thought that he had other evidence to support his thesis. For example, he could have based his theory on the passage in which Pliny described the meridian instrument set up by Augustus in the Campus Martius. The Egyptian obelisk that Augustus had brought to Rome played the central role of a gnomon in this apparatus. But by Pliny's account, the Romans who brought the obelisk to Rome, not the Egyptians who first hewed it from the rock, were the ones who thought of using it in this way. Nothing in Pliny's description directly suggests that Augustus saw himself as imitating Egyptian precedents.[29] In the end, Pliny's text contains no solid reason to think that obelisks were Egyptian observatories.

Perhaps Rheticus's search for financial support led him to leap, like a skater on cracking ice, from shaky evidence towards a striking conclusion. In the second half of the sixteenth century, a number of imperial princes competed to develop programs of astronomical innovation. They invited astronomers and craftsmen to develop more and more splendid and effective instruments, which served as both scientific and political instruments. More than one princely astronomer made important observations. William IV of Hesse-Kassel recorded the movements of the comet of 1572 so precisely, according to Bruce Moran, that he felt able to debate the comet's path and nature with Tycho Brahe.[30] Rheticus's plan, with its call for the erection of a rare, costly instrument and its evocation of the big science of Augustus and the Pharaohs, may have been calculated to catch some particular princely eye.

Rheticus deeply regretted that astronomers had ceased to use this perfect instrument. Some obelisks languished in captivity under the "Turkish despot" in Constantinople, "where this treasure serves no purpose except his vanity."[31] The Roman obelisk of Augustus had ceased to yield observations centuries before. But help was on the way. In Cracow, Rheticus had found a patron, who paid for him "to set up an obelisk, forty-five Roman feet high."[32] The Lutheran astronomer became, in his own eyes, the last in the great line of Egyptian priests.

In summer 1554, Rheticus described his obelisk and his plans for a new astronomy in a letter to Johannes Crato, to whom he confided that "with God's help I hope to use my obelisk to make new measurements of the entire sphere of the fixed stars."[33] Seven years later, Crato persuaded a student of Ramus's, Jacques Caloni, to visit Rheticus. He found both the astronomer and the obelisk fascinating. After Caloni returned to Paris, in fact, he managed to convince Ramus to invite Rheticus to Paris. In a letter of 17 August 1563 he described Ramus's plans to see to it that Rheticus received a chair in the Collège du Roi.[34] Eight days later Ramus sent the letter which Kepler would later cite, in which he praised the German astronomer's writings and accomplishments. He also pointed out the pressing need for a reformation in astronomy: "As to the form of astronomy I wanted to create, I could find no real aids either in logic or in books or among men when I considered how this subject has become so difficult and been thrown into such confusion by those hypotheses." Rheticus's obelisk impressed Ramus as just the right instrument for developing a kind of astrology "which is entirely set free not only from the efforts required by the devices of Pythagoras and Jabir, but also from those many hundreds of tables."[35]

The enthusiasm with which Ramus received and adopted Rheticus's theory reflects, in part at least, his reading of other texts, one of which he ascribed to the same author. In arguing that Copernicus had gone wrong, like his forerunners, by devising wild hypotheses, Ramus drew on the anonymous preface to *De revolutionibus*. This argued, as we have seen, that Copernicus meant only to propose a hypothesis, not what he considered a true system of the world. Ramus's letter shows that he thought Rheticus, not Osiander, had written this disclaimer: "Your letter—if I am not mistaken—which serves as the preface to Copernicus shows that the hypotheses regarding epicycles and eccentrics are false and ridiculous inventions."[36] Though the young Rheticus had revered Copernicus as his teacher and agreed with him, Ramus imagined that he had subtly tried to distance himself both from the Copernican theory and its Ptolemaic forerunner. Only this productive misunderstanding enabled him to draw from Rheticus's obelisk project the general conclusion that he rejected all hypotheses—an interpretation based on a false assumption which Rheticus, for reasons of his own, did not correct.

But the fuller set of historical and astronomical theories that Ramus connected with Rheticus's enterprise drew on other, late antique sources as well— sources that reflected the serious effort of scholars to probe the defects of their system of the world. Proclus had argued, in his commentary on the *Timaeus*, "that Plato applied no astronomical hypotheses." In his *Hypotyposes* he had gone further, claiming that the Greek philosophers had feared

> that they might find the movement of the stars irrational, undefined and irregular. They even feared that the heavens might collapse and dissolve. Therefore the most important astronomers developed the hypotheses, which assumed a reasonable, regular rhythm, which obeyed definite, ordered and identical causes. They hoped in this way to defend and support the eternal order of the heavens.[37]

Simplicius, finally, offered a vital clue in his commentary on Aristotle's *De caelo*. There he quoted a passage from Porphyry, according to which Callisthenes, who had accompanied Alexander the Great to Babylon, sent the astronomical observations of the Chaldeans, which he found in the conquered city, back to

Greece.³⁸ Unfortunately, these pure data, which had not been forced into the distorting mold of a hypothesis, never arrived. Instead, the Greek astronomers from Eudoxus to Aristotle, developed a series of increasingly elaborate geometrical models, which Simplicius discussed in detail, drawing on the lost history of mathematics by Eudemus. Even these earliest models, Ramus held, suffered from their hypothetical character. But they were simpler, and therefore better, than the more complex ones devised by later astronomers. Above all, Ramus claimed, drawing on Aristotle as well as Proclus, the Pythagoreans were at fault, since they had "rejected concentric motions and invented epicycles and eccentrics."³⁹ Ramus's history of the mathematical sciences rested on energetic study of a wide range of texts. It formed the context within which he encouraged Rheticus to develop an "astronomia sine hypothesibus" and reproached the Greeks for ruining astrology with their complex models. And it explains why, though Ramus knew Rheticus's obelisk only from the image of it printed on Rheticus' *Canon*, this inspired him with the conviction that he now had the key that could unlock the mysteries of history and astronomy alike. Years later, meeting Tycho in Strasbourg, Ramus tried to convince him that the Egyptian astronomers had possessed a perfectly empirical astronomy.⁴⁰

Ramus influenced Rheticus in his turn. As early as 1551, Rheticus printed what he claimed to be Copernicus's theory that the weaknesses of modern astronomy derived in large part from the lacunae and errors in its ancient sources. The Greek astronomers, who should have based their theories on the data, in fact did the reverse, and preserved only those observations which supported their models. Ptolemy's data in particular deserved little credence. Only a program of systematic observation, designed to test and verify the transmitted ancient materials, could restore modern astronomy. No wonder that Rheticus came to suspect that Ptolemy might have done serious harm to astronomy.⁴¹ In a fragmentarily preserved letter to Ramus of 1568, Rheticus proposed again to restore the astronomy "of the first creators of this science." He now appropriated Ramus's terminology. Instead of simply emphasizing the virtues of the obelisk as an observational tool, he insisted on the theoretical necessity of redeeming astronomy as a whole from its original sin of hypothesis framing: "I will now set to work on the project that occurred to you as well: to free the art of astronomy from hypothesis, contenting myself with the observations alone."⁴² Perhaps Rheticus still hoped for a call to Paris. In any event, this time the outside offer did not come. As to Rheticus's obelisk in Cracow and the data he compiled using it, both seem to have vanished without trace.

Cardano also found obelisks enchanting. In his *De rerum varietate* of 1557 he described in some detail the Vatican obelisk, which was one of Rome's wonders, and the still more astonishing one which the emperor Constantine had transported from Egypt to Constantinople. But he presented them only as imposing pieces of stonework, not as scientific instruments: they were no more and no less scientific than the forum of Hadrian, which he also described.⁴³ And the same relatively calm attitude towards Egyptian achievements that manifested itself in this passage also appeared in his more technical astronomical and astrological works of the same period.

At the beginning of Cardano's career, as we saw, he believed firmly in the near eastern origins of Greek science. In his *De temporum et motuum erraticarum restitutione libellus* of 1543, for example, he argued that the Jews had developed a lunar calendar before the Flood, and that the Egyptians had devised a solar year shortly afterwards.[44] The Greek astronomers, Eudoxus and Hipparchus, came later and "imitated the methods of the Egyptians." Cardano portrayed matters in the same way in his short *Encomium astrologiae*, which appeared in the same collection. He called Moses the first astronomer, because he used the equinoxes and new moons to set the dates of holidays for the Jewish people. And he praised the later astronomical achievements of the Egyptian Hermes Trismegistus and the Chaldean Berosus.[45] For the most part, the *Encomium* followed the doxographic approach and exhibited the enthusiastic style normal for the genre to which it belonged.

Already in this early work, however, Cardano showed that he did not share his German friend's Egyptomania. His *Encomium* treated astronomy as an art that developed over time. He evoked a golden chain of astronomical learning, which extended chronologically and geographically from Israel to Italy, over Ethiopia, Egypt, Africa, Babylon, Phoenicia and Greece. Apparently, he saw the later astronomers as perfectly competent. At one point he even remarked that Greeks, Egyptians and Phoenicians, in that order, studied the heavens "more curiously" than the Jews—a tactful but unmistakable effort to correct the widely held belief that astronomy had declined over time.[46]

In his commentary on Ptolemy's *Tetrabiblos*, which appeared in 1554, Cardano made a more systematic effort than before to reconstruct the early development of a Greek science: astrology. He began by suggesting a historical analogy. Hippocrates, the founder of Greek medicine, developed a sophisticated set of medical theories, with which Cardano came into contact as early as the 1520s, when Fabio Calvo printed both the first full Latin translation and the Greek texts of his works.[47] The Hippocratic method, though speculative in its conclusions, seemed entirely empirical in its foundations. For Hippocrates derived the components of his system from direct observation: in large part, from case histories of the sort that he had recorded in the *Epidemics*. The modern medical man who, like Cardano, wished to enlarge and improve Greek medicine, must do so on the basis of its originally empirical method.

The modern astrologer, Cardano argued, should work in the same way—just as his ancient counterparts had. He assumed that Ptolemy had had at his disposal a collection of examples, a series of horoscopes. These he drew from now lost texts, perhaps those of the Egyptians and Chaldeans. They provided the empirical basis of Ptolemy's astrology.[48] Cardano held that he could reconstruct this lost body of empirical data by a combination of philological and astrological research. His argument forms a rich mixture of precise textual interpretation and bold historical speculation.

Cardano drew on many texts, taking account of the historical background as well as of the preserved astrological sources. Reading the Roman historians, above all Suetonius and Tacitus, he produced a list of the astrologers who had worked in the Roman imperial court just before and during Ptolemy's time. He also portrayed, with admiration, their scientific and political accomplishments, assuming that those described as "Chaldeans" really came from the Near East.[49] And he boldly conjectured that these professionals had supplied Ptolemy with his data. Taking as his occasion a passage in which Ptolemy discussed Egyptian medicine, Cardano summarized his theory:

> When [Ptolemy] says that the Egyptians developed this art, he means not only them, but also the neighboring peoples, for example the Chaldeans. Suetonius describes in his life of Nero how they prophesied both his rule and his murder of his mother. When the mother learned of this, she said simply 'occidat cum imperet.' One of these men was Thrasyllus, who predicted the rule of Tiberius during his exile and enjoyed enormous authority ... Ascletarion, who lived under Diocletian, was less successful ... It seems, then, as though the art of the Chaldeans and Egyptians made them incredibly famous. After Ptolemy had collected their writings, he summarized them in this book. Unfortunately, he did so with such brevity that the bulk of the art is missing. It would have been better if he had followed the example of the great Hippocrates. After he had transformed the data into a systematic art, he should also have written a book of individual examples on the model of the *Epidemics*.[50]

The Roman historians, in other words, seemed to Cardano to prove that the tradition of Near Eastern astrology had survived as late as the second century A.D., and that Ptolemy's achievement lay in his synthesis of data and principles that had previously been scattered.

Cardano did not lack boldness. Because Ptolemy in fact wrote no *Epidemics*, he tried to fill this lacuna by compiling his own horoscope collections, even though the task involved great technical difficulties.[51] After Cardano drew up and interpreted the horoscope of Nero, for example, he saw that he had assumed a wrong birth date. This error, however, did not bother him. He computed the positions of the planets for the correct date and produced a second horoscope, which he not only included in the second edition of his collection, but also interpreted in exactly the same way (and even in the same words). He proudly announced that the second horoscope and the emperor's character resembled one another as precisely as two snowdrops, saying nothing about the first horoscope.[52] Even more curious were his dealings with another horoscope, that of Jesus. Cardano found that the details of Jesus's horoscope—considered as that of an ideal human being—corresponded with startling precision to the requirements laid out by Ptolemy. He went so far as to suggest that this one horoscope could have formed the entire foundation for Ptolemaic astrology on its own— and commented on it at great length, shocking many readers. Cardano's horoscope collections took a digressive and sometimes bizarre form that their austere Hippocratic prototype lacked. But their central inspiration remained clear: he compiled them both to confirm and to exemplify his view that "the *experimenta* of the Caesars Tiberius, Claudius, Nero, Domitian, Hadrian and Gordian, as well as the things that befell Picinnino in our time, show the truth of the art."[53]

Cardano's admiration for Chaldean and Egyptian court astrologers had limits. He rarely emphasized the age of their observations, and took more interest in the well-documented achievements of the astrologers of imperial Rome than in the legendary ones of their older predecessors. Gradually he came to see that he could not even reconstruct the doctrines of the last Chaldean astrologers in detail. The humanist Pietro Crinito and others had pointed out that Firmicus Maternus, writing in the fourth century, compiled rich materials on the powers of the degrees of the zodiac, which he described as Egyptian in origin. Gaurico used Firmicus to identify Hermes Trismegistus, along with his pupils Petosiris, Necepso, and Aesculapius, as early practitioners of astrology.[54] Cardano, however, remarked disappointedly that Firmicus was a grammarian, not an astrologer. From Firmicus's scattered and unsystematic data, the sources for which he did not cite, Cardano felt able to draw no firm conclusions. The Greek commentaries on the *Tetrabiblos*, which Hieronymus Wolf published in 1559, offered little more than a few bibliographical details about Ptolemy's predecessors.[55]

In so far as Cardano insisted on reconstructing ancient astrology from preserved sources, through precise inferences, he resembled some of the astronomers who had attacked similar problems before him. More than one of his Renaissance predecessors had, in fact, devoted some serious attention to the early history of their art, both because they still relied so heavily on ancient data and methods and because they were also well-schooled humanists to whom a historical and philological approach came readily. As early as the mid-fifteenth century, Regiomontanus produced the first modern, critical interpretation of the basic handbook of ancient astronomy, Ptolemy's *Almagest*. Here and in other, shorter texts, he devoted some attention to the lost sources Ptolemy drew on, and drew from the sources sober reconstructions of ancient practices and results. Instead of mocking the pitiful instruments the Greeks had used, Regiomontanus offered a detailed reconstruction of the armillary sphere, "which Ptolemy used, above all, and Hipparchus before him, to investigate the motions of the stars."[56] This instrument, Regiomontanus held, enabled the Greek astronomers and their Islamic successors to found scientific astronomy. The modern astronomer who wished to improve on them must use it as well. Egyptians and Chaldeans played almost no role in Regiomontanus's history of astronomy. Cardano's work on the astrological tradition had far more in common with Regiomontanus than with Rheticus. The decline and fall that Ramus and Rheticus sketched out found its complement, or opposite, in Cardano's sketch of cultural and technical progress.

## Last Stop: Back to Kepler

The period of intellectual transition around 1600 saw many intellectuals show their willingness to work from contradictory premises without feeling paralyzed. Kepler read both Rheticus and Cardano, with care. In his first book, the *Mysterium Cosmographicum* of 1596, he cited Rheticus's attack on Ptolemy.[57] For years to come he would continue, as he did in the *Mysterium*, to account for Ptolemy's mistakes by historical research and to rebuild astronomy on a new empirical basis, just as Rheticus had recommended.

Reading Cardano left a deeper mark on Kepler. True, he believed that Cardano was "not the best in astrology," and explained that the Italian had not practised his art empirically, in Kepler's sense: "in his *aphorisms* one often sees that he writes them on the basis of one example."[58] Still, Cardano was the only modern astrologer who really continued to interest Kepler. Throughout his life, Kepler worked on the relationship between the astrological aspects and the musical consonances. He found the inspiration for this, as he said, in Cardano's commentary on Ptolemy, where Cardano tried to use the principles of musical harmony to explain the aspects. From *De varietate* Kepler learned that Ptolemy had devoted an entire, as yet unpublished book to harmonics. The interest in the music of the spheres that led Kepler to hunt for manuscripts of Ptolemy and prepare a critical edition of the text was thus awakened by Cardano.[59] In his own *Harmonice mundi* he went so far as to speak of the "correspondence of the aspects with musical consonances, as set out by Ptolemy and explicated by Cardano."[60] The treatments of the history of astronomy and astrology that Kepler laid out in the preface to his *Tabulae Rudolfinae* and elsewhere gave Ptolemy the starring role and emphasized progress, much as Cardano had long before.

Towards the end of Kepler's years in Prague, he encountered serious intellectual and political difficulties. The emperor Rudolf II and others pressed him to evaluate the current situation of the Empire astrologically and to derive practical advice for them from what he observed. Kepler, however, no longer believed that astrology could predict individual fates or events. He cited examples from the history of Roman astrology, hoping to convince his masters that the political influence of astrologers had led to crises and revolutions. Tacitus, for example, had rightly condemned the astrologers as disloyal to those who held power and deceitful to those who sought it. Thrasyllus and other evil counsellors had used their prestige as astrologers to exercise an evil influence on more than one emperor.[61] Kepler, in short, followed Cardano in investigating the social and political functions of astrology above all in the context of the Roman empire—just as he had when he treated astrology itself as a relatively modern discipline, the development of which could be reconstructed only from late sources and for the historical period. Even for the later Kepler, the opponent of classical astrology, Cardano still provided essential hints and guidance.

The divergent genealogies of science that Rheticus and Cardano traced were in fact the two complementary strands of a double helix. Each reflected its creator's effort to put ancient science back into its historical context and to trace its development in some detail. Each provoked and stimulated Kepler, and each supplied him with vital materials. When Kepler gave his account of ancient science visual form on the title page of the *Rudolfine Tables*, it took the form of a splendid round temple of astronomy, designed to indicate how and where the art had developed.[62] Both Rheticus and Cardano contributed their modest pillars to this splendid new historical structure.[63]

## Notes

1. Johannes Kepler, *Gesammelte Werke*, ed. Max Caspar (München, 1937–), 3:6. See E.J. Aiton, "Johannes Kepler and the Astronomy without Hypotheses," *Japanese Studies in the History of Science* 14 (1975): 49–71.

2. Kepler, *Gesammelte Werke*, 3:6.

3. For a good case in point, written by an astronomer and astrologer of considerable technical competence, see L. Gaurico, "Oratio de laudibus astrologiae, habita in Ferrariensi achademia," in *Spherae tractatus*, ed. Gaurico (Venice, 1531), sigs. a ii recto–[a vi verso].

4. N. Jardine, *The Birth of History and Philosophy of Science* (Cambridge, MA, 1984); A. Grafton, *Defenders of the Text* (Cambridge, MA, 1991), ch. 7.

5. Jardine offers suggestive remarks on Rheticus; so do H. Hugonnard-Roche et al. in their edition of and commentary on Rheticus' *Narratio prima* (Wrocław, 1982).

6. On Rheticus see the *Narratio prima*, ed. Hugonnard-Roche et al.; K.H. Burmeister, *Georg Joachim Rhetikus 1514–1574: Eine Bio-Bibliographie* (Wiesbaden, 1967–68), and *Dictionary of Scientific Biography*, s.v. "Rheticus, George Joachim," by E. Rosen.

7. Kepler, *Gesammelte Werke*, 3:6.

8. On Cardano the best short biography remains O. Ore, *Cardano, The Gambling Scholar* (Princeton, 1953). See also A. Ingegno, *Saggio sulla filosofia di Cardano* (Florence, 1980); E. Kessler, ed., *Girolamo Cardano: Philosoph, Naturforscher, Arzt* (Wiesbaden, 1994); and N. Siraisi, *The Clock and the Mirror* (Princeton, forthcoming). His *Opera omnia* appeared at Lyons in 1663 (reprint, Stuttgart-Bad Canstatt, 1965). For Naudé's judgment of him see the sketch of his character in Cardano, *De vita propria liber* (Amsterdam, 1654), sig. *5 recto-verso.

9. J.C. Scaliger, *Exotericarum exercitationum liber quintus decimus, de subtilitate, ad Hieronymum Cardanum* (Paris, 1557). See I. Maclean, "The Interpretation of Natural Signs: Cardano's *De subtilitate* versus Scaliger's *Exercitationes*," in *Occult and Scientific Mentalities in the Renaissance*, ed. B. Vickers (Cambridge, 1984), 231–52.

10. Kepler, *Gesammelte Werke*, 8:15, 113.

11. Both Petreius's eagerness to publish astrological materials and his collaboration with Rheticus are documented by his letter to Rheticus of 1 August 1540, in Antonius de Montulmo, *De iudiciis nativitatum liber praeclarissimus* (Nuremberg, 1540), sig. A ij verso.

12. Cardano, *De supplemento Almanach* (Milan, 1538), sig. [A vii recto–A viij recto) = *Libelli duo* (Nuremberg, 1543), sig. D ij verso (*Opera omnia*, 5:584–85).

13. Rheticus, *Narratio prima*, 47–50; Burmeister, *Rhetikus*, 3:133–34.

14. Cardano, horoscope 64, in *Libelli duo* (Nuremberg, 1543), sig. cc ij recto-verso.

15. Cardano, *Aphorismi astronomici*, vii, *Libelli quinque* (Nuremberg, 1547), fol. 299 recto (*Opera omnia*, 5:85–86).

16. G. Ernst, *Religione, ragione e natura* (Milan, 1991), 213–14; P. McNair, "Politian's Horoscope," *Cultural Aspects of the Italian Renaissance*, ed. C.H. Clough (Manchester and New York, 1974).

17. Ptolemy, *Quadripartitum*, ed. G. Cardano (Basle, 1554), 245, 251 (*Opera omnia*, 5:278, 284–85).

18. Cardano, *Libelli quinque* (Nuremberg, 1547), horoscope 89, fol. 175 recto. Cardano may have been embarrassed at having described the horoscope of Regiomontanus's teacher, Georg Peurbach, as that of Regiomontanus himself, in his *Libelli duo* (Nuremberg, 1543), horoscope 66,

sig. cc iij verso. Rheticus, in turn, may well have earned Cardano's irritation by pointing out this obvious error, which he was particularly well equipped to notice. (Some readers of Cardano's *Libelli duo* corrected it in their copies: e.g., Oesterreichische Nationalbibliothek 72 J 123 and 72 X 5.)

19. Burmeister, *Rhetikus*, 3:121, 123.
20. Ibid., 181, 186.
21. Cardano, *Libelli duo* (Nuremberg, 1543), sig. G recto (*Opera omnia*, 5:1).
22. Cardano, *Libelli quinque* (Nuremberg, 1547), fol. 215 verso (*Opera omnia*, 5:35).
23. Rheticus, *Narratio prima*, 222–23.
24. Pliny, *Naturalis historia*, 36.14.64; 36.35.74.
25. Burmeister, *Rhetikus*, 3:139; cf. Rheticus, *Narratio prima*, 234.
26. Burmeister, *Rhetikus*, 3:139.
27. Ibid., 139–40.
28. Pliny, *Naturalis historia*, 36.14.71: "Inscripti ambo rerum naturae interpretationem Aegyptiorum philosophia continent."
29. Ibid., 36.15.72; cf. M. Schütz, "Zur Sonnenuhr des Augustus auf dem Marsfeld," *Gymnasium* 97 (1990): 432–57.
30. B. Moran, "Princes, Machines and the Valuation of Precision in the Sixteenth Century," *Sudhoffs Archiv* 61 (1977): 209–28; "German Prince-Practitioners: Aspects of the Development of Courtly Science, Technology and Procedures in the Sixteenth Century," *Technology and Culture* 22 (1981): 253–74; "Christoph Rothmann, the Copernican Theory, and Institutional and Technical Influences on the Criticism of Aristotelian Cosmology," *Sixteenth Century Journal* 13 (1982): 85–108.
31. Burmeister, *Rhetikus*, 3:140.
32. Ibid., 139.
33. Ibid., 123.
34. Ibid., 170.
35. Ibid., 173.
36. Ibid., 175. See E. Rosen, "The Ramus-Rheticus Correspondence," *Journal of the History of Ideas* 1 (1940): 363–68.
37. Burmeister, *Rhetikus*, 3:174–75; for the sources see the detailed commentary in *Narratio prima*, 242–45.
38. A. Grafton, *Joseph Scaliger* (Oxford, 1983–93), 2:264–67.
39. Rheticus, *Narratio prima*, 238.
40. See T. Brahe, *Opera omnia*, ed. J.L.E. Dreyer, (Copenhagen, 1919), 6:88–89.
41. Rheticus, *Narratio prima*, 221–27.
42. Burmeister, *Rhetikus*, 3:188.
43. Cardano, *De rerum varietate* 17.98; *Opera omnia*, 3:345.
44. Cardano, *Libelli duo* (Nuremberg, 1543). sig. G recto (*Opera omnia*, 5:1).
45. Cardano, *Libelli duo* (Nuremberg, 1543), sig. dd iii verso; cf. ibid., sig. dd iiii verso.
46. Ibid., sig. dd iii verso.
47. See N. Siraisi, *The Mirror and the Clock* (Princeton, forthcoming), for a full account of Cardano's medical thought and scholarship.
48. Ptolemy, *Quadripartitum*, 53.
49. On astrology in imperial Rome, see M.T. Fögen, *Die Enteignung der Wahrsager* (Frankfurt am Main, 1993), D. Potter, *Prophets and Emperors* (Cambridge, MA, 1994), and T.S. Barton, *Power and Knowledge* (Ann Arbor, MI, 1994).
50. Ptolemy, *Quadripartitum*, 33–34 (*Opera omnia*, 5:117).
51. Cardano, *Aphorismi astronomici*, fol. 270 recto (*Opera omnia*, 5:69).
52. Cardano, horoscope 40, *Libelli duo* (Nuremberg, 1543), sig. x verso–x ii recto; (ed. Nuremberg, 1547), fol. 143 verso (*Opera omnia*, 5:480–81); *Aphorismi*, fol. 270 recto (*Opera omnia*, 5:69).
53. Cardano, *Aphorismi astronomici*, 1; *Libelli quinque* (Nuremberg, 1547), sig. 211 verso.
54. Gaurico, "Oratio," sig. [a v recto].

55. Ptolemy, *Quadripartitum*, 33–34 (*Opera omnia*, 5:117–18); 194 (*Opera omnia*, 5:158–59).
56. J. Regiomontanus, *Scripta* (Nuremberg, 1544; repr. Frankfurt am Main, 1976), fols. 21 recto, 22 verso.
57. Kepler, *Gesammelte Werke*, 1:63–64.
58. Ibid., 4:256, 258.
59. U. Klein, "Johannes Keplers Bemühungen um die Harmonieschriften des Ptolemaios und Porphyrios," *Johannes Kepler: Werk und Leistung* (Linz, 1971), 51–60.
60. Kepler, *Gesammelte Werke*, 6:267.
61. J. Kepler, *Opera omnia*, ed. C. Frisch, (Frankfurt and Erlangen, 1858): 1:632; see the fine analysis by B. Bauer, "Die Rolle des Hofastrologen und Hofmathematikus als fürstlicher Berater," in *Höfischer Humanismus*, ed. A. Buck (Weinheim, 1989), 93–117.
62. Grafton, *Defenders of the Text*, ch. 7.
63. The original text of this article, written in German, will appear in a collection tentatively entitled *Philologia perennis*, ed. W. Schmidt-Biggemann (Frankfurt, forthcoming). The present translation is abridged and slightly revised.

# LEGITIMIZING A DISCIPLINE
## James Mackenzie's History of Health (1758)

### Heikki Mikkeli

Preservation of health and the prevention of disease, is a kind of neutral ground, between the several branches of medicine, and the common sense and daily observation of well informed men, and of course is open to everyone.[1]

From the above words of Sir John Sinclair it becomes clear why the field of hygiene can be considered an interesting topic for medical historians. Hygiene was understood as one of the five principal parts of academic medicine, but at the same time the part of medicine that was most closely associated with the interests of the laity. Prevention was already in the eighteenth century at the periphery of the physician's professional interests and therefore hygiene was the very field where academic and popular medicine most easily ended in confrontations.[2]

The history of hygiene (or dietetics)[3] in early modern Europe has hitherto mostly been written of as a part of popular medicine, or the popularization of academic ideas in vernacular medical literature.[4] Almost no attention has been paid to preventive medicine as an academic discipline or to the various writings where the scientific nature of hygiene was compared with other branches of the medical discipline. The popular medical healthbooks and manuals constitute, however, only one part of the medical literature that deals with dietetical problems.

If we turn our gaze to academic or learned writings in general, John Sinclair's *Code of Health and Longevity* is, perhaps, not the most interesting example. In 1758 James Mackenzie (1680–1761), a Scottish physician, wrote a book called *The History of Health and the Art of Preserving It*.[5] The work consists of three parts: an introduction (15 pages), a history of dietetics entitled "The History of Health" (310 pages), and Mackenzie's own "Rules of Health" (140 pages). These three parts together amount to almost 500 pages. As a work, it differs totally from other contemporary histories of medicine such as Daniel LeClerc's *Histoire de la médicine* (1st ed. 1696, Engl. trans. in 1699) or John Freind's *History of Physick* (1725–26).[6]

In his introduction Mackenzie sets forth his intentions. First he introduces himself as a writer: he was a retired country physician, who due to his age could no longer "pursue the painful practice of a country physician." If Mackenzie could no longer "ride long journies to remove distempers," he could nevertheless try to prevent them by presenting "the most effectual rules to preserve health." He saw clearly the meaning of history for his pursuit: "That I might add a greater weight and authority to these rules, I resolved to trace them from their sources, by giving the history of the whole art of preserving health, from the most remote antiquity down to the present time."[7] It would be a hasty conclusion, however, to treat Mackenzie's book exclusively as an old man's hobby-horse. In fact, he had in mind more ambitious plans in trying to defend hygiene as a part of medicine by using history as a means of legitimation.[8] He also wanted to stress the measurable nature of medicine in general, and hygiene especially, which was partly due to instruments that had been invented in this field. For Mackenzie, this was a further proof for the scientific status of hygiene.[9]

In the introductory part of his treatise, Mackenzie criticized the prevailing situation in the field of dietetics, and especially the use of the term "non-naturals," which was a central concept in the preventive part of medicine. "The very sound of the epithet non-natural, when applied to aliment, air, sleep, etc., so essential to the subsistence of mankind, is extremely shocking."[10] He attributed the term to the peripatetic school and, according to him, non-naturals cannot be understood without commentaries. In the latter part of his book he tried to lay the rules for a new kind of hygiene that would no more exclusively depend on the principles of Galenist humoral pathology, namely on the six non-naturals.

Mackenzie's criticism leads us to several questions concerning the status of preventive medicine in early modern Europe. What were these non-naturals? How were they used in the organization of dietetical literature? And above all, why did Mackenzie not approve of using the term? Thus my present study consists of two parts. First I am going to give an overview of the history of preventive medicine, and how it was established in the medical curriculum at the time of the Renaissance. Second I am going to introduce Mackenzie's own ideas and show what he meant by his criticism, and in what way he thought hygiene could be legitimized as a part of medicine in the mid-eighteenth century.

## Natural, Non-natural, and Contra-natural

In Western medical tradition the factors influencing health and disease were divided into the natural, the non-natural and the contra-natural. The division had its origins in ancient medical thinking, but it took shape in medieval commentaries.[11] The Latin translation of Johannitius's introduction to Galen, the *Isagoge*, clearly divided material into "res naturales," "res non naturales," and "res contra naturam." The same idea of non-naturals as well as the term itself are found in the Latin translation of Haly Abbas's "Royal Book," the *Liber pantegni*. Haly Abbas directly connected the "res non naturales" with the number six.[12]

Natural things form the basic constitution of an individual human body. There are seven of these, such as elements, qualities, humors, and spirits. They provide a basic set of factors that are essential for life, but which also are subject to great variation in order to produce either a healthy or a diseased state of life. The non-naturals, by contrast, are those things which are able to cause variation in the natural constituents and functions. The traditional list of the six non-naturals usually included the ambient air, food and drink, exercise and rest, sleep and waking, evacuation and repletion, and the passions of the soul. The non-naturals are also essential aspects of life, but the individual has more or less control over the particular ways in which they are used—for example, how much and what kind of food one eats—and they can be either helpful, harmful, or indifferent, depending upon how one lives life. If carried beyond a certain point they can have more long-lasting effects in the form of disease. This brings us to the things contrary to nature, which include diseases, their causes, and their symptoms.[13]

In early modern commentaries the whole field of medicine was usually divided into five subdisciplines: anatomy (physiology), semeiology, pathology, hygiene, and therapeutics. The fivefold division had its origin in ancient medical literature, but it gained new popularity in the Renaissance when the classical texts became better known.[14] The six non-naturals had in fact a double role in this division. They could either be the conceptual disposition of the preventive part of medicine (i.e., hygiene) or they could also belong to the curative part, namely to therapeutics, which was usually further divided into three sections: the use of diet through the right use of the non-naturals, surgery and pharmacy.

As to the term "non-natural" itself, there were usually two different explanations given to it from the Middle Ages on. Either the six non-naturals were considered to be things that were not ordered by nature, like natural things were, but by a human being him- or herself, or these factors were considered to be outside the human body and in this sense external to her or him, in contrast to the natural things, which were thought to be internal.[15]

## Preventive Medicine in the Middle Ages

In medieval universities the three main medical textbooks utilized were Galen's *Tegni* Avicenna's *Canon* and Hippocratic *Aphorisms*, all of which dealt also with the preventive part of medicine and the six non-naturals.[16] In the beginning of the third book of his *Tegni* Galen discussed the causes of diseases and there he introduces the familiar list of non-naturals, even if he does not mention the term. The third part of the first book of Avicenna's *Canon* deals with the preservation of health. Moreover, Avicenna speaks about causes instead of things. By calling the non-naturals "efficient causes" when they harm the body, and "necessary causes" when they help maintain a person's health, Avicenna gives a good description of the ambivalent nature of the non-naturals. When used in excess, they are causes of sickness, but when used in moderation, they prevent sickness.

In medieval commentaries the exact number of the non-naturals was questioned. For example, baths were usually listed, but they appeared under the term "exercise and rest" or "things excreted and retained." At the beginning of the fourteenth century in his commentary of Galen's *Tegni*, Torrigiano de Torrigiani asks whether there are more than six non-naturals. He considers whether geographical location, the altitude of the stars, as well as human occupations and customs should not also be included among them, since these conditions unavoidably affect the environment in which the human body flourishes or fails to flourish. Like most of his contemporaries, he concludes, however, by deciding that the traditional six categories in fact provide an adequate scheme of division that should properly be understood to include the additional environmental conditions mentioned above.[17]

In the Middle Ages, however, dietetics was not just a part of academic medical literature, but already constituted a central part of popular medical literature as well. Dietetic information is also to be found in the related genres of the *consilium* (i.e., a physician's report prescribing care and treatment for an individual patient on a specific occasion) and in calendars (i.e., *Volkskalender* in German).[18] *Regimen sanitatis salernitanum*, for example, was a health regimen from Salerno's medical school written in verse, which became extremely popular in late Middle Ages and early modern Europe, especially as it contained a commentary which was thought to have been written by Arnaldo da Villanova in the thirteenth century. In Salernitan regimens all the six non-naturals are mentioned, even if the treatment of food and drink is given a central place.[19]

## Medical Humanism and the Recovery of Ancient Hygienic Literature

In the beginning of the sixteenth century, there arose in Italy a movement that has been labelled as medical humanism. Medical humanists translated ancient medical treatises into Latin from the "uncorrupted" sources in order to reform medical practices.[20] With regard to the preventive part of medicine, Thomas Linacre translated Galen's *De sanitate tuenda* in 1517, Erasmus made a new edition of Plutarch's work on the same subject, and Junio Paulo Crasso translated Galen's small treatise called *Ad Thrasybulum* in 1538.

The above two of Galen's works are probably the most significant classical contributions to the genre under investigation here. Perhaps more clearly than any other author before him, Galen defined the nature of dietetics as an autonomous art. In *Ad Thrasybulum* he distinguishes the special office of the hygienist from that of the gymnast and the physician.[21] This was well understood in the sixteenth century when preventive medicine was considered to be the other major, but hitherto neglected, part of the art of medicine. In this respect, hygiene (or dietetics) can be compared to anatomy: during the century anatomical writers also were eager to emphasize the independent nature of anatomy and make it, instead of natural philosophy, the basis of all medical studies.[22]

In the sixteenth century medicine was usually divided into preventive and curative parts. Most writers considered the preventive part to be primary and more valuable than the curative one. Regarding this claim, they sought evidence from ancient sources, and usually found it, too. In *Ad Thrasybulum* Galen had distinguished four different stages of health, two of which could also be called neutral ones (*neutrum*). Distinguishing the various types was an important task for the physician, because he had to choose between preventive and curative measures. If the body was diagnosed to be in a healthy or neutral state, he should strengthen the balance of the temperaments (the principle *similia similibus*). If the patient was thought to be ill, however, the disease ought to be cured by using the opposite effects (the principle *contrario contariis*).[23]

In the first part of the sixteenth century most authors on preventive medicine just made translations of newly found ancient texts, but in the second half there emerged a new tradition of commentary based on these translations. The ancient interpretations were not taken for granted anymore, but the need for new contributions was also increasingly understood in this field of medicine. As a result, two major commentaries on preventive medicine were written during the latter part of the century.

Girolamo Cardano wrote his *De sanitate tuenda* in 1560.[24] Cardano's work consists of four books, the first of which is a general introduction to the subject. In this book the six non-naturals are also dealt with. In the second and third books, Cardano deals more profoundly with different foodstuffs and drinks. The last book is dedicated to the themes of old age and longevity, where he tries to identify general symptoms of longevity, which can be considered, for the most part, as true indications of long life, when they occur in the same person. Cardano named three things. Firstly, a long-lived person must descend from a long-lived family, at least by one of the parents. Secondly, he or she must be a cheerful person, and thirdly, she or he ought to be a naturally good and sound sleeper.

Fifteen years later another Italian scholar, Giulio Alexandrini, finished his large treatise entitled *Salubrium sive de sanitate tuenda* (1575). He divided his work into thirty-three books which consist of over six hundred columns. In Alexandrini's view, the preventive part of the medical art should be considered to be most important, but nonetheless it was the most neglected part of medicine. He touched on all the six non-naturals, but dealt most profoundly with food and drink, as did many other sixteenth-century writers. Some of the non-naturals were also treated in separate studies, such as Girolamo Mercuriale's *Artis gymnasticae* (1559) or Andrea Bacci's work on baths, *De thermis* (1571).[25]

## Two Sixteenth-Century Popular Writers: Cornaro and Elyot

In the fifteenth and especially in the sixteenth century, medical advice books, most of which were written in the vernacular, grew extremely popular. There were several reasons for this growth. In the Italian city-states there arose a new civilized group, the bourgeoisie, which was willing to read different kinds of advice books. At the same time, numerous advice books for princes were published and were often called mirrors, such as "The Mirror of Health," or "Fürstenspiegel."[26]

The invention of the printing press made this kind of material more available, and some of these health mirrors reached several editions. For example, Thomas Elyot's *The Castle of Health* and Thomas Moulton's *The Mirror of Health* both reached over twenty editions during the sixteenth century.[27] The fear of plague was another factor that increased the demand for health manuals. Many citizens were able and also willing to hear or read advice on how to avoid plague, which raged several times in Europe during the sixteenth century.

A Venetian nobleman Luigi Cornaro wrote in the middle of the sixteenth century a book called *Discorsi della vita sobria*, which became one of the most popular books ever on longevity.[28] In his discourses Cornaro takes his own life as an example of what a moderate and regulated life can result in. Cornaro's book concentrates on alimentation, but all the other non-naturals are also mentioned. Longevity was his primary aim and a regular, moderate diet the means to attain this end. Part of the attraction of the book resulted from the fact that Cornaro was claimed to have been almost a hundred years old while writing it. It was, therefore, easy to consider him as an authority on matters dealing with longevity.[29]

Cornaro's *Discorsi della vita sobria* was, moreover, noted approvingly in the academic world, for example, in Girolamo Cardano's *De sanitate tuenda*. His book was also translated into Latin at the beginning of seventeenth century by Leonard Lessius, a Belgian Jesuit who stressed the religious aspects of temperate life. At the turn of the eighteenth century (1714), Bernardino Ramazzini, a professor of medicine at Padua, wrote his own annotations to Cornaro's treatise. Thus it can be seen that the barriers between academic and nonacademic writings in this field of medicine were not insuperable. This was partly due to the fact that Galenist humoralism and the idea of the six non-naturals was the organizing principle both in the academic and popular medical traditions.[30]

In England the most influential of sixteenth-century popular health books was Thomas Elyot's (ca.1490–1546) *The Castle of Health*, which first appeared in 1536.[31] It was also the first health book written in the English language, except for the translation *Regimen sanitatis salernitanum*, which had appeared eight years earlier. *The Castle of Health* is divided into four books. The first of these introduces the general principles of Galenist humoral pathology concerning the four humors of the human body and their balance. The second book consists mainly of an introduction to over a hundred different foodstuffs and cooking recipes, and this also contains a short introduction on exercise. The topic of the third book is the recovery of the humoral balance through purgatives and vomiting. It also contains a description of central human passions such as anger, grief, and joy. The fourth book is dedicated to poor digestion and to analyzing the symptoms of different diseases by examining the urine.

The whole treatise ends in an introduction on the ideal diet which should be followed, especially during pestilence.

## Francis Bacon and the Idea of Longevity

The theoretical background of the popular medical advice books was the same as that of the academic commentaries: in both the idea of health was based on the theory of the regulation of individual humoral balance through the use of the six non-naturals. In these popular books, however, the emphasis was not on the doctrine of causes of diseases. Instead they concentrated on practical medical advice. The main question was: Is it possible to prolong your life, and if the answer is positive, then by what means?[32] Most of the sixteenth-century authors did not see any kind of conflict between the prevention of diseases and the prolongation of life. In this sense, however, Francis Bacon was an interesting exception.

In the second chapter of the fourth book of *The Advancement of Learning*, Bacon mentions the three parts of medicine: the preservation of life, the cure of diseases, and the prolongation of life. He explicitly notes that the last of these ought to be kept separate from the other two. According to Bacon, the last part of medicine, the prolongation of life, is new, deficient, and the most noble of all.

> Men should rightly observe and distinguish between those things which conduce to a healthy life, and those which conduce to a long life. For there are some things which tend to exhilarate the spirits, strengthen the bodily functions, and keep off diseases, which yet shorten the sum of life, and without sickness hasten on the decay of old age. There are others also which are of service to prolong life and retard decay, which yet cannot be used without danger to health, so that they who use them for the prolongation of life should at the same time provide against such inconveniences as may arise from their use.[33]

As we saw in the case of Cornaro, most of the classical and Renaissance texts on preventive medicine were based on the doctrine of moderation. In Bacon's view, however, things that are so strong that they "turn back the course of nature" cannot be mixed with common food or compounded with any medicine. Instead, they ought to be "used in series, and regularly, and at set times recurring at certain intervals."[34] Unfortunately Bacon himself did not give a detailed consideration of the things by which the prolongation of life could be attained, but was content with outlining some general principles.

After the Counter-Reformation in the seventeenth century, popular health books acquired more religious and moral tones, especially in Germany. It was the duty of a human being to remain healthy; not just a personal duty, but also a duty towards the state.[35] On the whole, until the mid-eighteenth century these popular books on health and longevity for laymen fell into two broad categories. To the first of these belongs books on regimen and long life, which were clearly written for the leisured and educated. Such texts were often as much moral treatises as medical ones. Thus books like Cornaro's *Discorsi* or George Cheyne's *An Essay of Health and Long Life* (1725) were to be read and contemplated.[36] To the second category belongs more pragmatic health manuals, consisting essentially of recipes or lists of medicaments and their application in the home treatment of perceived illness. These books were not so much to be contemplated as to be used in kitchens and sickrooms.[37]

## Hygiene in the Medical Curriculum in Early Modern Europe

It can be said that at least from the mid-seventeenth century, one can trace in European universities the development of a self-consciously scientific medicine, represented in the teaching of those faculties that sought to break free of the Aristotelian heritage, to create a rationally founded pathology and therapeutics, and to critically re-examine the empirical tradition. There developed in parallel a medical profession that strove to establish clearer boundaries between itself and nonprofessionals and to base its claim not only on experience but also on the mastery of a scientific medicine not accessible even to educated laymen.[38]

As regards the preventive part, in the introductions to the whole discipline of medicine which were written during the seventeenth and early eighteenth century such as Hoffman's *Fundamenta medicinae*, Hermann Conring's *Introductio ad medicinam artem*, and Boerhaave's *Institutiones medicae*, it is treated more and more briefly. A German physician Günther Christopher Schelhammer wrote additions to the second edition of Conring's *Introductio* (1726) where he tackles the question of whether hygiene or dietetics really is a part of medicine or not. Schelhammer nonetheless gives a positive answer to the question, as he does also in his posthumously published *Ars medendi universa* (vol.3, 1752), but the importance of this discussion lies in the fact that the question was ever raised.

The fiercest attack on the preventive part of medicine, however, had already been published in the previous century. In 1630 Petrus Lauremberg, a physician from Rostock, wrote a book called *Porticus Aesculapi*, which was a general introduction to the field of medicine. In this book Lauremberg severely attacks the idea of prevention and wants to exclude it from the canon of medical studies completely.[39] Lauremberg's extended argumentation is briefly summarized below.

His basic idea was that all cases in medicine where perfect health is broken can be described as due to some kind of deficiency. Consequently, all such cases do no longer belong to the preventive, but to the curative part, where the aim of the physician is to restore health in the patient.[40] On the other hand, the case of perfect health is a natural state which does not belong to medicine at all. According to Lauremberg, even Celsus had confirmed this in his *De re medica*, where he had stated that a healthy person should not go to see a physician. This was already a common proverb at the time of Lauremberg, and also

Mackenzie notes it disapprovingly in his book when he deals with the seven rules invented by Friedrich Hoffman.[41]

In the first history of medicine in English, *The History of Physick* (1723–24) by John Freind, dietetics is more or less separated from medicine and connected with other arts such as the art of cooking.[42] For Freind, as for many of his contemporaries, anatomy and chemistry were the developing and most important parts of medicine. In academic circles medicine was more and more defined through diseases (ts. nosology and pathology) and the central question was how to cure some particular disease; not any more how to take care of one's healthy body. This was partly due to the fear that in modern times many new diseases were discovered that were totally unknown in classical times and therefore medicine should concentrate on curing instead of caring.

The emphasis on cure can also be noticed in some early eighteenth-century medical dictionaries. In G. Motherby's dictionary, which was later corrected by George Wallis, Motherby defined medicine as "the art of preserving present, and restoring lost health; more properly the last."[43] Thus in a way both the preventive and the curative parts belonged still in medicine, but the latter one was already considered to be far more important. The same observation can be made when looking at the discussion that arose when George Cheyne published his *An Essay of Health and Long Life* in 1724. In the preface of his book entitled *An Essay concerning the Nature of Aliments* (1731), John Arbuthnot described the situation in the following way:

> This Book was receiv'd by the Publick, with the Respect that was due to the Importance of its Contents; it became the Subject of Conversation, and produc'd even Sects in the dietetick Philosophy. In some of those symposiac Disputations amongst my Acquaintance, being appeal'd to; I happen'd to affirm that the dietetick Part of Medicine depended, as much as any of the rest, upon scientifick Principles.[44]

As a result of his observations, John Arbuthnot decided to write a book about all the non-naturals where the scientific principles of this part of medicine could be expounded. However, only the books concerning aliment and air ever appeared.

When James Mackenzie, therefore, took on the task of writing a complete history of health in the 1750s, the whole subdiscipline of hygiene was increasingly considered to be something of less importance to the progress of medicine, or even something that was wholly outside the realm of scientific medicine. Thus this can clearly be seen as an attempt to answer the claims against hygiene. In short, Mackenzie tried to lay a basis for a new scientific hygiene that could meet these challenges.

## Legitimation by the Use of History

In the short historical sketch in his introduction (pages 5–12), Mackenzie expresses a clear notion of progress in medicine. He repeats several times in two or three pages words like "improve," "progress," and "improvement." Mackenzie had, however, a somewhat different idea of progress than other eighteenth-century writers on the history of medicine. Daniel LeClerc had noted in his history of medicine that dietetics is the part of the science where almost no progress had taken place since classical times. Mackenzie did not wholly agree with LeClerc, because he tried to include in medicine the new kind of dietetics that had been developed, at least during the previous hundred years.[45]

According to Mackenzie, of the six instruments of health, aliment is the only one of which mention is made before Pythagoras. And it was mere necessity that made human beings invent the first rudiments of the art of preserving health. Adam was obliged "to contrive some method of dressing the fruits of the earth, in such a manner as to make them agree better with him, than they had done quite rude and unprepared."[46] Hippocrates, however, was the first author to deal with these "six articles necessary of life," but those lay scattered throughout his works. Thus Mackenzie considers himself the first author to have attempted to collect them together and build up a canon for this branch of medicine.

In spite of his low estimation of past authors, Mackenzie did not see himself as the first modern author in this field. In his view, the modern foundation of dietetics was laid in the works of Santorio Santorio (1561–1636), a professor of theoretical medicine at the University of Padua in the early seventeenth century.[47] What was important in Santorio's work, according to Mackenzie, was the fact that he had developed the idea of insensible perspiration through the human skin in his book called *Ars de statica medicina* published first in 1614.[48]

Mackenzie believed that the ancient writers had made right observations in the field of dietetics but, unfortunately, they relied on a very confused medical theory. Thus Santorio was the person who had confirmed the observations of the ancients, but who had also added many valuable observations of his own.

> It is true that Hippocrates, Galen, and others among the ancients, by diligently observing the operations of nature, and following her steps, have given us excellent practical rules concerning health; but their knowledge of the animal machine was defective, and their reasoning obscure.
>
> The nature and quantity of insensible perspiration, discovered by Sanctorius, opened to physicians a much clearer view into the reasons and grounds of the rules of health established by the ancients than they had before.[49]

## Legitimation by the Measurable Method

Even more important was the fact that Santorio's method was based on measurement and calculation, and thus was mathematical at least with regard to weighing. Santorio had considered that the amount of insensible perspiration through the skin and lungs could be measured by a scale invented by himself. Over the course of many years he had weighed himself many times a day and as a result of these calculations he had formed some basic principles. Firstly, the *perspiratio insensibilis* that had been known since Erasistratus but which had been considered imponderable, could be determined by systematic weighing. Secondly, insensible perspiration was greater than all other forms of sensible bodily excretions combined. And finally, that it is not constant but varies considerably as a function of several internal and external factors, such as cold and sleep, which decrease it, and fever, which increases it.

The mathematical method was considered one of the basic guarantees of the scientific presentation and status of a discipline during the first half of the eighteenth century.[50] For example, in the introduction to the English translation of Santorio's *De statica medicina* James Keill wrote that "physical writers of late, have with a great deal of Industry and Success introduced Geometry into their Studies, and endeavoured to account for all that concerns the Animal Oeconomy upon Mechanical Principles."[51] Within a few decades, however, the applicability of mathematics to medicine was increasingly brought into question. In 1738 John Burton wrote in his book dealing with the non-naturals:

> For my Part, I can't agree with a very great Man, who says, that a thorough Knowledge of the Mathematics, ought to be made the distinguishing Characteristic of a Physician from a Quack; for the Mathematics can give us no more Help in the Cure of Diseases, than they can in explaining the Mysteries of reveal'd Religion.[52]

Neither was Mackenzie very keen on adapting mathematics to the field of medicine, or hygiene especially. "Method" and "rules" were keywords for him, too, but he did not believe in the force of mathematical methods in this branch of knowledge. Instead, Santorio's measurable method was a guarantee for Mackenzie that dietetics could be claimed to be scientific in the terms of the eighteenth-century view of science. He placed the work of Santorio even before Harvey's ideas of the circulation of the blood. He admitted that the theory of health was greatly improved by the knowledge of blood circulation, but the practical rules for preserving health, however, underwent few alterations, having been founded in nature and confirmed by the experience of ages long before Harvey's discovery.[53] For this reason, Mackenzie criticized Frederick Hoffman and other "moderns" who "address the public with such an air of superiority, as if themselves had invented the rules which they only transcribe."[54]

Interestingly, for Mackenzie the founding father of modern medicine was not Vesalius or Harvey, as most of us would be inclined to think today, but Santorio Santorio. In fact, during the eighteenth century there did not yet prevail any unanimity about the founders of modern medicine. Only a few decades earlier John Freind in his *History of Physick* (1725–26) had praised Thomas Linacre as the founder of the Royal College of Physicians and, by implication, as the initiator of modern progressive medicine as well.[55]

Mackenzie expounded in full his sources. He regretted that he has not found all the texts he had been looking for, even though he had searched "the immense libraries of Oxford" and written to his friends in London and Holland as well. In this context Mackenzie makes an interesting remark about his relationship to the writers of the other fields of medicine. He states that "systematical writers in physic I seldom take notice of, as most of them touch but very slightly on my subject."[56]

The gap between hygiene and other parts of medicine seemed to be a reality for Mackenzie. He stated that he wanted to preserve the spirit and sense of past authors rather than give a close translation of their words, and thus tried to reduce different precepts "to a proper method." In the end of his introduction, while dealing with his dedication to the Bishop of Worcester, Mackenzie connects the preservation of health with philosophical, namely moral, issues. In his view, this branch of medicine "is a philosophical as well as a medical subject."[57] Mackenzie even saw it as a duty to take care of one's own health. He wanted to upgrade the status of dietetics also by referring to its high moral principles, and to the common good which is attainable, if certain rules are followed.

## Laying Out a New Basis for Hygiene: Mackenzie's Rules of Health

In the last part of his *History of Health* James Mackenzie introduces his own rules of health as the basis of a new dietetics. In his own words he had "collected into a narrow compass those general and particular rules which are most conducive to health in the several periods and circumstances of life."[58] These rules of health, however, cannot be understood without some basic facts about the functions of the human body, the animal economy, as it was termed in those days.[59]

> And here we may, with pleasure, remark a surprising agreement and harmony between the successful practice of the antients, directed only by their assiduous observation of nature, and the mechanical theory of the moderns, founded upon the wonderful structure of our solids, and the perpetual rotation of our fluids, with which the ancients were unacquainted.
>
> Anatomy discovers ten thousand beauties in the human fabrick, which I have no room to mention here; nor is it possible, in a performance of this kind, to describe the geometrical accuracy with which the author of nature has formed every part of the body to carry on the animal oeconomy, and answer the various purposes of life. All I propose in this place is, by touching upon a few particulars, to give those, who are unacquainted with our profession, a general idea of the structure of their own bodies, from which they will easily apprehend, that intemperance, sloth, and several other vices and errors, have a necessary and mechanical tendency to destroy health.[60]

In Mackenzie's view, Hermann Boerhaave was the man who had given the most complete view of the animal economy. His book, however, was written for physicians only, "and no man, probably, of any other profession will ever take the pains to understand it perfectly."[61]

As regards the rules of health, some of them "are general or common to all ages and conditions of men; and some are particular, or adapted to different periods and circumstances of life."

> Under the general rules are comprehended those which relate to the six instruments of life, as air, aliment, etc. together with some other useful maxims. Under the particular rules are reckoned, first, Those which are peculiar to different temperaments, namely, the bilious, sanguine, melancholic and phlegmatic. Secondly, Those rules that belong to different periods of life, as infancy, youth, manhood and old age. Thirdly, Those that are appropriated to different conditions and circumstances of men, considered as active and indolent, wealthy or indigent, free or servile.[62]

The last group of rules mentioned by Mackenzie deals with the social circumstances of his environment. During the previous centuries the dietetical literature was mainly directed to the upper middle class. In the treatises the leisured and unhealthy courtly life is compared with the active and healthy life in the countryside. The life of urban intellectuals, who did not have enough exercise, was often considered dangerous for one's health, too.[63]

Also, in his *History of Health* James Mackenzie held the opinion that even poor people have always some leisured hours during the day, which may be employed for the purpose of health. He came to the conclusion that "the poor, if they are virtuous and cleanly, have great advantages over the rich, with respect to health and long life, as the narrowness of their circumstances prompts them to labour and withdraws all temptations to luxury."[64] The quotation well confirms the claim made recently that the equation of health with the poorest sections of society can be interpreted as an implied justification for doing nothing about their condition of life.[65]

Mackenzie wanted to replace the six non-naturals with the term "six instruments of life."[66] According to our writer, the term non-naturals is extremely shocking, even if the six things themselves are necessary to the life of man. Mackenzie thus followed the Galenist distinction of the six non-naturals, even if he condemned the actual phrase, and he even stuck to the temperaments as the four basic human constitutions. Thus Mackenzie was not able to consider the discipline of hygiene in a new light, even if his division between the general and particular rules is reminiscent of the division that later came to be called public and private hygiene.

In the ancient world the human constitution had been highly individual and thus the regimens which were to be followed were always written for some particular person, never for a group or a class of people. During the eighteenth century, however, the idea of common hygienic rules was introduced and it gained acceptance rapidly.[67] The other important change in the late eighteenth-century hygienic literature was the new emphasis put on environment and its impact on the individual. In the Galenist framework an individual could affect his own health by following the regimen, but in the more and more popular

Hippocratic emphasis, the environment increasingly determined human society, manners, and health.[68]

## The Aftermath of Mackenzie

Mackenzie's *History of Health* gained wide popularity after its publication and in two years it went to three editions and was translated into French and German.[69] The book was considered as a general introduction to the history and practice of hygiene which, in spite of the critique mentioned above, was in general still held to be a part of medical studies. However, when the public health movement spread rapidly at the end of the eighteenth century, authors like Mackenzie, who in spite of some reformatory ideas, had approved the classical idea of hygiene as a personal program, began to look hopelessly outdated.

In any event, Mackenzie's ideas lived an interesting afterlife in John Sinclair's (1754–1835) book *The Code of Health and Longevity*.[70] Sir John Sinclair, first president of the board of agriculture, was a typical encyclopedist of his time who tried to build up a system he called the "Codean system of literature," in which all knowledge was to classified into four departments, comprising agriculture, health, political economy, and religion. After the *Code of Health* he published a similar code of agriculture (1817), but the other two works were never completed. *The Code of Health and Longevity* was directly influenced by Mackenzie's book, even to the extent that some considered it to be a plagiary of it. Sinclair's intention was to lay out in similar vein the rules of hygiene, and he also included in his work a list of all dietetical literature published until his own time, amounting to 1878 items![71]

In the last volume of the *Code* Sinclair also included some translations of central dietetical texts such as Santorio's book *De statica medicina*, which had been a source of inspiration for Mackenzie, too. However, Sinclair also admired the recent work of Jean Nöel Halle, where the idea of public health dominated the field of hygiene, and Halle's article on hygiene was translated in the last volume of Sinclair's *Code*.[72] In spite of these additions, Sinclair's book was already considered to be too old-fashioned to be of any use in contemporary practice.

Already by the 1790s modified hygienic humoralism had become outdated; partly because it no longer provided an adequate terminology for the late eighteenth century physiology of the nervous system and sense impressions.[73] In other words, mechanical explanations had turned to vitalist ones.[74] This does not mean, however, that hygiene as a discipline would have definitively been discarded from scientific medicine. Rather it had once more to be scientifically justified. This was the situation already in the 1820s and 1830s when the professionalization of public hygiene was established.[75]

## Notes

1. John Sinclair, *The Code of Health and Longevity*, vol.1 (Edinburgh, 1807), 170.

2. Ginnie Smith, "Prescribing the Rules of Health: Self-help and Advice in the Late Eighteenth Century," in Roy Porter (ed.), *Patients and Practioners. Lay Perceptions of Medicine in Preindustrial Society* (Cambridge, 1985), 253–82, esp. 254.

3. In the Renaissance the word "dietetics" was often used as a synonym for the word "hygiene," which is a word of Greek origin. Later dietetics referred most often to one of the three parts of therapeutics, the other parts being surgery and pharmaceutics. On the other hand, at that time, hygiene referred to the whole preventive part of medicine. In this article both these terms refer to the preventive part of medical studies.

4. Angel González de Pablo, "La dietetica para el hombre sano en el pensamiento medico del mundo moderno," *Asclepio* 42 (1990): 69–117; Richard Palmer, "Health, Hygiene and Longevity in Medieval and Renaissance Europe," in Y. Kawakita, S. Sakai, and Y. Otsuka, eds., *History of Hygiene* (Tokyo, 1991), 75–98; Andrew Wear, "The Popularization of Medicine in Early Modern England," in Roy Porter, ed., *The Popularization of Medicine 1650–1850* (London and New York, 1992), 17–41; A. Wear, "The History of Personal Hygiene," in W.F. Bynum and Roy Porter, eds., *Companion Encyclopedia of the History of Medicine*, vol.2 (London and New York, 1993), 1283–1308.

5. Mackenzie was educated at Edinburgh University and entered the University of Leiden in 1700. After completing his studies he became a member of the Royal College of Physicians, in Edinburgh, and practiced as a country physician in Worcester. Mackenzie retired in 1751, settled in Kidderminster, and died at Sutton Coldfield, Warwickshire, in 1761. He wrote *The History of Health* only three years before his death and dedicated it to Isaac Maddox, the Lord Bishop of Worcester, who initially had encouraged him to write it. For more biographical information on Mackenzie, see the *Dictionary of National Biography*, vol. 35 (ed. by Sidney Lee, London, 1893), 154–55.

6. LeClerc's and Freind's histories are usually considered to be in the field of medicine the two best examples of the history writing that gained wide popularity in the late seventeenth and early eighteenth centuries. In spite of the differences in their approaches, they at least shared the idea of the importance of physiological, anatomical and pathological studies as the basis of medical studies. Therefore, neither of them approved very much of hygienic literature and the preventive part of medicine. On Freind's book and its political connections, see Julian Martin, "Explaining John Freind's History of Physick," *Studies in History and Philosophy of Science* 19 (1988): 399–418; on LeClerc, see P.Roethlisberger, "Daniel Le Clerc und seine 'Histoire de la médecine,'" *Gesnerus* 21 (1964): 126–41. On the eighteenth-century histories of medicine, see Edith Heischkel, *Die Medizinhistoriographie im XVIII. Jahrhundert* (Leiden, 1931); and on the idea of writing disciplinary histories, see Peter Burke, "Reflections on the Origins of Cultural History." in Joan H.Pittock and Andrew Wear, eds., *Interpretation and Cultural History* (London, 1991), 5–24.

7. James Mackenzie, *The History of Health and the Art of Preserving It* (3rd ed., Edinburgh, 1760), 3. I have used the Arno Press facsimile reprint of the third edition of Mackenzie's book (New York, 1979).

8. On the various ways of legitimating a discipline, see Nicholas Jardine, "Legitimation and History," ch. 6 in *Scenes of Inquiry* (Oxford, 1991), 121–45.

9. On a parallel case of raising the status of a discipline, see Lissa Roberts, "Filling the Space of Possibilities: Eighteenth-century Chemistry's Transition from Art to Science," *Science in Context* 6 (1993): 511–53.

10. Mackenzie, *History of Health*, 4.

11. Even if Mackenzie called these factors "Galen's Six Non-naturals," the term, in fact, dates from the Middle Ages. However, the list of six factors can be found in texts attributed to Galen, for

example, in the third book of *Tegni*. On the six non-naturals in classical time, see Luis García-Ballester, "On the Origin of the 'Six Non-natural Things' in Galen". In Jutta Kollesch und Diethard Nickel eds., "Galen und das hellenistische Erbe," *Sudhoffs Archiv, Beiheft* 32 (1993): 105–15.

12. On the conceptual history of the term "non-naturals," see García-Ballester, "On the Origin of the Six Non-natural Things"; L.J. Rather, "The 'Six things non-natural': a Note on the Origins and Fate of a Doctrine and a Phrase," *Clio Medica* 3 (1968): 337–47; Saul Jarcho, "Galen's Six Non-naturals: a Bibliographical Note and Translation," *Bulletin of the History of Medicine* 44 (1970): 372–77; Jerome J. Bylebyl, "Galen on the Non-natural Causes of Variation in the Pulse," *Bulletin of the History of Medicine* 45 (1971): 482–85; Peter H. Niebyl, "The Non-naturals," *Bulletin of the History of Medicine* 45 (1971): 486–92.

13. On humoralism, see Vivian Nutton, "Humoralism," in *Companion Encyclopedia of the History of Medicine*, vol.1 (London and New York, 1993), 281–91; Jerome Bylebyl, "The Medical Side of Harvey's Discovery: The Normal and the Abnormal," in Jerome J. Bylebyl, ed., *William Harvey and his Age* (Baltimore and London, 1979), 28–102, esp. 28–32.

14. Nancy G. Siraisi, *Avicenna in Renaissance Italy* (Princeton, 1987), 101–3, and Jerome J. Bylebyl, "Teaching 'Methodus medendi' in the Renaissance," in F. Kudlien and R.J. Durling, eds., *Galen's Method of Healing* (Leiden, 1991), 157–89, esp. 165.

15. Mackenzie himself, referring to Friedrich Hoffman, emphasizes the latter alternative. See Mackenzie, *History of Health*, 4.

16. On the medical curriculum in Renaissance universities, see Bylebyl, "The School of Padua: Humanistic Medicine in the Sixteenth Century," in C. Webster, ed., *Health, Medicine and Mortality in the Sixteenth Century* (Cambridge, 1979), 335–70; Bylebyl, "Medicine, Philosophy and Humanism in Renaissance Italy," in J.W. Shirley and F.D. Hoeniger, eds., *Science and the Arts in the Renaissance* (Washington D.C. and London, 1985), 27–49.

17. Siraisi, *Taddeo Alderotti and His Pupils* (Princeton, 1981), 138, 226–27.

18. For an example of medieval *consilia*, see Taddeo Alderotti, *I consilia*, ed. G.M. Nardi (Turin, 1937).

19. P.W. Cummins, "Introduction," in *A Critical Edition of 'Le regime tresutile et tres proufitable pour conserver et garder la santé du corps humain'* (Chapel Hill, 1976); Melitta Weiss-Amer, "Food and Drink in Medical-dietetic *Fachliteratur* from 1050 to 1400," thesis (University of Toronto, 1989).

20. On medical humanism, see Nutton, "John Caius and the Linacre Tradition," *Medical History* 23 (1979): 373–91; Nutton, "John Caius and the Eton Galen: Medical Philology in the Renaissance," *Medizinhistorisches Journal* 20 (1985): 227–53; Nutton, "Hippocrates in the Renaissance," *Sudhoffs Archiv, Beiheft* 27 (1989): 420–39; Siraisi, *Avicenna in Renaissance Italy*, 61–76; Daniela Mugnai Carrara, *La biblioteca di Nicolò Leoniceno. Tra Aristotele e Galeno: cultura e libri di un medico umanista* (Florence, 1991).

21. On Galen's text called *Trasybulos*, see Ludwig Englert, *Untersuchungen zu Galens Schrift Trasybulos* (Leipzig, 1929).

22. On the intentions of the anatomical writers, see Roger French, "Berengario da Carpi and the Use of Commentary in Anatomical Teaching," in A. Wear, R.K. French, and I.M. Lonie, eds., *The Medical Renaissance of the Sixteenth Century* (Cambridge, 1985), 42–74; Andrew Cunningham, "Fabricius and the 'Aristotle Project' in Antomical Teaching and Research at Padua," ibid., 195–222; and Heikki Mikkeli, *An Aristotelian Response to Renaissance Humanism. Jacopo Zabarella on the Nature of Arts and Sciences* (Helsinki, 1992), 148–59.

23. Per-Gunnar Ottosson, *Scholastic medicine and philosophy* (Naples, 1984), 277–80. On the organization and principles of the preventive part of medicine in Renaissance universities, see Heikki Mikkeli, *Preventive Medicine as a Discipline in Early Modern Europe* (forthcoming).

24. Eleanor B. English, "Girolamo Cardano and 'De sanitate tuenda': A Renaissance Physician's Perspective on Exercise," *Research Quarterly for Exercise and Sport* 53 (1982): 282–90. On Cardano and his work, see Siraisi, "Girolamo Cardano and the Art of Medical Narrative," *Journal of the History of Ideas* 52 (1991): 581–602.

Legitimizing a Discipline 293

25. Nutton, "Les Exercices et la santé: Hieronymus Mercurialis et la gymnastique médicale," in J. Céard, M.M. Fontaine, and J-C. Margolin, eds., *Le Corps à la Renaissance* (Paris, 1990), 295–308.

26. On the metaphor of mirror in the titles of these advice books, see Herbert Grabes, *The Mutable Glass. Mirror-imagery in Titles and Texts of the Middle Ages and English Renaissance* (Cambridge, 1982).

27. Paul Slack, "Mirrors of Health and Treasures of Poor Men: The Uses of the Vernacular Medical Literature of Tudor England," in C. Webster, ed., *Health, Medicine and Mortality in the Sixteenth Century* (Cambridge, 1979), 237–73.

28. Alvise Cornaro, *Scritti sulla vita sobria*, ed. M. Milani (Venice, 1983). See esp. the thorough introduction by Marisa Milani (5–60).

29. As a matter of fact Cornaro was "only" 82 years old when his work was published. He did, however, make later additions to his book at the age of 89. See M. Milani's "Introduzione," 12.

30. On the idea of doctrinal differentiation as a precondition of medical professionalization, see John Henry, "Doctors and Healers: Popular Culture and the Medical Profession," in S. Pumfrey, P.L. Rossi, and M. Slawinski, eds., *Science, Culture and Popular Belief in Renaissance Europe* (Manchester, 1991), 191–221.

31. Like many other writers of these health books, Sir Thomas Elyot was not an academic teacher, but a humanist who acted as a civil servant and a diplomat. His most famous book was *The Book Named The Gouvernour* (1531), which was an advice-book for rulers and other governors. On *The Castle of Health*, see John Villads Skov, "The First Edition of Sir Thomas Elyot's 'Castell of Helthe' with Introduction and Critical Notes," unpublished dissertation (University of California, Los Angeles, 1970). On English health manuals in general, see Jane O'Hara-May, *Elizabethan Dyetary of Health* (Lawrence, 1977).

32. Gerald J. Gruman, "The Rise and Fall of Prolongevity Hygiene 1558–1873," *Bulletin of the History of Medicine* 35 (1961), 221–29.

33. Francis Bacon, Translation of the *De augmentis scientiarum*, in *The Works of Francis Bacon*, vol. 4, ed. J. Spedding (London, 1858), 391.

34. Ibid., 393.

35. Alfons Labisch, *Homo Hygienicus. Gesundheit und Medizin in der Neuzeit* (Frankfurt and New York, 1992), 69–104.

36. On Cheyne's work, see Paul W. Child, "Discourse and Practice in Eighteenth Century Medical Literature: The Case of George Cheyne," unpublished diss. (University of Notre Dame, 1992).

37. Charles E. Rosenberg, "Medical Text and Social Context: Explaining William Buchan's 'Domestic medicine,'" 23–24. *Bulletin of the History of Medicine* 57 (1983): 22–42.

38. Matthew Ramsey, "The Popularization of Medicine in France, 1650–1900," in Porter, ed., *The Popularization of Medicine 1650–1850*, 97–99.

39. Petrus Lauremberg, *Porticus Aesculapi, seu generalis artis medicae constitutio* (Rostochi, 1630)., esp. ch. 13: "Sanitatis conservatio, an pertineat ad medicinam?" (fol.67–78).

40. The critique against hygiene, introduced by William Cullen 150 years later, rested basically on the same principle. Cullen wrote, "The common language is that 'Medicine is the art of preserving health and of curing diseases', but I have said 'the art of preventing diseases'; for although I do not deny that the preserving of health is the object of a physician's care, yet I maintain that there is truly no other means of preserving health but what consists in preventing disease. Every other idea is false, and has led to a superfluous, very often a dangerous practice. I say, that health properly understood, we cannot add to it, nor increase its powers. There is never room for our art, but when there is some defect in the constitution—some bias and tendency towards disease; and it is only by preventing this tendency, by correcting these defects, that is by preventing disease, that we preserve health." William Cullen, MSS, *Lectures on physiology*, 5 vols. (ca.1770), cited from Rosalie Stott, "Health and virtue: or, how to keep out of harm's way. Lectures on pathology and therapeutics by William Cullen, ca.1770," *Medical History* 31 (1987): 123–142, esp. 128.

41. According to Mackenzie, Hoffman had borrowed five of the seven rules from Galen and one from Hippocrates. At the end of his list Hoffman had added a "curious rule of his own": "Avoid physic and physicians, if you 'have any value for your health.'" (Fuge medicos et medicamenta, si vis esse salvus.) See Mackenzie, History of Health, 11–12.

42. John Freind, The History of Physick, vol. 2 (London, 1750), 202–3, 290–91. I have used the AMS Press facsimile reprint (New York, 1973).

43. G. Motherby, A New Medical Dictionary, or General Repository of Physic, 3rd rev. ed. with additions by George Wallis (London, 1791), 505. On other authors attacking the preventive part of medicine, see Harold J. Cook, "Physick and Natural History in Seventeenth-century England," in P. Barker and R. Ariew, eds., Revolution and Continuity. Essays in the History and Philosophy of Early Modern Science (Washington, DC, 1991), 63–80.

44. John Arbuthnot, An Essay concerning the Nature of Aliments (London, 1731), preface, iii.

45. Mackenzie, History of Health, 251–52.

46. Ibid., 28.

47. For biographical information on Santorio, see M.D. Grmek, "Santorio, Santorio," in Dictionary of Scientific Biography, vol. 12, ed. by C.C. Gillispie (New York, 1975), 101–4.

48. On the history of the idea of insensible perspiration, see E.T. Renbourn, "The Natural History of Insensible Perspiration: A Forgotten Doctrine of Health and Disease," Medical History 4 (1960): 135–52.

49. Mackenzie, History of Health, 288–89.

50. On the importance of the mathematical model for the eighteenth century concept of science, see T. Frängsmyr, J.L. Heilbron, and R.E. Rider, eds., The Quantifying Spirit in the Eighteenth Century (Berkeley, 1990).

51. James Keill, Medicina statica: Being the Aphorisms of Sanctorius, Translated into English with Large Explanations (London, 1720), 1. On Keill and the adoption of Newtonian principles in medicine, see Anita Guerrini, "James Keill, George Cheyne, and Newtonian Physiology," Journal of the History of Biology 18 (1985): 247–66.

52. John Burton, A Treatise on the Non-naturals (York, 1738), 10. The "very great" man Burton refers to seems to be Boerhaave, under whom he had studied in Leiden, and to whom the whole book is dedicated.

53. Mackenzie, History of Health, 290.

54. Ibid., 10.

55. John Freind, The History of Physick, 400–7.

56. Mackenzie, History of Health, 12.

57. Ibid., 15.

58. Ibid., 14.

59. John P. Wright, "Metaphysics and Physiology. Mind, Body, and Animal Economy in Eighteenth-century Scotland," in M.A. Stewart, ed., Studies in the Philosophy of the Scottish Enlightenment (Oxford, 1990), 251–301. On the early conceptual history of the term "animal economy," see B. Balan, "Premières recherches sur l'origine et la formation du concept d'économie animale," Revue d'histoire des sciences 28 (1975): 289–326.

60. Mackenzie, History of Health, 330–31.

61. Ibid., 333 (note). On the importance of Boerhaave for eighteenth-century Scottish medicine, see the illuminating article by Andrew Cunningham, "Medicine to Calm the Mind. Boerhaave's Medical System, and Why It Was Adopted in Edinburgh," in A. Cunningham and R. French, eds., The Medical Enlightenment of the Eighteenth Century (Cambridge, 1990), 40–66. Mackenzie shared with Boerhaave the view that history was really important for progress in medicine, and that Hippocrates had been the most prominent of the ancient authors in this field (ibid., 47, 54).

62. Mackenzie, History of Health, 366–67.

63. For example, Giovanni Argenterio, in his commentary on Galen's Tegni (1566), while discussing the causes of diseases, notes that country people leading a simple life are more healthy than princes or priests in towns. Giovanni Argenterio, In artem medicinalem Galeni commentarii (Monte Regali, 1566), 366A.

64. Macenzie, *History of Health*, 419.
65. A. Wear, "Making Sense of Health and the Environment in Early Modern England," in A. Wear, ed., *Medicine in Society. Historical Essays* (Cambridge, 1992), 119–47, esp. 132. On the "diseases of civilization," see Roy Porter, "Civilisation and Disease: Medical Ideology in the Enlightenment," in J. Black and J. Gregory, eds., *Culture, Politics and Society in Britain, 1660–1800* (Manchester and New York, 1991), 154–83.
66. Mackenzie, *History of Health*, 367.
67. On the idea of hygiene as a part of public health, see Ann F. La Berge, *Mission and Method. The Early Nineteenth-century French Public Health Movement* (Cambridge, 1992); L.J. Jordanova, "Policing Public Health in France 1780–1815," in T. Ogawa, ed., *Public Health* (Tokyo, 1981), 12–32; William Coleman, "Health and Hygiene in the *Encyclopédie*: A Medical Doctrine for the Bourgeoisie," *Journal of the History of Medicine* 29 (1974): 399–421.
68. Harold J. Cook, "The New Philosophy and Medicine in Seventeenth-century England," in D. Lindberg and R. Westman, eds., *Reappraisals of the Scientific Revolution* (Cambridge, 1990), 397–436, esp. 423; and L.J. Jordanova, "Earth Science and Environmental Medicine: The Synthesis of the Late Enlightenment," in L. Jordanova and R. Porter, eds., *Images of the Earth* (Chalfont St. Giles, 1979), 119–46.
69. Jaques Mackenzie, *Histoire de la santé et de l'art de la conserver* (The Hague, 1759); James Mackenzie, *Die Geschichte der Gesundheit und die Kunst dieselbe zu erhalten* (Altenburg, 1762).
70. For biographical information on John Sinclair, see *Dictionary of National Biography*, vol. 52, ed. by Sidney Lee (London, 1897), 301–5.
71. Sinclair *Code of Health and Longevity*, 2:301.
72. Jean Noël Hallé, "Hygiène," in *Encyclopédie méthodique médecine* vol. 7 (Paris, 1798), 373–437.
73. Virginia Smith, "Physical Puritanism and Sanitary Science: Material and Immaterial Beliefs in Popular Physiology, 1650–1840," in W.F. Bynum and R. Porter eds., *Medical Fringe and Medical Orthodoxy 1750–1850* (Kent, 1987), 174–97, esp. 186; Guenter B. Risse, "Medicine in the Age of Enlightenment," in A. Wear, ed., *Medicine in Society* (Cambridge, 1992), 149–95.
74. Theodore M. Brown, "From Mechanism to Vitalism in Eighteenth Century English Physiology," *Journal of the History of Biology* 7 (1974): 179–216.
75. On this development, see Ann F. La Berge, "The Early Nineteenth Century French Public Health Movement: The Disciplinary Development and Institutionalization of 'hygiène publique,'" *Bulletin of the History of Medicine* 58 (1984), 363–79; Caroline Hannaway, "From Private Hygiene to Public Health: A Transformation in Western Medicine in the Eighteenth and Nineteenth Centuries," in T. Ogawa, ed., *Public Health*, 108–28.

# THE MANTLE OF MÜLLER AND THE GHOST OF GOETHE
## Interactions between the Sciences and Their Histories

### Nicholas Jardine

### Introduction

In much of the recent literature the interactions between the scientific disciplines and their histories are treated as inessential, undesirable, and avoidable. Exceptionally, when data collected long ago are of current relevance, scientists may have to concern themselves with the histories of their disciplines; but otherwise pursuit of the history of the sciences is no part of the practice of the sciences themselves. Likewise, to the extent that historians of the sciences show themselves prejudiced by current scientific beliefs or involved in current scientific debate they depart from properly historical research and writing. Further, it is often implied, such entanglements are unnecessary. Thus the persuasive and polemical uses of history by medics and scientists are treated as merely rhetorical moves, as if it were always open to them to forswear such shabby tactics in favor of wholesomely rational argument and objective presentation of information.[1] Moreover, imposition by historians on past actions and beliefs of presuppositions, concepts, and values derived from present-day science and medicine is castigated as a vice, curable through honest reflection on their prejudices and sympathetic attention to past agents' own categories.[2]

The study which follows does not directly address general philosophical and historiographical issues.[3] Rather it sets out to show by example that these views about the separability of disciplines from their histories are too easy and simplistic: that at least in some fields and in some periods historical work is to be seen not as rhetorical embellishment but as inextricable from the processes of scientific and medical discipline formation; and conversely, that historians' concerns with the nature and goals of medicine and the sciences may be sources of interest and critical power.

After some historical stage-setting I shall tell how Emil Du Bois-Reymond and Rudolf Virchow used history to articulate and promote their own very

---

I thank Professor Robert Brain, Dr. Soraya de Chadarevian, Dr. Michael Hagner, Dr. Harmke Kamminga, Professor Donald Kelley, Dr. Marina Frasca-Spada, Professor Piyo Rattansi, and Dr. Ulrich Schneider for valuable advice. I owe special debts to Professor David Cahan and Dr. Keith Anderton, who offered constructive comments on a draft of this article and kindly rescued me from some outright errors.

different conceptions of science and medicine, their rival plans for the reform of scientific and medical education, and their sharply contrasting visions of the proper roles of science and medicine in German society. I shall conclude with some reflections on the implications of such interactions between disciplines and their histories for the current historiography of medicine and the sciences.

## Medicine, Science, and Revolution

Let us start the story in Berlin in the revolutionary decade, the decade which ends with the insurrection of March 1848 and the reactionary suppression of the liberal reforms at the end of 1849.[4] In this period Du Bois-Reymond and Virchow shared certain aspirations. Both were taught by Johannes Müller, Professor of Anatomy and Physiology at the University of Berlin, and in 1843 both received their doctorates for dissertations on physiological topics from Müller acting in his capacity as Dean of the Medical Faculty. Both were members of the Berlin *Physikalische Gesellschaft*. From Virchow's numerous articles in the radical journal *Medicinische Reform* it is clear that he endorsed the basic reformist goals of the *Gesellschaft*. And from Du Bois-Reymond's sentimental correspondence with the radical Eduard Hallmann, it appears that he too was in sympathy with them. First and foremost amongst these goals was the creation of a new kind of teaching and practice of medicine and the sciences, centered on laboratory research. They envisaged a natural science that would unite theory and practice in a new way, opposed on the one hand to blind empiricism, on the other hand to the vain speculations of *Naturphilosophie*. It was to be a strictly mechanistic science with a mechanistic physiology as its standard bearer—thus it was substantially opposed to the vitalistic and teleological physiology of their master Johannes Müller. It was to be a unified science, united under mechanistic principles, and a unifying science, fostering the union of the German lands by breaking down the barriers of superstition.

The *Physikalische Gesellschaft* had more specifically institutional goals as well. Its members wanted participation in the government of the university by all university teachers, including *Privatdozenten* and extraordinary professors, the highest academic ranks to which most of the still rare breed of natural scientists could aspire. These plans were to be crushed by a state-dictated decree of 1849, pronounced by the conservative Müller, reluctantly serving as Rector of the University. They wanted a standardization of medical education in which natural scientific, clinical, and German language training would be greatly increased. These plans came to partial fruition in 1861 with the replacement of the *Tentamen Philosophicum* by the *Tentamen Physicum*. They wanted abolition of the state licensing boards with their power to interfere in all aspects of the lives of physicians—in some states, for example, the boards permitted a physician to marry only after his fiancée had been vetted. The abolition of the licensing boards was achieved in 1869.

Even at this early stage in their meteoric careers, however, there are apparent differences of orientation between the confident, emotionally mature Virchow and the insecure, excitable, and opportunistic Du Bois-Reymond.[5] For all his occasional fits of revolutionary enthusiasm and thrills at the sight of Republican troops, Du Bois-Reymond played no active role in revolutionary activities, and stayed at home on the day of the *Marzkampf*.[6] His primary commitment was already to the laboratory researches through which he dreamed of achieving scientific fame and a prominent place in Berlin society. Worldly affairs he found a distraction from work—he had a particular horror of ending up in medical practice.[7] He was, and was to remain, a university figure. It was through his electrophysiological researches that he sought international recognition as a leading German man of science. And it was in the often ferocious infighting of the University of Berlin and the Berlin Academy of Sciences that he was to exercise his formidable oratorical talents. Virchow, by contrast, was active on a far wider stage—a political radical who fought at the *Marzkampf* barricades, a practicing clinician and surgeon at the Charité military hospital, he was already editing *Medicinische Reform* and agitating for the formation of the democratically elected free associations that would govern medical and social practice once the petty tyrannies of state regulation were swept away.[8]

There is, as far as I know, no evidence that Du Bois-Reymond supported Virchow's more radical proposals and much evidence to suggest his lack of sympathy with them. The new medicine envisaged by Virchow was to be centred on a new pathological physiology and a reformed clinical practice. With the help of statistical methods, it would reveal the laws relating material conditions to society and culture; thus it would provide the foundation for social legislation and for the unification and regeneration of the German *Volk*. As one of the slogans of the medical reform movement put it: "medicine is a social science and politics is nothing but medicine on a large scale."[9] Further, Virchow was already speaking out fiercely against interference in medicine by outsiders: "we dispute the right of any discipline not itself rooted in the contemplation of diseased life to share in the interpretation of its phenomena."[10] By contrast, Du Bois-Reymond, like Hermann Helmholtz and Ernst Brücke, endorsed a reformed mechanistic physiology or "organic physics" centered on study of the healthy functions of living bodies. Indeed, he already dreamed, so his friend Carl Ludwig later testified, of precisely the reform of medicine from without that Virchow feared, of "going through the medical kraals, burning and destroying, the fiery sword of physics in his hand."[11]

In the decades of reaction after 1849, Virchow's and Du Bois-Reymond's positions on academic and medical reform diverged. For the rest of his life Virchow battled tirelessly for the reform of medical practice as part of a sweeping program of reformist social legislation. While banished to Würzburg from 1849 to 1856, then in Berlin as Professor of Pathology, head of a clinical section at the Charité, city counselor, and leader of the Progressive Liberals in the Reichstag, at every stage, Virchow agitated with outstanding effect for social reforms—of medical education, of clinical practice, of hospital building and design, of public hygiene, of factory and housing conditions, of prisons, of elementary schools, of social insurance, and so on and on, the list seems endless.[12]

After 1849 there is, to my knowledge, no evidence of Du Bois-Reymond's having shown radical sympathies. On the contrary, like many academic liberals of his generation, he committed himself to the Humboldtian ideals of his master, Müller, ideals of *Bildung*, of disinterested research, of detachment of the university from party squabbles and the commercial world of "practical interests."[13] He was, indeed, keen to give natural science the credit for the technical advances that had bettered man's estate. But his enthusiasm was increasingly dampened by cultural pessimism. The "disease of socialism" could, he hoped, be defeated; not so the barbarous Americanization of Germany that was the inevitable result of utilitarian attitudes and the commercial exploitation of technology.[14] Nationalistic, chauvinistic, and ever more deeply involved in academic administration, he consistently pleaded the virtues of an academically free university, promoting a balance of arts and sciences, as a vital organ and servant of the state.[15] In his notorious Rectoral address of 1870, "Der deutsche Krieg," he even went so far as to describe the University as "the intellectual bodyguard of the House of Hohenzollern," receiving a personal letter of congratulations from Bismarck and, a couple of months later, state funding for the palatial new Physiology Institute for which he had so long petitioned.[16] One of the few contentious public issues on which he did see eye to eye with Virchow was that of the need for a total secularization of the state and society. Indeed, he and Virchow briefly joined forces with Bismarck in the *Kulturkampf* against the Catholics, a move which Virchow later saw as a "miscalculation."

What of the crucial field of medical education?[17] This was crucial for Virchow because his whole vision of a future German liberal-democratic federation was centred on a reformed medicine. And it was crucial for both of them in their careers because the personal incomes they derived from capitation fees were conditional on recruitment of pre-clinical medical students as was much of the funding of their research institutes. In this field their divergence was especially sharp. Virchow's primary focus was on clinical reform, in particular on the clinical applications of his new discipline of cellular pathology.[18] With respect to teaching his emphasis was on the need for more and better instruction in the sciences both in the schools and in the preparation for the university *Tentamen Physicum*. Virchow opposed moves to make the university Doctorate a licensing condition for physicians. And like most others in the medical reform movement he pinned his hopes on the reform of the *Realschulen*, despairing of the elite *Gymnasia* and opposing their exclusive right to university places. Du Bois-Reymond, on the other hand, affirmed the priority of physiology over pathology in medical training; and he ardently promoted the pre-clinical university Doctorate in experimental physiology as a precondition of sound medical practice.[19] Moreover, though initially sympathetic to the cause of the *Realschüler*, as Rector of the University he eventually had a hand in the defeat of Virchow's movement for their admission to the universities.[20] In medical practice, as in other practical realms, Virchow showed a constant respect for tradition, in this case the tradition of "clinical tact" based on long experience.[21] Here it was not the radical Virchow but the conservative-liberal Du Bois-Reymond who spoke the language of revolution, publicly denouncing the entire medical profession as obscurantist and bigoted.[22]

## The Mantle of Müller and the Ghost of Goethe

Du Bois-Reymond and Virchow repeatedly voiced their commitment to history, both endorsing the slogan that things can be properly known only by their origins.[23] Both were convinced of the value for students of medicine and the natural sciences of instruction in the history of medicine and the sciences.[24] Both wrote extensively on historical topics, and did so in ways which measured up to the high standards of precise and critical scholarship prevalent in the Berlin of Ranke and Mommsen (a personal friend of Du Bois-Reymond). And, as we shall see, in the history of medicine and the life sciences their writings proved remarkably influential—Du Bois-Reymond establishing a framework for the interpretation of the history of physiology that remained unchallenged until very recently; Virchow pioneering a type of social history of medicine that still flourishes. But in their writings about history, as in the history they wrote, they differed very sharply indeed.

Du Bois-Reymond promoted and wrote what he called "inductive history," that is, history of the sciences premised on the thesis that "the historical path of the inductive sciences is for the most part almost the same as the path of induction itself."[25] Such a history, intended to inspire the students to independent research, tells how progress has been achieved by the heroes of the empirical sciences through rigorous experimental testing of hypotheses. From the Archimedean standpoint of the cosmos revealed by the sciences their history appears as "the absolute organ of culture and the only real history of mankind."[26] (Du Bois-Reymond here acknowledges Hegel's philosophy of history as his inspiration, remarking that the crucial difference lies in Hegel's commitment to *a priori* as opposed to *a posteriori* knowledge.) He contrasts this noble and edifying history of progressive understanding of the cosmos with the futile and pointless "bourgeois history" of kings and queens, bloodthirsty generals, plots, and assassinations.[27] World history he divides into epochs which end in revolutions, the latest (and, he strongly implies, the last) being the "technical-inductive age," the age of the experimental method. Opening with Galileo and Bacon, this age has only recently come to its wonderful but dangerous maturity with the fulfilment in industrial technology of Bacon's prophetic "knowledge is power."[28]

Virchow's historicism is more thoroughgoing, permeating all levels of his discourse.[29] In his central research fields his methods are declaredly historical or genetic, tracing the etiologies of diseases and of epidemics, the lineages of cells, the genealogies of races, cultures, and institutions. In his extensive works on local, popular, and disciplinary history, he is above all at pains to show how culture, including the culture of medicine and the sciences, is determined by material and social conditions. Thus, for example, he explores the relations of feudal society and feudal religious institutions to medieval medical practice; and he relates the formation of modern scientific medicine to the institution of teaching in specialized clinical hospitals in France.[30] In a manner reminiscent of the early Marx he is constantly alert to the ways in which theoretical systems reflect the political regimes under which they arise and whose interests they covertly serve—how Stahlian vitalism reflected and served absolute monarchy; how his own cellular pathology reflects the harmonious cooperation of members of an ideal liberal democracy.[31]

Both in public and in their correspondence from 1864 to 1894 (largely of a formal nature arising from their joint involvment in the sponsorship of scientific expeditions by the Humboldt Stifftung) Du Bois-Reymond and Virchow maintained cordial relations—perhaps surprisingly, given their dissonant views and interests, their competition for students and posts, and their aggressive oratory and polemic in other contexts.[32] From the moment of Virchow's triumphant recall from Würzburg to Berlin in 1856, however, there are copious—if coded—signs of conflict in their historical speeches and publications.

At the end of April 1858 Johannes Müller died. He had students in place in well over half the thirty or so chairs of physiology and anatomy in the German lands.[33] He had been, at least in the academic realm, one of the two or three most powerful German "men of science" of his time.[34] And for many, right across the political spectrum, he was the living embodiment of the Humboldtian ideals of the *wissenschaftlich Gebildeter* and of the disinterested *Forscher*. His mantle was well worth a fight. In this period memorial addresses provided a conventional forum for such struggles.[35] Du Bois-Reymond, Müller's successor in the Chair of Physiology, got in first with a blockbuster delivered to a crowded session of the Berlin Academy of Sciences. Virchow had his say more briefly before a yet larger audience a few days later in the Aula Magna of the University. Reading the two addresses together is a strange experience. It is hard to believe that Virchow and Du Bois-Reymond are talking of the same person. They agree on Müller's good looks, his conservatism, his exemplary disinterestedness, and his standing as a reformer of physiology, but there is little further common ground.

Du Bois-Reymond's Müller owed his successes to a disposition formed by nature and nurture. Born in Coblenz under the French tyranny over the Rhineland, he was inspired to heroic deeds by the warlike demeanor of the Napoleonic troops.[36] He combined his father's powerful physique and fine looks with his mother's will-power, perseverance, and sense of order.[37] In his early studies he became sunk in the *Traummeer* of *Naturphilosophie*, and was led by Goethe into a false valuation of observation over experiment.[38] During his stay in Berlin in 1823–24 he attended Hegel's lectures, but his anatomical studies under Rudolphi rescued his from *Naturphilosophie*.[39] As a professor of physiology and anatomy at Bonn he was led to the verge of mental collapse by is auto-experimental studies of sensation and consciousness.[40] However, following the ministerial call to Rudolphi's chair of anatomy and physiology at Berlin, Müller became less speculative; indeed, his subsequent career was a model of disinterested pursuit of empirical knowledge.[41]

Virchow's Müller too was formed by blood and soil. But how very differently. Like Cuvier, he was stimulated by the mixture of French and German culture in his education.[42] His origins were petty-bourgeois—his father was a cobbler—and Virchow implies, though he does not actually state it, that it was to this that he owed his practicality and talent for *Forschung*. It was from his mother that he acquired his imaginative religiosity.[43] Far from having corrupted him, Goethe saved him from nature-philosophical speculation, providing an example of *Enthaltsamkeit* (self-possession or moderation) in a frivolous age, inspiring him with a dedication to precise observation.[44]

In their treatments of Müller's career at Berlin the discrepancies are no less striking. Du Bois-Reymond's Müller belonged to a bygone era, his achievements transcended by the new experimental physiology of his students. Müller's careful observation in comparative anatomy and physiology, his polymathy, and his systematizing powers, made him a supreme completor and synthesizer of the work of others—a latter-day von Haller.[45] But because of his commitment to "plastic observation" rather than theoretical analysis and controlled experiment he made no discoveries of the first magnitude.[46] Du Bois-Reymond characterizes Müller as the Erasmus of the Reformation of Physiology.[47] Not one to hide his light under a bushel, he implies that he himself is the Luther (having in mind Luther's famous, and probably mythical, remark "Erasmus laid the egg of the Reformation, but I hatched it"). For he insinuates that it is he who has surpassed Müller's achievements by introducing the experimental-analytic method into physiology. Müller, he suggests, took on the colours of those he had vanquished, combining in his mature career a metaphysical vitalism, derived from the conquered *Naturphilosophen*, with an observational empiricism at odds with metaphysical speculation.[48] He himself, Du Bois-Reymond asserts, played the central role in the refutation of Müller's appeal to vital force.[49]

For Virchow, in sharp and, in this instance, surely deliberate contrast, it is precisely Müller's exemplary mastery of the true inductive method, that is, of hypothesis controlled by minute observation and a "genetic" approach based on comparison and cautious analogy, that enabled him in the course of his triumphant tenure at Berlin to unify and reform the entire field of the life sciences.[50]

In connection with their rival views on the reform of medical education, particular interest attaches to the discrepancy between their accounts of Müller's attitude to vivisection. Virchow himself performed many experiments in which damage was inflicted on live animals and the pathological anatomical effects were posthumously examined, but he showed little concern with the physiological monitoring of vivisected animals.[51] He reports with apparent approval

Müller's reluctance to vivisect and his belief that little is to be learned about vital functions from the abnormal responses of tormented animals.[52]

Du Bois-Reymond, by contrast, is at pains to rebut the view that Müller undervalued vivisection, portraying him as an active vivisector in his early career and remarking with Bernardian gusto that "it is a necessary attribute of a physiologist that his hands daily steam with blood."[53] At first sight, this is very odd indeed. For Du Bois-Reymond himself did not think of vivisection per se as an important research tool (though, of course, animals had to be sacrificed to provide the living preparations on which he performed his experiments). Indeed, a letter from Carl Ludwig reveals that he had started to learn vivisectional techniques only a few months previously. What is going on here? Ludwig's letter provides a clue:

> You feel like passing the time away for a while with vivisection: how happy I would be if I could be of nay use to you in this. Everything I can show you is trivial, absolutely commonplace, but that is why it has to be shown to you, else you would never hit upon it. It is necessary to you because the doctors want to see blood; they have to get used to it in good time if they ever want to be efficient in their profession as destroying angels. So come, you must crush Müller and can only do so if you step down a few thousand coils from your high region. I shall be glad to help you in this; you will go back to Berlin completely demoralised.[54]

It seems that Du Bois-Reymond had reluctantly acknowledged that his electrophysiological experiments were not enough, that performance of demonstration vivisections was crucial if he was to make his experimental physiology sufficiently attractive to compete for pre-clinical medical students with Müller's anatomy-based teaching. Only by recruiting and disciplining medics through the initiation rites of vivisection could he hope to achieve his reformist goals. In 1877, at the inauguration of his giant new Physiology Institute, Du Bois-Reymond was to declare his position more explicitly. Comparing the present state of physiology with that of his youth, he tells how, in the absence of teaching laboratories, "those of us who wanted to do physiological research" had to vivisect frogs and rabbits in their lodgings.[55] Later in the address he embarks on a bombastic defence of vivisection against the hypocritical objections of the "fox-hunting British," who want vivisection subjected to restrictive legislation.[56] He pleads instead for regulation of vivisection by the conscience of the physiologists themselves, and emphasises the indispensability of vivisection for experimental physiology.[57] Such a vivisection-based experimental physiology is, he insists, a better training for medics than pathology (an obvious poke at Virchow's adjacent Pathology Institute).[58] Indeed, he concludes, it is the primary task of the Institute to provide such teaching in an experimental physiology able to serve as a "torch and a weapon in the questionable twilight of medicine."[59] No wonder the squeamish Du Bois-Reymond felt compelled to feign belief in the power of vivisection as a tool of research.

Let us return to Goethe. In 1861 Virchow followed up his positive assessment of Goethe's influence on Müller and German physiology with a semi-popular booklet, *Goethe als Naturforscher*.[60] Here Goethe appears as an ideal of humanity, at once supreme artist, enlightened administrator, and distinguished empirical scientist. He is presented as the co-founder with Caspar Friedrich Wolff of the genetic method, generally acknowledged as the key to the empirical study of living beings.[61] Tribute is paid to his major discoveries in comparative anatomy, comparative physiology, and optics.[62]

Two decades later Du Bois-Reymond responded with his vitriolic "Goethe und kein Ende," delivered before the Berlin Academy of Sciences. It opens engagingly enough, playfully contrasting Professor Heinrich Faust in his "Gothic cell" with the professoriate of the new German universities.[63] But the tone rapidly becomes aggressive and contemptuous. Du Bois-Reymond presents Goethe as an amateur, one who "entirely lacked the organ for theoretical knowledge," namely quantitative and controlled experimentation.[64] He denounces Goethe's purely observational methods as incapable of revealing the mechanical causes of phenomena, as having given a disastrously wrong direction to German science, already lead astray by speculative nature-philosophy.[65] His alleged anatomical discoveries—here the reader is referred to Virchow's booklet—were largely anticipated by others.[66] As for his famous theory of colours, it was "the stillborn foolery of an autodidact dilettante."[67]

Du Bois-Reymond was by this time widely admired for his theatrical oratory, his *Prunkvortrag*.[68] But on this occasion he went too far, questioning the literary quality of Goethe's *Faust*. In particular, he claimed that Faust himself—at once lover, adventurer, man of action, and seeker after contemplative wisdom—lacks "psychological truth."[69] In his youth Goethe wavered in his commitment to his poetic vocation, admiring the life of heroic action symbolized by Napoleon and wasting much time in administrative activities in Weimar; and the Protean Faust's psychological untruth is but a projection of Goethe's confused mental state.[70] So too, he implies, are the many other "logical inconsistencies and ethical monstrosities" that mar the work.[71]

A *Goethestreit* ensued in the journals and newspapers—as Du Bois-Reymond later reported, the furor was second only to that occasioned by the address in the presence of the King in which he had compared Darwin's achievement with that of Copernicus.[72]

Virchow's motives in his tribute to Goethe are transparent—he is using Goethe to promote his own ideals of humanity and his own program in medicine and the life sciences. Du Bois-Reymond's long-delayed counter assessment is harder to explain, and the following remarks are conjectural. Others of his writings of the period show that Du Bois-Reymond was becoming seriously concerned about what he saw as a widespread and dangerous reversion to the speculative excesses of *Naturphilosophie*—the psychologist and physiologist Gustav Fechner, for example, had strayed from the path of empirical science into the swamps of metaphysics, as had the botanist Karl von Nägeli and the zoologist Ernst Haeckel.[73] And he was involved in a heated confrontation with an internationally renowned, ultra-reactionary (and eventually insane) astrophysicist, Karl Friedrich Zöllner, who had responded to Du Bois-Reymond's "Über die Grenzen des Naturerkennens" of 1872 by holding him up as an example of the way in which arrogance, materialism, and narrow specialization lead men of science into scepticism, agnosticism, and irresponsibility.[74] Further, Du Bois-Reymond and Virchow were by now in head-on competition for control of pre-clinical teaching at the University, and on opposite sides in the long-running and bitter public controversy over the admission of *Realschüler* to the universities.[75] Small wonder then that Du Bois-Reymond decided to make an example of Goethe, so denouncing him as to implicate all who prostitute ideas through their involvement in worldly affairs, and all who betray science by straying from the path of analytical experiment.

To this it is tempting to add an explanation more in Du Bois-Reymond's own style. Given the evidence from his youthful correspondence of his identification with Goethe and of his daydreaming of heroic deeds, together with his adult tributes to such heroes as Frederick the Great, one may well suspect that Du Bois-Reymond is denouncing himself: that in describing Goethe's projection of himself in Faust he describes his own projection onto Goethe; that it is he who yearns to be not just an academic man of science, but also adventurer and lover; that it is his own timid fascination with men of action that he attributes to Goethe; that the psychological untruth he detects in Goethe is his own.[76] Perhaps it is this undertone of empathy with his victim that makes the attack so unpleasantly effective.

## The Entanglement of Medicine in Its Histories

The historical writings I have discussed have preserved their reputation right up to the present day as important scholarly contributions to the history of medicine and the sciences. And it is hard to challenge this assessment, given that Du Bois-Reymond's interpretation of the rise of experimental physiology in Germany has provided the framework for most subsequent accounts; and that Virchow's position as a pioneer of the social and cultural history of medicine is unquestioned. Yet on my reading, these works are to be read also as moves in a power struggle. Both Du Bois-Reymond and Virchow are out to advance their personal positions, taking on the mantle and mystique of Müller by portraying him in their own image.[77] Both use their selectively interpreted Müller to legitimate their own programs for reform of medicine and the sciences, and Virchow does the same with Goethe. And each wields history as a weapon against the type of medicine and science espoused by the other—Du Bois-Reymond associating Goethe and hence, by implication, Virchow and his program with the speculative excesses of *Naturphilosophie*, Virchow finding in Müller one who had doubts about an analytical, vivisection-based physiology of the type so ardently promoted by Du Bois-Reymond.

Du Bois-Reymond's attack on Goethe can, I think, be read as a defence of a fully articulated position against established disciplinary rivals; not so the memorial addresses for Müller and Virchow's tribute to Goethe. They appear so only if one reads them with hindsight, attributing to Virchow and Du Bois-Reymond in the 1850s and 1860s the research programs and ideologies set out in their publications and embodied in the institutions they controlled in the 1870s and 1880s. Set in their proper periods these writings appear rather as stories in the "stipulative mode," to borrow Rachel Laudan's apt phrase: histories in whose very writing new methods and disciplinary agendas are articulated.[78] Thus, the section of Du Bois-Reymond's memorial address in which he criticises Müller's methodology contains an account of the proper methods of physiology that is more general than anything in his earlier writings.[79] And in Virchow's tribute to Goethe's holistic and poetic vision of nature we find some of the earliest indications of the vitalist holism that dominated the researches of his later years, but was substantially at odds with the mechanism and materialism of his pre-revolutionary propaganda. Moreover, for all their differences about the nature of reform in physiology and medicine, these works took a common stand in presenting the reform of physiology as a *fait accompli* even though their revolutionary programs were far from securely established even within the confines of the University of Berlin.

In this conflict Du Bois-Reymond's plan for the reform of medicine was, on balance, a winner. At the hands of his students the experimental physiology of his Institute, preserving substantial aspects of his program, remained powerful in these fields far into the present century; whereas Virchow's far fewer students were dispersed, his Institute moving away from his cellular pathological program even within his lifetime.[80] However, Du Bois-Reymond's victory was incomplete. It was not his biophysical program in physiology, but chemical physiology and its successor, biochemistry, that were eventually to dominate academic physiology and medicine.[81] And though defeated in Berlin, substantial parts of Virchow's morphological-genetic program and quest for a pathology-centred social medicine were long preserved in Germany in the fields of social hygiene and public health.[82]

At the ideological level Du Bois-Reymond's victory seems more clear-cut. For the general image of physiology and medicine that he conveyed in his historical writings lives on in the self-images of physiology and medicine—the image of a laboratory-based, objective, experimental physiology that has transformed medicine through its discoveries; the complementary image of a medicine that has become scientific and achieved unprecedented progress through its alliance with laboratory physiology. The vision that Virchow projected in his extensive medical historical writings, that of a social and political medicine centered on a general pathology and creating a new society and culture, has vanished almost without trace.

Even at the ideological level, however, Du Bois-Reymond's victory was less than total. Despite the "laboratory revolution" in medical education and practice at the end of the nineteenth century, substantial minorities of medics have continued to protest at the resultant scientism that devalues clinical tact, experience, and humanity.[83] And though few would now endorse Virchowian visions of a general cultural pathology and social medicine, there remain many who insist that it is only in the context of comprehensive social, educational, and political reform that the goals of medicine can be realized. The angry recent debates between those who blame the rising incidence of tuberculosis in Britain on social deprivation and those who blame it on mismanagement and professional incompetence in the health services shows the persistence of the contrast between official scientistic and unofficial Virchowian perspectives.

Explanation of the fates of Du Bois-Reymond's and Virchow's medical reform programs would be a massive undertaking, far beyond the modest scope of this article. I do, however, venture the suggestion that, at least in the local context of the University of Berlin, Du Bois-Reymond's disciplinary victory owed much to his oratorical performance, of which his historical addresses form a substantial part.[84] One may well suspect that after the foundation of the Prussian Reich Du Bois-Reymond's Bismarkian bombastics were more in tune with the public mood than Virchow's parades of a reasonable, if sometimes quietly malicious, humanity.[85] Certainly they were more in line with the mood at the University of Berlin after 1870—to see this, one has only to contrast the cautious, humane, and rational public pronouncements of von Ranke and Mommsen with the fiery chauvinism of their successors, the liberal nationalist historians, von Sybel and Treitschke.[86]

In the fields of physiology and medicine Du Bois-Reymond was, if not an outright victor, at least winner on points. What of the domain of historiography?

If one looks just at the standard recent histories of physiology one may indeed conclude that Du Bois-Reymond won hands down. For they endorse his historiography at several levels. His memorial address is regularly mined for information about Müller; and the scientific achievements attributed to Müller are by and large those that Du Bois-Reymond attributed to him.[87] Though recently challenged on certain points, it is Du Bois-Reymond's general picture of the nineteenth-century history of German physiology that has prevailed, a picture in which the speculative excesses of Romantic *Naturphilosophie* were swept away by a new laboratory-oriented analytical and mechanistic physiology at the hands of Du Bois-Reymond, Helmholtz, Brücke, Ludwig, and the schools of experimental physiology that they founded; a picture in which experimental physiology brought medicine into the fold of the sciences.[88] And at the most general level we find that for much of the present century the historiography of the sciences has been dominated by "positivistic" narratives of the kind promoted by Du Bois-Reymond, stories of triumphant scientific progress mediated by the experimental method.

When we turn to medical historiography, however, a rather different picture emerges. Let us consider the historiography of medicine at its major institutional sites from the 1930s to the 1950s, the heyday of positivistic historiography of the sciences.[89] First at the Leipzig Institute, then as emigrés at the Johns Hopkins Institute, Henry Sigerist and his associates, Walter Pagel, Erwin Ackerknecht, and Oswei Temkin, promoted a historiography far removed from Du Bois-Reymond's, a Virchowian historiography which attempted to integrate the history of medicine into social and cultural history, to present each age and culture in its own terms avoiding judgement of past ideas and practices by the standards of present medicine and science.

The clash between the positivistic historiography of the sciences and the sociocultural historiography of medicine of Sigerist's school is dramatically displayed in a sharp exchange between Sigerist and George Sarton, a key figure in the formation of history of science as an academic discipline in America.[90]

Sarton opened hostilities in his journal *Isis* with an editorial aggressively entitled "The History of Science Versus the History of Medicine." He started with a lukewarm welcome to the Johns Hopkins Institute and its newly appointed Director, Henry Sigerist.[91] He moved on to contrast the "popular," but richly funded history of medicine with the poorly endowed history of science. The history of medicine, while bound to mention the scientific discoveries that have so benefited medicine, can deal with them only "in a very superficial way."[92] The history of science, however, can do full justice to them. Indeed, Sarton declares, echoing Du Bois-Reymond's grandiloquent claims:

> The history of science should form the core of every history of humanity because the elements which it describes and discusses are the ones which reveal most clearly the rational, progressive and cumulative tendencies and explain in the best manner the function and purpose of man in the scheme of things.[93]

Sigerist responded at once with an editorial in the *Bulletin* of his Institute, conciliatorily entitled "The History of Science *and* the History of Medicine." There he resisted Sarton's claim that history of science should form the core of medical history.

> Science is one aspect of medicine and there are a great many other aspects the history of which has to be investigated. The history of medicine in a very large sense is the history of the relation between physician and patient, between the medical profession in the largest sense of the word (including administrators, public health officers, scientists, nurses, priests, quacks, etc.) and society. We therefore have to study the position of physician and patient in a given society, their attitude towards the human body, the valuation of health and disease at a given time.[94]

He went on to deflate Sarton's grander vision of a history of science at the heart of the history of humankind. History of science, he declared, is just one component of a history of civilisation whose study requires the cooperation of a host of historical disciplines centered not on history of science but on political, social, and economic history.[95]

Triumphalist and positivistic historiography of the type promoted by Du Bois-Reymond and embraced by Sarton and his associates came to dominate the history of science in the phase of its post-World War II establishment in university curricula.[96] By contrast, in the history of medicine, though this kind of historiography became common enough, it existed alongside flourishing schools of social and cultural history of medicine.[97] How are we to account for the diversity and richness of twentieth-century historiography of medicine in comparison with twentieth-century historiography of the sciences?

Part of the answer is straightforward. History of medicine became fully institutionalized within the French and the German medical faculties in the course of the nineteenth century.[98] The reasons for this early embodiment of history of medicine in medical institutions are complex. Doubtless it is attributable in part to the widespread view that the study of history complemented the study of the classical languages in the moral and humane education of physicians. More specifically it may be related to the fact that many fields of medicine, most notably those such as epidemiology involving study of past case histories and their provenances, were perceived as intrinsically historical.[99] And, at least in the German lands, this was aided and abetted by the massive impact of Romantic *Naturphilosophie* on medical teaching; for it was a central tenet of *Naturphilosophie* that the history of medicine and the sciences is integral to medicine and the sciences themselves, because it reveals the primordial ideas of antiquity from which they have developed.[100] By contrast, as we may gather from Sarton's complaints, even in the 1930s the academic institutionalization of the history of science as an ancillary of the sciences had barely begun. Thus the history of medicine was established as a discipline allied both to medicine and to other branches of history well before the "great divide" between arts and sciences of the latter decades of the nineteenth century, whereas the history of science was established in virtual isolation from other branches of history long after the Divide.[101]

But this cannot be the whole story. For the history of historiography shows many instances of rapid shift in style and fashion. Given the evident links between the histories of science and medicine, how are we to explain the resistance of the history of medicine to conquest by positivistic historiography, and how are we to account for the long neglect (from the 1930s to the 1960s) by historians of science of the issues in social and cultural history explored by medical historians?

It is fashionable to assume that it is precisely through separation from the presuppositions and values of our science and medicine that the histories of those disciplines can achieve historiographical maturity and sophistication. At the same time, few would deny that current history of medicine is indebted for much of its richness and diversity to the medical historians of the Leipzig school. It comes, therefore, as something of a shock to learn that the leading figures of the Leipzig school, Walter Pagel, Erwin Ackerknecht, and Henry Sigerist, were deeply involved in medical politics and explicitly presented the history of medicine as an instrument of medical education and reform.[102]

The issue of morally and politically engaged versus disinterested historiography has itself a long history. Consider, for example, the famous attack of Johann Gustav Droysen on Leopold von Ranke and his school.[103] Here, as in the slightly later confrontation between Du Bois-Reymond and Virchow, the fundamental issue was that of Humboldtian academic detachment from worldly affairs versus reformist engagement. The radical liberal Droysen's target was the conservative von Ranke's program for a scientific historiography. Ranke's critical method can yield only a "eunuch-like objectivity," Droysen insisted, a heartless and pointless accumulation of facts.[104] True history, Droysen maintained, is an instrument of moral education and reform. Moreover, to the extent that objectivity is attainable in history, it is to be reached not through the impossible attempt to write without a standpoint but through a critical awareness of one's standpoint and moral aims.[105]

Droysen's paradoxical thesis that historical objectivity, insofar as it is attainable, is to be reached through the right kind of partisanship had theological and teleological underpinnings that few would now accept.[106] This is not the occasion for a secular defence of Droysen's thesis. Instead, I merely declare my belief that we should side with Droysen, if not with respect to all branches of history at least with respect to the history of medicine and the sciences. The historiographical poverty of much twentieth-century history of science has resulted from a false objectivity, from a disengagement that disguises an uncritical complicity in the positivistic and triumphalist ideologies of science. Conversely, the historiographic richness and diversity of the history of medicine have, I suggest, been products not of distance from medicine, but of critical involvement in the problems and politics of medical education, practice and reform.

## Notes

1. On persuasive and polemical uses of history of medicine and the sciences, see, for example, L. Graham, W. Lepenies, and P. Weingart, eds., *Functions and Uses of Disciplinary Histories* (Dordrecht, 1983); N. Jardine, *The Scenes of Inquiry: On the Reality of Questions in the Sciences* (Oxford, 1991), ch. 6, "Legitimation and History"; R. Laudan, "Histories of the Sciences and Their Uses: A Review to 1913," *History of Science* 31 (1993): 1–33; also works cited in notes 35, 77, and 84.

2. For opposed points of view on this issue, see, for example, Q.R.D. Skinner, "Meaning and Understanding in the History of Ideas," *History and Theory* 8 (1969): 3–53; D.L. Hull, "In Defense of Presentism," *History and Theory* 18 (1979): 1–15; A. Wilson and T. Ashplant, "Whig History and Present Centred History," *The Historical Journal* 31 (1988): 1–16.

3. In *Scenes of Inquiry* I argue on more general grounds for the inextricability of disciplines and their histories.

4. The bulk of the information in this and the following paragraph is derived from T. Lenoir, "Social Interests and the Organic Physics of 1847," in E. Ullmann-Margalit, ed., *Science in Reflection* (Dordrecht, 1988), 169–91, and "Laboratories, Medicine and Public Life in Germany 1830–1849: Ideological Roots of the Institutional Revolution," in A. Cunningham and P. Williams, *The Laboratory Revolution in Medicine* (Cambridge, 1992), 14–71; P. Cranefield, "The Organic Physics of 1847 and the Biophysics of Today," *Journal for the History of Medicine and Allied Sciences* 12 (1957): 407–23.

5. For Du Bois-Reymond's youthful hopes and fears, see *Jugendbriefe an Eduard Hallmann* (Berlin, 1918), esp. 75–76. A comprehensive, but differently angled, account of Du Bois-Reymond's and Virchow's attitudes is in K.M. Anderton, "The Limits of Science: A Social, Political and Moral Agenda for Epistemology in Nineteenth-Century Germany," thesis (Harvard, 1993).

6. See, for example, ibid., 128–29, and *Two Great Scientists of the Nineteenth Century: Correspondence of Emil Du Bois-Reymond and Carl Ludwig*, trans. S. Lichtner-Ayer (Baltimore, 1982) 9–14.

7. Ibid., 104.

8. Except where indicated, the following information about Virchow is from E. Ackerknecht, *Rudolf Virchow. Doctor, Statesman and Anthropologist* (Madison, 1953, reprinted New York, 1981).

9. The slogan was appropriated, with acknowledgment, by Virchow from his collaborator in the medical reform movement, Salomon Neumann: Ackerknecht, *Rudolf Virchow*, 46, 247.

10. "On standpoints in scientific medicine" (1847), trans. L.J. Rather, *Disease, Life, and Man: Selected Essays by Rudolf Virchow* (Stanford, 1959), 26–39, esp. 32.

11. *Two Great Scientists*, 88. On the "organic physics" envisaged by Du Bois-Reymond, Brücke, and Helmholtz, see Lenoir, "Social Interests"; Cranefield, "The Organic Physics"; and A. Tuchman, "Helmholtz and the German Medical Community", in D. Cahan, ed., *Hermann von Helmholtz and the Foundations of Nineteenth-Century Science* (Berkeley, 1993), 17–49.

12. On Virchow's political career, see B.A. Boyd, *Rudolf Virchow: The Scientist as Citizen* (New York, 1991).

13. Du Bois-Reymond defends academic freedom and disinterestedness in "Über Universitätseinrichtungen" (1869) and "Kulturgeschichte und Naturwissenschaft" (1877), in *Reden von Emil Du Bois-Reymond*, ed. Estelle Du Bois-Reymond, 2nd ed. (Leipzig, 1912), (hereafter cited as *Reden*), 1:356–69 and 567–629, respectively. On Prussian academic ideals, see C.E. McClelland, "Zur Professionalisierung der akademischen Berufe in Deutschland," in J. Kocka and W. Conze, *Bildungsbürgertum im 19. Jahrhundert*, 1 (Stuttgart, 1985), 234–37. On German liberal ideals of disinterestedness, see J.L. Sheehan, *German Liberalism in the Nineteenth Century* (Chicago, 1978). On the ideological role of illusions of disinterestedness and autonomy of the professoriate, see P. Bourdieu and J.-C. Passeron, *La réproduction: éléments pour une théorie du système d'enseignement* (Paris, 1970).

14. *Reden*, 602–7.

15. Du Bois-Reymond's enthusiasm over the virtues of German academic freedom and his disparagement of French university education are closely paralleled in Helmholtz's more moderate writings: compare, for example, his "Über Universitätseinrichtungen" (1869), *Reden*, 1:356–69, with Helmholtz's "Ueber die akademische Freiheit der deutschen Universitäten" (1877), in *Vorträge und Reden*, 3rd ed., vol. 2 (Brunswick, 1884).

16. "Der deutsche Krieg", *Reden*, 1:393–420. The "bodyguard" remark was widely criticized and mocked, for example, by Karl Friedrich Zöllner (see n. 74) and Fritz Mauthner (see n. 85). Du Bois-Reymond looked back with misgivings on his war mania: *Reden*, 1:419–20.

As David Cahan has noted (personal communication), the timing of the approval of the new physiology institute may also have been due to the parallel negotiations to bring Helmholtz to Berlin and to build a physics institute there for him: the two institutes occupied one and the same building. On the foundation of the physiology and physics institutes, see G. Fritsch, "Das physiologische Institut," and H. Rubens, "Das physikalische Institut," in M. Lenz, ed., *Geschichte der königlichen Friedrich-Wilhelms-Universität zu Berlin*, vol. 3 (Halle, 1910), 154–63 and 278–96, respectively; also T. Lenoir, "Social Interests and the Organic Physics of 1847," 183–86.

17. On the medical reform movement and its outcomes, see K. Finkenkrath, *Die Medizinalreform. Die Geschichte der ersten deutschen ärztlichen Standesbewegung* (Leipzig, 1929); H.H. Simmer, "Principles and Problems of Medical Undergraduate Teaching in Germany during the Nineteenth and Early Twentieth Centuries," in C.D. O'Malley, ed., *The History of Medical Education* (Berkeley, 1970), 173–200; C. Huerkamp, *Der Aufstieg der Ärtze im 19. Jahrhundert* (Göttingen, 1985); U. Frevert, *Krankheit als politisches Problem* (Göttingen, 1985), and "The Making of the Modern Medical Profession: Prussian Doctors in the Nineteenth Century," in G. Cocks and K.H. Jarausch, eds., *German Professions 1800–1900* (Oxford, 1990), 66–84.

18. See, for example, "Standpoints in Scientific Medicine" (1847), "Scientific Medicine and Therapeutic Standpoints" (1849), "Cellular Pathology" (1855), *Disease, Life, and Man*, 26–39, 40–70, and 70–101, respectively.

19. See, for example, "Der physiologische Unterricht sonst und jetzt" (1877), *Reden*, 1:630–53, esp. 648–51.

20. *Reden*, 1:607–19.

21. See, for example, his "Scientific Medicine and Therapeutic Standpoints," in *Disease, Life, and Man*.

22. "Der physiologische Unterricht," in *Reden*.

23. See, for example, Ackerknecht, *Rudolf Virchow*, 44; Du Bois-Reymond, *Reden*, 1:435.

24. Virchow, however, opposed the creation of a chair of History of Medicine at Berlin, evidently supposing his own lectures on the subject to suffice: Ackerknecht, *Rudolf Virchow*, 38.

25. *Reden*, 1:435.

26. "Kulturgeschichte und Naturwissenschaft," *Reden*, 1:595. Du Bois-Reymond's accounts of the culture of science have much in common with Helmholtz's public pronouncements on science's civilizing powers. However, Helmholtz's notion of *Bildung* allowed a far greater role for the arts. For a fine account of Helmholtz as scientific *Kulturträger* see D. Cahan, "Helmholtz and the Civilising Power of Science," in *Hermann von Helmholtz*, 559–601.

In a masterly article Christoph Gradmann has related Du Bois-Reymond's scientist *Kulturgeschichte* and its negative reception (notably by Theodor Mommsen) to the ambivalent attitude of the new brand of natural scientists to the cultural ideals of the *Bildungsbürgertum*: "Naturwissenschaft, Kulturgeschichte und Bildungsbegriff bei Emil Du Bois-Reymond," *Tractrix* 5 (1993): 1–16. See also A. Demandt, "Natur- und Geisteswissenschaft im 19. Jahrhundert," *Berichte zur Wissenschaftsgeschichte* 6 (1983): 59–78. (I thank Dr Keith Anderton for drawing my attention to this article.)

27. "Kulturgeschichte und Naturwissenschaft," in *Reden*, 1:596.

28. Ibid., 1:592ff.

29. On Virchow's medical historical writings, see Ackerknecht, *Rudolf Virchow*, 146–55.

30. "On Hospitals and Military Hospitals" (1866) and "The Hospital Order of the Holy Ghost, Especially in Germany" (1877), in *Rudolf Virchow: Collected Essays on Public Health and Epidemiology*, ed. and trans. L.J. Rather, vol. 2 (Canton, MA, 1985), 7–21 and 22–45, respectively; "Morgagni and the Anatomic Concept" (1882), trans. R.E. Schluter and J. Auer, *Bulletin of the History of Medicine* 7 (1939): 975–89.

31. On Stahlian vitalism and monarchy, see "Über die Reform der pathologischen und therapeutischen Anschauungen durch die mikroskopischen Untersuchungen," *Virchows Archiv* 1 (1847): 207–55; for cellular analogies with the state, see, for example, "Cellular-Pathologie," *Virchows Archiv* 8 (1855): 3–39; "Die Kritiker der Cellularpathologie," *Virchows Archiv* 18 (1860): 1–14. On the Virchowian cell state, see O. Temkin, "Metaphors of Human Biology," in R.C. Stauffer, ed., *Science and Civilization* (Madison, 1949), 169–94; G. Mann, "Medizinisch-biologische Ideen und Modelle in der Gesellschaftslehre des 19. Jahrhunderts," *Medizinhistorisches Journal* 4 (1969): 1–23; P. Weindling, "Theories of the Cell State in Imperial Germany," in C. Webster, ed., *Biology, Medicine and Society 1840–1940* (Cambridge, 1981), 99–155, esp. 117–19; R.G. Mazzolini, *Politisch-biologische Analogien im Frühwerk Rudolf Virchows* (Marburg, 1988).

32. See Klaus Wenig, ed. and intro., *Rudolf Virchow und Emil du Bois-Reymond, Briefe 1864-1894* (Marburg, 1994). In 1894, however, du Bois-Reymond ventured cautious criticism of Virchow's vitalism: see his "Über Neo-vitalismus" (1894), *Reden*, 2:492–515.

33. On Müller's students, see K. Rothschuh, *History of Physiology* (1953), trans. G.B. Risse (Huntington, NY, 1973); on chairs of physiology, see H.-H. Eulner, *Die Entwicklung der medizinischen Spezialfächer an der Universitäten des deutschen Sprachgebietes* (Stuttgart, 1970).

34. Du Bois-Reymond's memorial address remains the most substantial assessment of Müller's career. See also W. Haberling, *Johannes Müller. Das Leben des rheinischen Naturforschers* (Leipzig, 1924); Rothschuh, *History of Physiology*; G. Koller, *Das Leben des Biologen Johannes Müller* (Stuttgart, 1958); J. Steudel, *Le Physiologiste Johannes Müller* (Paris, 1963).

35. On memorial addresses, see D. Outram, "The Language of Natural Power: The Funeral Éloges of Georges Cuvier," History of Science 16 (1978): 153–78; C.B. Paul, Science and Immortality (Berkeley, 1980).
36. Reden, 1:135–317, esp. 139: first publ. Abhandlungen der königlichen Preussischen Akademie der Wissenschaften (1860), 25–191.
37. Ibid., 138.
38. Ibid., 143ff.
39. Ibid., 147ff.
40. Ibid., 156–57.
41. Ibid., 172ff.
42. Johannes Müller: An Eloge Pronounced in the Hall of the University of Berlin by Professor Rudolf Virchow, trans. A. Mercer (Edinburgh, 1859), 4; first publ. Johannes Müller: Gedächtnisrede (Berlin, 1858).
43. Ibid., 6–7.
44. Ibid., 12.
45. Reden, 1:270.
46. Ibid., 1:203, 225, 263–71. Du Bois-Reymond even challenges Müller's account of the clicking of the auditory ossicles he could produce by waggling his ears: 198, 300.
47. Ibid., 204–5.
48. Ibid., 205–9.
49. Ibid., 208–9.
50. Johannes Müller: An Eloge, 14ff.
51. On Müller's attitude toward vivisection, see B. Lohff, Die Suche nach der Wissenschaftlichkeit der Physiologie in der Zeit der Romantik (Stuttgart, 1990), 120ff. On Virchow's conception and practice of experiment, see H.-P. Schmiedebach, "Pathologie bei Virchow und Traube. Experimentalstrategien in unterschiedlichem Kontext," in H.-J. Rheinberger and M. Hagner, eds., Die Experimentalisierung des Lebens. Experimentalsysteme in den biologischen Wissenschaften 1850/1950 (Berlin, 1993), 116–34.
52. Reden, 1:20.
53. Ibid., 1:211.
54. Letter of August 5, 1857, Two Great Scientists, 96.
55. "Der physiologische Unterricht," 634.
56. Ibid., 646.
57. Ibid., 646–47.
58. Ibid., 648–50.
59. Ibid., 651.
60. Goethe als Naturforscher und in besonderer Beziehung auf Schiller (Berlin, 1861; facsimile reprint, Darmstadt, 1962).
61. Ibid., 48.
62. Ibid., 17ff., 60ff.
63. Reden, 2:157–58.
64. Ibid., 2:168–69, 173.
65. Ibid., 172–74. Du Bois-Reymond's critique of Goethe is here close to Helmholtz's diagnosis of the methodological weaknesses of Goethe's Farbenlehre, "Ueber Goethes naturwissenschaftlichen Arbeiten" (1853), in Vorträge und Reden, 3rd ed., vol. 1 (Brunswick, 1896); on Helmholtz's assessments of Goethe, see G. Hatfield, "Helmholtz and Classicism. The Science of Aesthetics and the Aesthetics of Science," in D. Cahan, ed., Hermann von Helmholtz, 522–58.
66. Ibid., 174.
67. Ibid., 173.
68. See W. Kloppe, "Du Bois-Reymonds Rhetorik im Urteil einiger seiner Zeitgenossen," Deutsches medizinisches Journal 9 (1958): 80–82. (I thank Rebecca Bertoloni Meli and the librarian of the Royal Society of Medicine, London, for obtaining for me a copy of this article.)
69. Reden, 2:160–64.

70. Ibid., 2:162–63.
71. Ibid., 2:165.
72. Ibid., 2:180–82.
73. In particular Du Bois-Reymond's "Die sieben Welträtsel" of 1880, *Reden*, 2:65–98, is directed at all who address questions beyond the bounds of empirical sense.
74. Zöllner, "Über Emil Du Bois-Reymonds Grenzen des Naturerkennens," *Wissenschaftliche Abhandlungen* 1 (1878): 189–416; Zöllner also notes the discrepancy between Du Bois-Reymond's call to arms in "Der deutsche Krieg" and his championship of academic freedom and disinterestedness. On Zöllner's attacks on Du Bois-Reymond, see C. Meinel, *Karl Friedrich Zöllner und die Wissenschaftskultur der Gründerzeit. Eine Fallstudie zur Genese konservativer Zivilisationskritik* (Berlin, 1991).
75. On the battle for admission of *Realschüler*, see D.K. Müller, *Sozialstruktur und Schulsystem. Aspekte zum Strukturwandel des Schulwesens im 19. Jahrhundert* (Göttingen, 1977).
76. For Du Bois-Reymond's heroic daydreaming see, for example, *Jugendbriefe*, 19, 71–72; for his identification with Goethe see, for example, ibid., 23–25 (his account of his pilgrimage to Goethe's house) and 76 (where he mocks his own "goethesirende Sprache"). A reading along the lines suggested here is adumbrated in A. von Berger's delightful pamphlet *Goethes Faust und die Grenzen des Naturerkennens. Wider "Goethe und kein Ende" von Emil Du Bois-Reymond* (Vienna, 1883). Von Berger defends the psychological truth of Faust as the "Type of all types," including university professors, teasingly intimating that had Faust followed Du Bois-Reymond's advice to settle down, marry Gretchen, and make real scientific discoveries, he would have turned into Du Bois-Reymond.
77. On construction of disciplinary father figures, see L. S. Jacyna, "Images of John Hunter in the Nineteenth Century," *History of Science* 21 (1983): 85–108; M.G. Ash, "The Self-presentation of a Discipline: History and Psychology in the United States between Pedagogy and Scholarship," in Graham et al., eds., *Functions and Uses of Disciplinary Histories*, 143–89.
78. "Redefinitions of a Discipline: Histories of Geology and Geological History," ibid., 79–104, esp. 80–81.
79. Certain of the methodological points are, however, prefigured in his "Über die Lebenskraft" of 1848: *Reden*, 1:1–22.
80. Rothschuh, *History of Physiology*.
81. R.E. Kohler, *From Medical Chemistry to Biochemistry. The Making of a Biomedical Discipline* (Cambridge, 1982).
82. See P. Weindling, *Health, Race and German Politics between National Unification and Nazism 1870–1945* (Cambridge, 1989).
83. On clinicians' resistance to laboratory medicine, see C.J. Lawrence, "Incommunicable Knowledge: Science, Technology and the Clinical Art in Britain 1850–1914," *Journal of Contemporary History* 20 (1985): 503–20; G.L. Geison, "Divided We Stand: Physiologists and Clinicians in the American Context," in M.J. Vogel and C.E. Rosenberg, eds., *The Therapeutic Revolution: Essays in the Social History of American Medicine* (Philadelphia, 1979), 67–90.
84. I have argued elsewhere that in this period oratory played a major role in the establishment of new agendas in medicine and the sciences: "The Laboratory Revolution in Medicine as Rhetorical and Aesthetic Accomplishment," in A. Cunningham and P. Williams, eds., *The Laboratory Revolution in Medicine* (Cambridge, 1992), 304–23.
85. For a fine parody see Fritz Mauthner, "Friedrich der Grosse und der Gymnotus electricus," in his *Nach berühmten Mustern. Parodistischen Studien* (Stuttgart, 1898), 31–36. The climax of this spoof oration splendidly captures Du Bois-Reymond's bombastic and imperialist scientism: "once German science has forced the herring to spawn only in German waters, the buffalo to graze only on German prairies, the grape to ripen only under a German sun, once the rainbow has sacrificed its illogical and gaily shimmering colours on the altar of love of the fatherland and stands black-white-red on its fabulous bowl of German gold, then will I reply, 'so be it.' For the German is not chauvinistic."
86. On the attitudes of von Sybel and Treitschke, see Sheehan, *German Liberalism*, chs. 9 and 10.

87. Thus the standard account of Müller's achievements, those of Rothschuh (*History of Physiology*) and Steudel (*Le Physiologiste Johannes Müller*), make substantial use of Du Bois-Reymond's *Gedächtnisrede*. Steudel, however, questions the view that Müller's methodology was transformed by his "conversion" from *Naturphilosophie*: "Wissenschaftslehre und Forschungsmethodik Johannes Müllers," *Deutsche medizinische Wochenschrift* 77 (1951): 115–18; and this alleged discontinuity is questioned also by several of the contributors to M. Hagner and B. Wahrig-Schmidt, eds., *Johannes Müller und die Philosophie* (Berlin, 1992). (I am grateful to Michael Hagner for these references.)

88. An important challenge to this scenario is C. A. Culotta, "German Biophysics, Objective Knowledge and Romanticism," *Historical Studies in the Physical Sciences* 4 (1975–76): 3–38, who argues that the *Naturphilosophen* played a major role in forming the agenda of the new scientific physiology; cf. A R. Cunningham and N. Jardine, "Introduction: The Age of Reflexion," in *Romanticism and the Sciences* (Cambridge, 1990), 1–9. B. Lohff (*Die Suche*), provides extensive evidence against the standard view of the *Naturphilosophen* as unconcerned with experimentation in the sciences.

89. The following remarks draw on my reading of the works of Sigerist and his associates (guided by O. Temkin, "The Double Face of Janus," in *The Double Face of Janus and Other Essays in the History of Medicine* [Baltimore, 1977], 3–37) and of W. Pagel, "Erinnerungen und Forschungen," in K. Mauel, *Wege zur Wissenschaftsgeschichte*, 2 (Wiesbaden, 1982), 45–66. (I thank Professor Piyo Rattansi for sending me a copy of Pagel's article.)

90. On Sigerist see O. Temkin, "In Memory of Henry E. Sigerist," *Bulletin of the History of Medicine*, 31 (1957): 295–99; E.H. Ackerknecht, "Introduction," in G. Miller, ed., *A Bibliography of the Writings of Henry E. Sigerist* (Montreal, 1966), 1–7. On Sarton see A. Thackray, "On Discipline-building: The Paradoxes of George Sarton," *Isis* 53 (1972): 463–95; also the articles in the 1957 memorial issue of *Isis*.

91. *Isis*, 23 (1935): 313–20, esp. 313–14. Sarton's disgruntlement is understandable given the frustration of his own long-pursued plans for an institute for the history of science; see Thackray, "On Discipline-building:."

92. *Isis* 23 (1935): 315–16.

93. Ibid., 317.

94. *Bulletin of the Institute of the History of Medicine* 4 (1936): 1–13, esp. 5.

95. Ibid., 11–12.

96. See A. Thackray, "The Pre-history of an Academic Discipline: The Study of the History of Science in the United States, 1891–1941," *Minerva* 18 (1980): 448–73; and "History of Science," in P.T. Durbin, ed., *A Guide to the Culture of Science, Technology and Medicine* (New York, 1980), 3–69; S.A. Jayawardene, "Mathematical Sciences," and E.H. Beardsley, "The History of Americal Science and Medicine," in P. Corsi and P. Weindling, *Information Sources in the History of Science and Medicine* (London, 1983), 259–84 and 411–35, respectively.

97. See C. Webster, "The Historiography of Medicine," in Corsi and Weindling, *Information Sources*, 29–43; G.H. Brieger, "History of Medicine," in Durbin, ed., *A Guide to the Culture of Science*, 121–94.

98. Webster, "The Historiography of Medicine"; E. Heischkel, "Die Geschichte der Medizingeschichtsschreibung," in W. Artelt, ed., *Einführung in die Medizinhistorik* (Stuttgart, 1949), 202–37.

99. Webster, "The Historiography of Medicine."

100. D. von Engelhardt, *Historisches Bewußtsein in der Naturwissenschaft von der Aufklärung bis zum Positivismus* (Freiburg, 1979), 3, "Romantische Naturforschung"; H. von Seemen, *Kenntnis der Medizinhistorie in der deutschen Romantik* (Zurich, 1926).

101. On the impact of the Divide on historiography of the sciences, see von Engelhardt, *Historische Bewußtsein*, 4, "Positivistische Naturwissenschaft"; Demandt, "Natur- und Geisteswissenschaft"; Gradmann, *Kulturgeschichte*.

102. See the articles by Temkin and Pagel cited in n.89; also P. Diepgen, "Das Schicksal der deutschen Medizingeschichte im Zeitalter der Naturwissenschaft und ihre Aufgaben in der Gegenwart," *Deutsche medizinische Wochenschrift* 60 (1934): 66–70; H.E. Sigerist, "The Medical

Student and the Social Problems confronting Medicine Today," *Bulletin of the Institute of the History of Medicine* 4 (1936): 411–22; O. Temkin, "An Essay on the Usefulness of Medical History for Medicine," *Bulletin of the History of Medicine* 19 (1982): 9–47.

103. For accounts of this confrontation see F. Gilbert, *Johann Gustav Droysen und die preussisch-deutsche Frage* (Munich, 1931), 74–78; M.J. Maclean, "Johann Gustav Droysen and the Development of Historical Hermeneutics," *History and Theory* 21 (1982): 347–65.

104. R. Hübner, ed., *Historik. Vorlesungen über Enzyklopädie und Methodologie der Geschichte* (Munich, 1937), 287.

105. Ibid., 302ff.

106. On the theological bases of Droysen's historiography, see R. Southard, "Theology in Droysen's Early Political Historiography: Free Will, Necessity, and the Historian," *History and Theory* 18 (1979): 378–96, and Maclean, "Johann Gustav Droysen."

# GENDER IN EARLY MODERN SCIENCE

## Londa Schiebinger

> Men have not only excluded the Women from partaking of the Sciences and Employ by long Prescription, but also pretend that this Exclusion is founded in their natural Inability. There is, however, nothing more chimerical.
> "By a Lady," 1763

In 1667 Margaret Cavendish made her famous visit to the Royal Society of London, England's primer institution for the new experimental science. Robert Boyle prepared his "experiments of . . . weighing of air in an exhausted receiver; [and] . . . dissolving of flesh with a certain liquor" and other entertainments fit for a lady of Cavendish's stature. The Duchess, accompanied by her ladies, was much impressed by the demonstrations and left (according to one observer) "full of admiration."[1]

Undoubtedly Cavendish's curiosity was not satisfied by this fleeting encounter with the Royal Society's men of science; the meeting was painfully formal. Nor did this short visit make up for a lifetime of intellectual isolation. Cavendish was never admitted to the Royal Society, though she was well qualified for that position (men above the rank of baron could become members without scientific qualifications). From its founding in 1660 until 1945, the only permanent female member of the Royal Society was the female skeleton in the Society's anatomical collection.[2]

Excluded from the public life of the mind, the Duchess of Newcastle tried to make contact with the learned world through her books. In addition to her numerous poems and plays, she wrote eight books touching on natural philosophy, where she ruthlessly critiqued both the rationalists and the experimentalists. She sent each of her beautifully published volumes to Oxford and to Cambridge, where her husband and two brothers had been educated, and a complete set of her philosophical works to Christian Huygens at the University of Leyden.[3] Her boldness was sharply reprimanded by Joseph Glanvill, one of the leading figures in the Royal Society. In explicit reference to her work, Glanvill warned that "he is a bold Man, who dares to attack the Physics of Aristotle himself, or of Democritus . . . or Descartes, or Mr. Hobbs." Apart from this attack, however, her work suffered that worst censure of all—neglect.

Cavendish was not alone in this regard. Historical memory is highly selective: books that are ignored are lost, their message forgotten. Despite their efforts, women of the seventeenth and eighteenth centuries remained excluded from universities and scientific societies. Even the memories of those struggles were eventually also lost, as the words of those who opposed orthodox views of women's nature were seldom preserved in libraries or taught in university lectures. In the preface to his daughter's 1742 defense of women's right to higher education, the German medical doctor Christian Leporin noted that the renowned Dutch scholar, Anna van Schurmann, had published a book on the education of women in the previous century, but that "despite all my efforts, it was not to be had."[4] With Schurmann's work misplaced by tradition, Dorothea Erxleben—Leporin's daughter—never had the opportunity to sharpen her young mind on Schurmann's mature thought. Erxleben could not have known that the same fate would befall her own work. Her countrywoman Amalia Holst some fifty years later noted that Erxleben's *Inquiry into the Causes Preventing the Female Sex from Studying* was "no longer available." Holst could not procure a copy, nor could Dorothea Erxleben's stepson—a professor.[5] Feminists such as Theodor von Hippel and Mary Wollstonecraft, known in their day for having contributed to vital social questions, also fell victim to neglect. In 1806, L. W. Weissenborn lamented that Hippel's book of only a decade before was nearly forgotten, and Wollstonecraft's almost completely ignored.[6] It is significant in this context that Anton Amo, the first African to receive a degree from a European university (Wittenberg, 1734), suffered the same fate. His dissertation, *De Jure Maurorum in Europa*, on the rights of Africans in Europe (one of his earliest works) has been lost, while his writings on traditional philosophical questions—the art of philosophizing and the mind/body distinction, for example—have been preserved in university libraries and archives.[7]

Knowledge is indeed, as Donald Kelley writes in this volume, "acquired, communicated, recalled, criticized, transformed . . . by human effort in the course of time."[8] Gender is a crucial dimension in this process. As revealed in Margaret Cavendish's experience, modern knowledge was shaped by a series of exclusions: all but a few exceptional women were excluded from European universities for over seven hundred years, from their inception in the twelfth century until the late nineteenth century; women were not admitted to scientific societies, nor were they regular members of the "invisible college" that fostered informal intellectual exchange across Western Europe. Modern knowledge has also been shaped by apologia—reasonings on the supposed causes of these exclusions. Cavendish, in her 1662 "Femal Orations," provides her own interrogation of women's marginal position in science: is it nature, something in the physical, psychological, and intellectual nature of women that prohibits them from producing great science? or is it nurture, their exclusion from institutions of learning? Are women less noble—culturally and intellectually—than men, or are feminine qualities—she mentioned grace and beauty—more noble and worthy of praise? Cavendish left her queries unresolved, assuming the posture of the men at the Royal Society who professed not to meddle in politics or religion. Cavendish remarked that she spoke freely in these orations—*pro* and *con*—but did not take sides.[9] More important than the consequences for women and other intellectually disenfranchised persons, however, are the consequences for human knowledge more generally. The contours of modern knowledge—the questions it asks, the methods it employs, its disciplinary demarcations—

have been shaped by the systematic exclusions of women and their concern. When Immanuel Kant enthroned pure reason, he dethroned women and something historically disparaged in Western intellectual life as the "feminine."

## The Exclusion of Women from Modern Science

The scientific revolution in early modern Europe marks a pivotal juncture for studying the problems women have faced in the natural sciences. The scientific revolution shook the foundations of science: Copernicus challenged the notion of a geocentric universe and suggested instead that the earth revolves around the sun; Galileo and Newton revised our notions of the basic laws of matter and motion; in most general terms, Aristotelian science was exploded. This was a time when the conceptual base of science was being refashioned, new institutions were being founded, and the question of participation was much discussed.

Historians, treating this profound and far reaching revolution, tend to scrutinize developments internal to the scientific community (new institutions, new methods, new foundations and conceptions of nature), external to it (struggles with church and state, spurs from technology and warfare), and intrinsic to it (cultural impulses pervading and formed by science). Seldom, however, do they bring another fundamental transformation in this period—the formal exclusion of women from science—to bear on the structuring of the sciences.[10]

In the reconfiguration of intellectual culture associated with the rise of modern Western science, women and also a set of values, qualities, and characteristics subsumed under the term "femininity" were banished from science. In defining why women could not do science, science theorists were not defining women so much as what was to be considered unscientific in the West. This was not the first time science had been set in a dualistic relationship with its perceived other. During the seventeenth and eighteenth centuries scientists had made efforts to distinguish their enterprise from religion, politics, poetic expression, and so forth. In each case, particular limits were set to scientific knowledge by defining science as epistemologically distinct from its supposed opposite. Women—as representatives of femininity—became repositories for all that was not scientific: in a scientific age women were to be religious; in a secular age they were to be the keepers of morals; in a contractual society they were to provide the bonds of love. At the same time, Western-style middleclass femininity was conceived as a necessary ballast to Western-style middleclass masculinity: according to the theory of sexual complementarity, each gender, though incomplete in itself, joined to the other to make a perfect whole.

As it emerged in the eighteenth century, the exclusion of women from science was made to seem natural and just, a cultural development that fair-minded people could embrace. Four key elements coalesced to render invisible the injustices of exclusion. The culture of science, stretched across these girders, continues today to hold women at a distance.

A first factor was the new authority invested in nature and its laws. The

Enlightenment posed a challenge to existing orthodoxies of inequality with its rally cry: "All men [often interpreted to include women] are by nature equal." Philosophes and politicians, attempting to build a just society, sought to ground the laws of men in what they considered a higher authority—the laws of nature. Natural law (as distinct from positive law of nations) was held to be immutable, given either by God or inherent in the material universe.[11]

This first factor is closely allied to a second: the growing authority of science as a privileged source of knowledge about nature. As claims to equality increasingly came to be considered matters not of ethics but of science, natural philosophers came to mediate between the laws of nature and of legislatures. In order for persisting social inequalities to be justified within this framework, scientific evidence would have to show that human nature is not uniform, but differs along the lines of ethnicity, race, and sex. The marquis de Condorcet wrote in reference to the equality of women: if women are to be excluded from the *polis*, one must demonstrate a "natural difference" between men and women in order to legitimate that exclusion.[12] This mindset made it seem reasonable for Auguste Comte to claim in his epic 1830s *Cours de philosophie positive* that the "sound philosophy of biology" could offer a resolution to the much acclaimed equality of the sexes.[13] Comte believed that one could expect the gradual liberation of subordinated men because there is "no organic difference between the dominant and the dominated," but he expected the subordination of women to continue forever "because it rests directly on a natural inferiority."[14]

A third factor buttressing the exclusion of women from science was the revolution in scientific views of sexual difference.[15] The revolution in sexual science grounded gender (both the social relations between the sexes and ideological renderings of those relations) unredeemably in biology (understood at the time as ahistorical and immutable). For many, physical differences underlay and determined intellectual and moral differences observed in men and women's character and daily lives. Jean-Jacques Rousseau typically asserted that differences in minds and morals were connected to sex, though he admitted "not by relations which we are in a position to perceive." Rousseau opened his chapter in *Emile*, entitled "Sophie, or the Woman" (perhaps the single most influential piece written about women in the eighteenth century), with a discussion of comparative anatomy.[16] Many agreed that "nature herself has drawn the boundary very exactly and correctly between feminine and masculine pursuits."[17]

The revolution in sexual science was marked by a methodological rupture in explanations of sexual difference (though the view of woman as a lesser being remained remarkably constant).[18] In Aristotelian and Galenic science, divergent sexual temperaments had been seen as driven by cosmic principles reduplicating the macrocosm within the microcosm of the individual body. Breaking with these traditions, anatomists and physiologists of the eighteenth century deployed empirical methods to weigh and measure sexual difference. The first representations of the female skeleton in Western anatomy epitomizes this scientific search for sexual difference.[19] Although drawn from nature with painstaking exactitude, great debate erupted over the distinctive features of the female skeleton. Political circumstances drew immediate attention to depictions of the skull as a measure of intelligence and the pelvis as a measure of womanliness. In her influential drawing of a female skeleton, the French anatomist Marie Thiroux d'Arconville (one of the few women to practice the art) portrayed the female skull as absolutely and proportionally smaller than the male; she also drew attention to the pelvis—remarkable, in her view, for its breadth and width. In contrast to Thiroux d'Arconville, the German anatomist Samuel Thomas von Soemmerring portrayed the skull of the female (correctly) as larger in proportion to the body than that of the male. Soemmerring drew the ribs smaller in proportion to the hips, but not remarkably so. As one of Soemmerring's students pointed out, the width of women's hips should not be overemphasized; they only appear larger than men's because their upper bodies are narrower, which by comparison makes the hips seem to protrude on both sides.

Despite (or perhaps because of) its exaggerations, the Thiroux d'Arconville skeleton became the favored drawing. Soemmerring's skeleton, by contrast, was attacked for its "inaccuracies." In subsequent years, when anatomists had to concede the truth of Soemmerring's depiction of the female skull as larger than the male skull in proportion to the rest of the body, they did not conclude that women's large skulls were loaded with heavy and high-powered brains. Rather than a mark of intelligence, women's large skulls signaled their incomplete growth. Anatomists in the nineteenth century used this as evidence that physiologically women resembled children and "primitives" more than they did European men. In 1829, Edinburgh physician John Barclay, using Thiroux d'Arconville's plates, presented a skeleton family in order to "shew," as he said, "that many of those characteristics described as peculiar to the female, are more obviously discernible in the foetal skeleton."[20] The seemingly irrevocably sexed body became the epistemic foundation for prescriptive claims about social divisions of labor, power, and privilege.[21]

A fourth factor sealing this self-reinforcing system excluding women from science was the belief that science is value and gender neutral or "impartial" (unparteyisch, as Soemmerring called it). Since the Enlightenment, science has stirred hearts and minds with its promise of a "neutral" and privileged viewpoint, above and beyond the rough and tumble of political life.[22] Soemmerring's self-proclaimed neutrality was premised on the absence of dissenting points of view. Those who might have criticized new scientific views were barred from the outset, and the findings of science (crafted largely in their absence) were used to justify their continued exclusion.

Value neutral science arose alongside sexual science: both were integral parts of the revolution that brought modern science to life in the West. Yet, inequalities in power made true impartiality impossible in several important ways. First, science cannot be considered value-neutral while certain groups are systematically excluded from its institutions. Second, asymmetries in social power lend authority to the voice of science. (In the modern intellectual world, "disinterest" carries with it the premium of "objectivity"; those, by contrast, without that cloak speak with a lesser voice.) Third, science cannot be considered neutral so long as systematic exclusions from its enterprise generate systematic neglect (or marginalization) of certain subject matters and problematics.

It is instructive in this context to look at the place of the woman question in the work of Immanuel Kant. Though often overlooked by experts in Kantian philosophy, he did in fact have quite a lot to say about women. Kant is now well known for his infamous remarks that creative work in the sciences lies beyond the natural capabilities of woman. He associated woman's "beautiful understanding" not with science, but with feeling: "her philosophy is not to reason, but to sense."[23] Of greater significance perhaps, Kant's discussions of women are relegated to his precritical work and anthropological writings, where he mentions "lady" or "woman" (*Frauenzimmer*, *Weib*, and so forth) more than three hundred times. These words appear only five times in all of his critical philosophy.[24]

Kant's uncritical views on women cannot be excused on the grounds that he was unfamiliar with these issues; he wrote from the 1760s through the 1790s at the peak of Enlightenment debates about women. Theodor von Hippel, a well-known figure in Kant's Königsberg, published his lengthy *On Improving the Status of Women* in 1792, six years before Kant published his *Anthropologie*. (Kant expressed his annoyance with Hippel for having published some of his ideas before he himself got them into print.)[25] Hippel took exception to Kant's proclamation that women who wished to engage in science "should have a beard."[26] Others joined the criticism of Kant's views. L. W. Weissenborn in 1806 pointed out that the German philosopher's views on women contradicted the axioms of his own system.[27] An anonymous "Henriette" described Kant's words as unjust: man, she wrote, is made for woman as much as woman is made for man; both are equally human.[28] Amalia Holst provided a possible explanation for Kant's unkindness to women: he simply did not have a wife.[29]

The four elements highlighted above—the authority given to nature and to science as the knower of nature, the grounding of gender in biology, and the belief that science is value-neutral—participated in the making and remaking of scientific culture throughout the eighteenth century. It underwrote the fall of the traditionally feminine image of science which Immanuel Kant banished, "mourning like Hecuba," from critical philosophy. It fueled battles over scholarly style that banished all "feminine" qualities—metaphysics, poetry, and rhetoric ornament—from science.[30] It fostered the systematic neglect (or marginalization) of certain subject matters and problematics. We are aware what these inequalities have meant for women; more importantly, these same inequalities held profound consequences for the style and substance of science.

## Gendered Knowledge

Gender was to become one potent principle organizing eighteenth-century revolutions in views of nature, a matter of consequence in an age that looked to nature as the guiding light for social reform. Let me sketch two concrete examples of how gender molded the results of science. The first is the gendering of Linnaean botanical taxonomy.

As extraordinary as it seems today, it was not until the late seventeenth century that European naturalists began recognizing that plants reproduce sexually.[31] The ancients Greeks, it is true, had some knowledge of sexual distinctions in plants: Theophrastus knew the age-old practice of fertilizing date palms by bringing male flowers to the female tree; Pliny tells us that peasants' agricultural practices recognized sexual distinctions in trees such as the pistachio.[32] Plant sexuality, however, was not a major focus of interest in the ancient world. In this era and throughout the medieval period, plant classification generally emphasized the usefulness of plants to human beings as foods and medicines.[33]

Plant sexuality exploded onto the scene in the seventeenth and eighteenth centuries for a variety of reasons including the general interest in sexual differentiation among humans. When sexuality in plants was recognized, everyone wanted to claim the honor for having discovered it. In France, Sébastien Vaillant and Claude-Joseph Geoffroy tussled over priority; in England, Robert Thornton complained that the honor was always given to the French, though properly it belonged to the English. Carl Linnaeus, always keen to reap his due reward for scientific innovation (not, in fact, the first to describe sexual reproduction in plants), claimed that it would be difficult and of no utility to decide who first discovered the sexes of plants.[34]

Even in this era, interest in assigning sex to plants ran ahead of any real understanding of fertilization, or the "coitus of vegetables," as it was sometimes called.[35] Botanists distinguished certain parts of plants as male and female, Claude Geoffroy reported, "without knowing well the reason."[36] The English naturalist, Nehemiah Grew, the first to identify the stamen as the male part in flowers, developed his notion of plant sexuality by analogy to animal sexuality. In his 1682 *Anatomy of Plants*, Grew reported that "the attire" (his term for the stamen) resembles "a small penis," the various coverings upon it appear to be "so many little testicles," and the globulets (or pollen) act as "the vegetable sperme." As soon as the plant penis is erected, Grew continues, "it falls down upon the seed-case or womb, and so touches it with a prolific virtue."[37]

By the early part of the eighteenth century, the analogy between animal and plant sexuality was fully developed. Linnaeus, in his *Praeludia sponsaliorum plantarum*, related the terms of comparison: in the male, the filaments of the stamens are the *vas deferens*, the anthers are the testes, the pollen that falls from them is the seminal fluid; in the female, the stigma is the vulva, the style becomes the vagina, the tube running the length of the pistil is the Fallopian tube, the pericarp is the impregnated ovary; and the seeds are the eggs.[38] Julien Offray de La Mettrie along with other naturalists even claimed that the honey reservoir found in the nectary is equivalent to mother's milk in humans.[39]

Sexual differential, built on the imperfect analogy between plant and animal life, led to the privileging of certain sexual types over others. Most flowers are hermaphroditic with both male and female organs in the same individual. As one eighteenth-century botanist put it, there are two sexes (male and female) but three kinds of flowers: males, females, and hermaphrodites or, as they were sometimes called, androgynes. While most eighteenth-century botanists enthusiastically embraced sexual dimorphism, conceiving of plants as hermaphroditic ran into resistance. William Smellie, chief compiler of the first edition of the *Encyclopaedia Britannica* (1771) and an zealous anti-sexualist, rejected the whole notion of sexuality in plants as the offspring of prurient imaginations, and strenuously disapproved of the use of the term hermaphrodite, noting when using the word that he merely spoke "the language of the system." Smellie denounced Linnaeus for taking his analogy "far beyond all decent limits," claiming that Linnaeus's metaphors were so indelicate as to exceed the those of the most "obscene romance-writer."[40]

The ardent sexing of plants coincided with what is commonly celebrated as the rise of modern botanical taxonomy. In the sixteenth and seventeenth centuries plant materials from the voyages of discovery and newly established colonies flooded Europe (increasing the number of known plants by a factor of four between 1550 and 1700). The search for new methods for organizing and understanding this explosion of materials proliferated new systems of classification: by 1799, when Robert Thornton published his popular version of the Linnaean system, he counted fifty-two different systems of botany.[41] Classification systems were based on different part of plants—John Ray based his on the flower, calyx, and seed-coat; Tournefort in Paris based his on the corolla and fruit; Albrecht von Haller, taking a very different approach, argued that geography was crucial to an understanding of plant life and that embryogenesis should also be represented in a system of classification. Despite the number and variety of systems, Linnaeus's taxonomy swept away these other systems and after in the 1740s (at least outside France) and until the first decades of the nineteenth century was generally considered the most convenient system of classification.

Linnaeus founded his renowned "Key to the Sexual System" on the *nuptiae plantarum* (the marriages of plants), that is, on the number of husbands (stamen) or wives (pistils) in a particular union. His famous *Systema naturae* divided the vegetable world (as he called it) into *classes* based on the number, relative proportions, and position of the male parts or stamens. These classes were then subdivided into some sixty-five *orders* based on the number, relative proportions, and positions of the female parts or pistils. These were further divided into *genera* (based on the calyx, flower, and other parts of the fruit), *species* (based on the leaves or some other characteristic of the plant), and *varieties*.

One might argue that Linnaeus founded his system on sexual difference because he was one of the first to recognize the biological importance of sexual reproduction in plants.[42] But the success of Linnaeus's system did not rest on the fact that it was "natural"; indeed Linnaeus readily acknowledged that it was highly artificial.[43] Though focused on reproductive organs, his system did not capture fundamental sexual functions. Rather it focused on purely morphological features (i. e., the number and mode of union)—exactly those characteristics of the male and female organs *least* important for their sexual function.[44]

Furthermore, Linnaeus devised his system in such a way that the number of a plant's stamens (or male parts) determined the *class* to which it was assigned, while the number of its pistils (the female parts) determined its *order*. In the taxonomic tree, class stands above order. In other words, Linnaeus gave male parts priority in determining the status of the organism in the plant kingdom. There is no empirical justification for this outcome; rather Linnaeus brought traditional notions of gender hierarchy whole-cloth into science. His new language of botany incorporated fundamental aspects of the social world as much as those of the natural world. Although today Linnaeus's classification of groups above the rank of genus has been abandoned, his binomial system of nomenclature remains, together with many of his genera and species.[45]

My second example comes from zoological nomenclature. In 1758, in the 10th edition of his *Systema naturae*, Carl Linnaeus coined the term "mammalia" (meaning literally "of the breast") to distinguish the class of animals embracing humans, apes, ungulates, sloths, sea-cows, elephants, bats, and all other organisms with hair, three ear bones, and a four-chambered heart.[46] In so doing, he idolized the female *mammae* as the icon of that class.

Historians of science have taken Linnaeus's nomenclature more or less for granted as part of his foundational work in zoological taxonomy.[47] There was, however, a complex gender politics informing Linnaean taxonomy and nomenclature. Why Linnaeus called mammals mammals had as much to do with the fact that there is something special about the female breast as with eighteenth-century politics of wet nursing and maternal breast-feeding, and the contested role of women in both science and the broader culture.[48]

For over two thousand years most of the animals we now designate as mammals (along with most reptiles and several amphibians) were called *quadrupeds*.[49] In coining his new term, "mammalia," Linnaeus did not draw from tradition, as was common in this period, but devised a wholly new term.

Were there good reasons for Linnaeus to call mammals mammals? Does the longevity of Linnaeus's term reflect the fact that he was simply right, that the mammae, indeed, represent a primary, universal, and unique character of mammals (as would have been the parlance of the eighteenth century)? Yes and no. Linnaeus chose this term even though naturalists in this period did not consider the mammae a universal characteristic of the class of animals he sought to identify (in the eighteenth century, it was commonly accepted that stallions lacked teats).[50] More importantly, the presence of milk-producing mammae is but one characteristic of mammals, as was commonly known to eighteenth-century European naturalists. Linnaeus could indeed have chosen a more gender neutral name, such as "pilosa" (the hairy ones—though hair, and especially beards—was also saturated with gender) or "aurecaviga" (the hollow-eared ones).

Or had he wished to emphasize the importance of mother's milk to the young, he could have chosen something like "lactentia" (the "sucking ones") which, like the German term *Säugetier* (suckling animals), nicely universalizes the term, since male as well as female young suckle at their mothers' breasts.

If Linnaeus had alternatives, if he could have chosen from a number of equally valid terms, what led him to the term mammal? Zoological nomenclature—like all language—is to some degree arbitrary; naturalists devise convenient terms to identify groups of animals. But nomenclature is also historic, growing out of specific contexts, conflicts, and circumstances.

Linnaeus created his term "mammalia" partly in response to the question of humans' place in nature.[51] In his quest to find an appropriate term for a taxon uniting humans and beasts, Linnaeus made the breast—and specifically the fully-developed female breast—the icon of the highest class of animals. In privileging a uniquely female characteristic in this way, it might be argued that Linnaeus broke with long-standing traditions that saw the male as the measure of all things. It is important to note, however, that in the same volume in which Linnaeus introduced the term "mammalia," he also introduced the term "homo sapiens." This term—man of wisdom—was used to distinguish humans from other primates (apes, lemurs, and bats, for example). In the language of taxonomy, *sapiens* is what is known as a "trivial" name.[52] From a historical point of view, however, the choice of the term *sapiens* is highly significant. Man had traditionally been distinguished from animals by his reason.[53] Thus, within Linnaean terminology, a traditionally masculine characteristic (reason) marks our separateness from brutes; while a female characteristic (the lactating *mammae*) ties humans to brutes.

Though Linnaeus may have introduced the term "mammalia" into zoology, long before Linnaeus the female breast had been a powerful icon within Western cultures, representing both the sublime and bestial in human nature, which may have played a role in his focus on the breast.[54] The grotesque withered breasts on witches and devils represented temptations of wanton lust, sins of the flesh, and humankind fallen from paradise. The firm spherical breasts of Aphrodite—the Greek ideal—represented an otherworldly beauty and virginity. In the French Revolution, the bared female breast—embodied in the strident Marianne—became a resilient symbol of freedom.[55] From the multibreasted Diana of Ephesus to the fecund-bosomed Nature, the breast symbolized generation, regeneration, and renewal.

Myths and legends also portrayed suckling as a point of intimate connection between humans and beasts, suggesting the interchangeability of the maternal breast in this respect. A nanny goat, Amaltheia, was said to have nursed the young Zeus.[56] A she-wolf served as the legendary nurse to Romulus and Remus, the founders of Rome. From the Middle Ages to the seventeenth and eighteenth centuries, bears and wolves were reported to have suckled abandoned children. Linnaeus himself believed that ancient heros, put to the breast of lionesses, absorbed their very great courage along with their milk.[57] In rarer instances, humans were reported even to have suckled animals. Veronica Giuliani, beatified by Pius II (1405–1464), took a real lamb to bed with her and suckled it at her breast in memory of the lamb of God.[58] Humans suckling animals was also standard eighteenth-century medical practice. As Mary Wollstonecraft lay dying after childbirth, the doctor forbade the child the breast and "procured puppies to draw off the milk."[59]

Linnaeus's fascination with female mammae arose alongside and in step with key political trends in the eighteenth century—the restructuring of child care (the decline of the wet nurse and midwife) and the restructuring of women's lives as mothers, wives, and citizens.[60] The portrait he painted of the naturalness of a mother giving suck to her young fed into movements to undermine the public power of women and to attach a new value to mothering.

Most directly, Linnaeus joined the ongoing campaign to abolish the ancient custom of wet nursing. Linnaeus—himself a practicing physician—prepared a dissertation against the evils of wet nursing in 1752. In this treatise, entitled "Step Nurse, or a Dissertation on the Fatal Results of Mercenary Nursing," he alluded to his own taxonomy by contrasting the barbarity of women who deprive their children of mother's milk with the gentle care of great beasts—the whale, the fearsome lioness, and fierce tigress—who willingly offer their young the breast.[61]

To champions of enlightenment, the breast became nature's sign that women belonged in the home. It is remarkable that in the heady days of the French Revolution, when revolutionaries marched behind the fierce and bare-breasted Liberty, the maternal breast figured in arguments against women exercising civic rights.[62] Delegates to the French National Convention, where many of these decisions were made, declared that Nature had removed women from the political arena. In this case, "the breasted ones" were to be confined to the home.[63]

Linnaeus's term "mammalia" helped legitimize the restructuring of European society by emphasizing how natural it was for females—both human and nonhuman—to suckle and rear their own children. Linnaean systematics sought to render nature universally comprehensible, yet the categories he devised infused nature with European notions of gender. Linnaeus saw females of all species as tender mothers, a parochial vision he (wittingly or unwittingly) imprinted on Europeans' understandings of nature.

## The Future of Gender in Science

I chose these examples because they are foundational to the modern life sciences. Linnaeus's taxonomy and nomenclature were validated as the starting point of modern biological nomenclature by the international scientific community in 1905.[64] Linnaeus may well have been innocent of the broader implications of his actions. There is no evidence, for instance, that Linnaeus intentionally chose a gender-charged term when he named mammals "mammals." He may have done so naively, but not arbitrarily. He was led to his innovation in response to the world of human interests and political tensions surrounding him. The fact that scientists might be unaware of the implications of their work does not make them any less mediators or marketeers of political ideas (for many this is a studied innocence). We need to begin to appreciate the contingencies of scientific knowledge, and especially what is foregone in the choice of one particular course rather than another.

Scholarship on women and gender in science has a history stretching back at least to Christine de Pizan's fifteenth-century query whether women have made original contributions in the arts and sciences.[65] Until recently, this debate was often sidelined in the history of science and intellectual history more generally. Nonetheless, gender studies of science have come into their own in the late twentieth century. In the 1960s, scholars working in this area tended to sketch the statistical dimensions of the problems women faced in science; throughout the 1970s and early 1980s, they identified subtle and not so subtle institutional barriers; in the last 1980s and early 1990s, they began to analyze knowledge itself, identifying as I have above how gender molded the very content of specific sciences. Today, scientists and science studies scholars concerned with gender are increasingly exploring how a critical awareness of gender can produce new knowledge and new disciplinary configurations. We are today beginning to chart the changes an awareness of gender has brought to the humanities, social sciences, and life sciences, especially medicine, primatology, and biology.[66] The ferment in knowledge of the past decade has in many instances reshaped what is known and knowable.

## Notes

1. Thomas Birch, *History of the Royal Society* (London, 1756–57), 2:175; and Samuel Pepys, *The Diary of Samuel Pepys*, ed. Robert Latham and William Matthews (Berkeley: University of California Press, 1970–83), 8:243. See also Douglas Grant, *Margaret the First, A Biography of Margaret Cavendish, Duchess of Newcastle 1623–1673* (London, 1957); Henry Ten Eyck Perry, *The First Duchess of Newcastle and Her Husband as Figures in Literary History* (Boston: Ginn, 1918); R.W. Goulding, *Margaret (Lucas) Duchess of Newcastle* (London, 1925); Virginia Woolf, "The Duchess of Newcastle," in *The Common Reader* (London: Hogarth Press, 1929), 98–109; Lisa Sarasohn, "A Science Turned Upside Down: Feminism and the Natural Philosophy of Margaret Cavendish," *The Hunting Library Quarterly* 47 (1984): 289–307; and Kathleen Jones, *A Glorious Fame: The Life of Margaret Cavendish, Duchess of Newcastle, 1623–1673* (London: Bloomsbury, 1988). This material on Margaret Cavendish is drawn from Londa Schie- binger, *The Mind Has No Sex? Women in the Origins of Modern Science* (Cambridge, MA: Harvard University Press, 1989).

2. "A Catalogue of the Natural and Artificial Rarities belonging to the Royal Society, and Preserved at Gresham College," in H. Curzon, *The Universal Library: Or, Compleat Summary of Science* (London, 1712), 1:439. Kathleen Lonsdale and Marjory Stephenson were elected to the Royal Society in 1945 (*Notes and Records of the Royal Society of London*, 4 [1946]: 39–40). See also Joan Mason, "The Admission of the First Women to the Royal Society of London," *Notes and Records of the Royal Society of London* 46 (1992): 279–300.

3. Grant, *Margaret the First*, 218.

4. Dorothea Leporin, *Gründliche Untersuchung der Ursachen, die das weibliche Geschlecht vom Studieren abhalten* (Berlin, 1742), sec. 9.

5. Amalia Holst, *Über die Bestimmung des Weibes zur höhern Geistesbildung* (Berlin, 1802), 80–83.

6. See L.W. Weissenborn, *Briefe über die bürgliche Selbstständigkeit der Weiber* (Gotha, 1806).

7. *De jure Maurorum in Europa* (1729), discussed in *Wöchentliche Hallische Nachrichten* 12 (1729), reprinted in *Antonius Guilielmus Amo, Afer aus Axim in Ghana: Dokumente, Autographe, Belege* (Halle: Martin Luther Universität, 1968), 5–6. His standard works, *De humanae mentis apatheia* (Wittenberg, 1734), *Ideam distinctam* (Wittenberg, 1734), and *Tractatus de arte sobrie et accurate philosophandi* (Halle, 1738) sound familiar themes: What is sensation? Does the faculty of sense belong to mind? On distinct ideas, intention, and learning. See also Paulin Hountondji, *African Philosophy: Myth and Reality*, trans. Henri Evans (London: Hutchinson University Library for Africa, 1983), 111.

8. Kelley, introduction to this volume.

9. Margaret Cavendish, CCXI. *Sociable Letters* (London, 1664), "The Preface."

10. On gender issues, see Carolyn Merchant, *The Death of Nature: Women, Ecology, and the Scientific Revolution* (San Francisco: Harper & Row, 1980); Schiebinger, *The Mind Has No Sex?*; Thomas Laqueur, *Making Sex: Body and Gender from the Greeks to Freud* (Cambridge, MA: Harvard University Press, 1990); Erica Harth, *Cartesian Women: Versions and Subversions of Rational Discourse in the Old Regime* (Ithaca: Cornell University Press, 1992); and David Noble, *A World Without Women: The Christian Clerical Culture of Western Science* (New York: Alfred A. Knopf, 1992).

11. For Enlightenment exclusions of women from the public sphere, see Elizabeth Blochman, "Das Frauenzimmer und die Gelehrsamkeit," *Anthropologie und Erziehung* 17 (1966): 10–75; Karin Hausen, "Die Polarisierung der 'Geschlechtscharaktere,'" in Werner Conze, ed., *Sozialgeschichte der Familie in der Neuzeit Europas* (Stuttgart, 1976); Susan Okin, "Women and the Making of the Sentimental Family," *Philosophy & Public Affairs* 11 (1982): 65–88; Maurice Bloch and Jean Bloch, "Women and the Dialectics of Nature in Eighteenth-century French Thought," in Carol P. MacCormack and Marilyn Strathern, eds., *Nature, Culture and Gender* (Cambridge: Cambridge University Press, 1980), 25–41; Joan Landes, *Women and the Public Sphere in the Age of the French Revolution* (Ithaca: Cornell University Press, 1988); Christine Fauré, *Democracy Without Women: Feminism and the Rise of Liberal Individualism in France*, trans. Claudia Gorbman and John Berks (Bloomington: Indiana University Press, 1991); Doris Alder, *Die Wurzel der Polaritäten: Geschlechtertheorie zwischen Naturrecht und Natur der Frau* (Frankfurt: Campus Verlag, 1992); and Geneviève Fraisse, *Reason's Muse: Sexual Difference and the Birth of Democracy*, trans. Jane Marie Todd (Chicago: University of Chicago Press, 1994).

12. Marie-Jean-Antoine-Nicolas de Caritat, marquis de Condorcet, "Sur l'admission des femmes au droit de cité," in *Oeuvres* (Stuttgart: F. Frommann, 1968), 10:129. See also Lynn Hunt, *Politics, Culture, and Class in the French Revolution* (Berkeley and Los Angeles: University of California Press, 1984); Joan Landes, *Women and the Public Sphere*; Christine Fauré, *Democracy Without Women*; Anne Phillips, *Engendering Democracy* (University Park: Pennsylvania State University Press, 1991); and Madelyn Gutwirth, *The Twilight of the Goddesses* (New Brunswick: Rutgers University Press, 1992).

13. Auguste Comte, *Cours de philosophie positive* (Paris, 1839), 4:569–70. On Comte, see Mary Pickering, *Auguste Comte: An Intellectual Biography, Volume 1* (Cambridge: Cambridge University Press, 1993).

14. Auguste Comte to J.S. Mill, 5 October 1843, in *Lettres inédites de J.S. Mill à A. Comte avec les résponses de Comte*, ed. L. Lévy-Bruhl (Paris, 1899), 256.

15. See Schiebinger, *The Mind Has No Sex?* chs. 6–8.

16. Jean-Jacques Rousseau, *Emile, ou De l'education* (1762) in Bernard Gagnebin and Marcel Raymond eds., *Oeuvres complètes* (Paris: Gallimard, 1959–69), vol. 4.

17. Holst, *Über die Bestimmung der Weibes*, 143.

18. For studies of sexual science, see also Ian Maclean, *The Renaissance Notion of Woman* (Cambridge: Cambridge University Press, 1980); Danielle Jacquart and Claude Thomasset, *Sexuality and Medicine in the Middle Ages*, trans. Matthew Adamson (Princeton: Princeton University Press, 1988); Monica Green, "Female Sexuality in the Medieval West," *Trends in History* 4 (1990): 127–58; Cynthia Russett, *Sexual Science: The Victorian Construction of Womanhood* (Cambridge, MA: Harvard University Press, 1989); Laqueur, *Making Sex*; Joan Cadden, *Meanings of Sex Difference in the Middle Ages: Medicine, Science, and Culture* (Cambridge: Cambridge University Press, 1993); Nancy Tuana, *The Less Noble Sex: Scientific, Religious, and Philosophical Conceptions of Woman's Nature* (Bloomington: Indiana University Press, 1993).

19. Schiebinger, *The Mind Has No Sex?* ch. 7.

20. John Barclay, *The Anatomy of the Bones of the Human Body* (Edinburgh, 1829), commentary to plate 32.

21. Environmentalism, a latter-day theory of humors, was still prominent among eighteenth-century medical men but was used more often to explain racial than sexual variation (European men and women were, after all, subject to similar environments).

22. On the origins of value-neutrality in science, see Robert N. Proctor, *Value-Free Science? Purity and Power in Modern Knowledge* (Cambridge, MA: Harvard University Press, 1991).

23. Immanuel Kant, "Beobachtungen über das Gefühl des Schönen und Erhabenen"(1766), in *Kants Werke*, ed. Wilhelm Dilthey (Berlin, 1900–19), 2:229.

24. Dieter Krallmann and Hans Martin, eds., *Wortindex zu Kants gesammelten Schriften* (Berlin, 1967), s.v. "Frau" and "Weib." On Kant's views on women, see Lawrence Blum, "Kant's and Hegel's Moral Rationalism: A Feminist Perspective," *Canadian Journal of Philosophy* 12 (1982): 287–302; Ellen Kennedy and Susan Mendus, eds., *Women in Western Political Philosophy: Kant to Nietzsche* (Brighton, Sussex: Wheatsheaf Books, 1987); and Robin Schott, *Cognition and Eros: A Critique of the Kantian Paradigm* (Boston: Beacon Press, 1988).

25. Timothy Sellner in Hippel, *On Improving the Status of Women* (1792), trans. Timothy Sellner (Detroit: Wayne State University Press, 1979), 29.

26. Theodor von Hippel, *Über die bürgerliche Verbesserung der Weiber* (1792), vol. 6 in *Sämmtliche Werke* (Berlin, 1828), 35.

27. Weissenborn, *Briefe über die bürgliche Selbstständigkeit der Weiber*, 22.

28. "Henriette," *Philosophie der Weiber* (Leipzig, 1802), 111.

29. Holst, *Über die Bestimmung des Weibes*, 95. Weissenborn held that only a businessman could fully appreciate the value of a learned wife (*Briefe über die bürgliche Selbstständigkeit der Weiber*, 27).

30. See Jean Le Rond d'Alembert, *Preliminary Discourse to the Encyclopedia of Diderot*, trans. Richard N. Schwab (Indianapolis: Bobbs-Merrill, 1963), 93–97.

31. These materials are drawn from Londa Schiebinger, *Nature's Body: Gender in the Making of Modern Science* (Boston: Beacon Press, 1993), ch. 1. See also François Delaporte, *Nature's Second Kingdom: Explorations of Vegetality in the Eighteenth Century*, trans. Arthur Goldhammer (1979; Cambridge, MA: MIT Press, 1982).

32. Pliny the Elder, *Natural History*, trans. H. Rackham (Cambridge, MA: Harvard University Press, 1942), XII, xxxii, 45. A.G. Morton, *History of Botanical Science: An Account of the Development of Botany from Ancient Times to the Present Day* (New York: Academic Press, 1981), 28, 38.

33. Karen Reeds, *Botany in Medieval and Renaissance Universities* (New York: Garland, 1991); and Agnes Arber, *Herbals: Their Origin and Evolution, A Chapter in the History of Botany 1470–1670*, 3rd ed. (Cambridge: Cambridge University Press, 1986), xxv and ch. 1.

34. Jacques Rousseau, "Sébastien Vaillant: An Outstanding Eighteenth-Century Botanist," *Regnum Vegetabile* 71 (1970): 195–228. "The Prize Dissertation of the Sexes of Plants by Carolus von Linnaeus," in Robert Thornton, *A New Illustration of the Sexual System of Carolus von Linnaeus* (London, 1799–1807).

35. William Smellie, "Botany," *Encyclopaedia Britannica* (Edinburgh, 1771), 20:646.

36. Claude Geoffroy, "Observations sur la structure et l'usage des principales parties des fleurs," *Memoires de l'Académie Royales des Sciences* (1711): 211.

37. Nehemiah Grew, *The Anatomy of Plants* (London, 1682), 170–72.

38. Linnaeus, *Praeludia sponsaliorum plantarum* (1729; reprinted Uppsala: Almquist & Wiksells, 1908), sec. 15. Linnaeus drew this information from Vaillant. See also Jean-Jacques Rousseau, *Dictionnaire des termes d'usage en botanique*, in *Botany: A Study of Pure Curiosity* (London: Michael Joseph, 1979), s.v. "pistil."

39. Julien Offray de la Mettrie, *L'Homme plante* (Potsdam, 1748).

40. Smellie, "Botany," *Encyclopaedia Britannica*, 1:653.

41. Robert Thornton, *A New Illustration of the Sexual System of Carolus von Linnaeus* (London, 1799–1807). See also David Allen, *The Naturalist in Britain: A Social History* (Harmondsworth Middlesex: Penguin Books, 1978) and Alice Stroup, *A Company of Scientists: Botany, Patronage, and the Community at the Seventeenth-century Parisian Royal Academy of Sciences* (Berkeley and Los Angeles: University of California Press, 1990).

42. Carl Linnaeus, *Systema naturae* (1735) ed. M.S.J. Engel-Ledeboer and H. Engel (Nieuwkoop: B. de Graaf, 1964), "Observationes in regnum vegetabile," no. 7.

43. Linnaeus, *Systema naturae* (1735), no. 12. See also Linnaeus to Haller of 3 April 1737, in *The Correspondence of Linnaeus*, James E. Smith, ed., 2 vols. (London, 1821), 2:229. Linnaeus's system was artificial, but it was neither arbitrary nor heterodox; see James Larson's discussion in his *Reason and Experience: The Representation of Natural Order in the Work of Carl von Linné* (Berkeley and Los Angeles: University of California Press, 1971), 61–62.

44. See Julius von Sachs, *Geschichte der Botanik vom XVI. Jahrhundert bis 1860* (Munich, 1876), 82–83.

45. For a history of binomial nomenclature, see John Heller, *Studies in Linnaean Method and Nomenclature* (Frankfurt: Verlag Peter Lang, 1983), 41–75.

46. The term "Mammalia" first appeared in a student dissertation, *Natura Pelagi*, in 1757, but was not published until 1760 (*Amoenitates academicae* [Erlangen, 1788], 5:68–77).

47. Linnaeus studies are voluminous. See *A Catalogue of the Works of Linnaeus* (London: The British Museum, 1933); Henri Daudin, *De Linné à Jussieu: Méthodes de la classification* (Paris: Félix Alcan, 1926); Ernst Mayr, *The Growth of Biological Thought, Diversity, Evolution, and Inheritance* (Cambridge, MA: Harvard University Press, 1982); Heinz Goerke, *Linnaeus* (New York: Charles Schribner, 1973); and Gunnar Broberg, ed. *Linnaeus: Progress and Prospects in Linnaean Research* (Stockholm: Almqvist & Wiksell International, 1980). Broberg's excellent *Homo Sapiens L.: Studier i Carl von Linnés naturuppfattning och människolära* (Stockholm: The Swedish History of Science Society, 1975) considers Linnaeus's work within its social contexts.

48. See Londa Schiebinger, "Why Mammals are Called Mammals: Gender Politics in Eighteenth-century Natural History," *American Historical Review* 98 (1993): 382–411 and *Nature's Body*, ch. 2.

49. Aristotle, *Historia animalium*, in *The Works of Aristotle*, trans. D'arcy Thompson (Oxford: Clarendon Press, 1910); G.E.R. Lloyd, *Science, Folklore and Ideology* (Cambridge: Cambridge University Press, 1983), 16; Aristotle, *Generation of Animals*, trans. A.L. Peck (Cambridge, MA: Harvard University Press, 1953), lxix; and Pierre Pellegrin, *Aristotle's Classification of Animals: Biology and the Conceptual Unity of the Aristotelian Corpus*, trans. Anthony Preus (Berkeley and Los Angeles: University of California Press, 1986).

50. Georges-Louis Leclerc, comte de Buffon, *Histoire naturelle* (Paris, 1749), 1:38–40. C. Prévost, the author of "Mammifères" in the *Dictionnaire classique d'histoire naturelle*, notes that in this period it was commonly thought that male horses had no teats and consequently that mammae were not a universal character of *Mammals* (Paris, 1826, 10:74).

51. See Gunnar Broberg, "Homo sapiens: Linnaeus's Classification of Man," in *Linnaeus: The Man and His Work*, ed. Tore Frängsmyr (Berkeley and Los Angeles: University of California Press, 1983), 156–94.

52. Broberg has shown that Linnaeus first used the term *sapiens* in 1753 to denote a species of monkey referred to as *Simia sapiens*—a species said to play a mean game of backgammon ("*Homo sapiens*," 176). Linnaeus wrote of "trivial names" in reference to botany: "I have put trivial names in the margin so that without more ado we can represent one plant by one name; these I have taken, it is true, without special choice, leaving this for another day. However, I would warn some solemnly all sensible botanists not to propose a trivial name without adequate specific distinction, lest the science fall back into its early crude state." Cited in Heller, *Studies in Linnaean Method and Nomenclature*, 278.

53. Linnaeus saw reason as the principal character distinguishing humans from other animals. In the preface to his *Fauna Suecica* (1746) he called reason "the most noble thing of all" that places man above all others. See also H.W. Janson, *Apes and Ape Lore in the Middle Ages and the Renaissance* (London: The Warburg Institute, 1952), 74–75.

54. The cultural significance of the breast and mother's milk is a large topic. See Marina Warner's *Alone of All Her Sex: The Myth and the Cult of the Virgin Mary* (New York: Alfred A. Knopf, 1976) and her *Monuments and Maidens: The Allegory of the Female Form* (New York: Atheneum, 1985) along with Caroline Bynum's *Jesus as Mother: Studies in the Spirituality of the High Middle Ages* (Berkeley and Los Angeles: University of California Press, 1982) and Marilyn Yalom's *A History of the Breast* (New York: Knopf, 1997). Heinz Kirchhoff's "Die künstlerische Darstellung der weiblichen

Brust als Attribut der Weiblichkeit und Fruchtbarkeit als auch der Spende der Lebenskraft und der Weisheit" (*Geburtshilfe und Frauenheilkunde*, 50 [1990]: 234-243) is rich and suggestive but written, like Erich Neumann's *Die Grosse Mutter* (Zurich: Rhein Verlag, 1956), from a Jungian perspective and without attention to historical context. Some helpful materials are also found in Anne Hollander, *Seeing Through Clothes* (New York: Penguin Books, 1975) and Françoise Borin, "Arrêt sur image," *Historie des femmes en Occident*, ed. Natalie Davis and Arlette Farge (Paris: Plon, 1991), 3:213–19. See also the anecdotes collected by Gustave-Jules Witkowski in *Les seins dans l'histoire* (Paris: A. Maloine, 1903).

55. Lynn Hunt, *Politics, Culture, and Class in the French Revolution* (Berkeley and Los Angeles: University of California Press, 1984), esp. part 1.

56. Warner, *Alone of All Her Sex*, 194.

57. Linnaeus also advocated having cows suckle children in order to improve fertility rates as was done in certain villages in France. Carl Linnaeus, "Nutrix Noverca," respondent F. Lindberg (1752) *Amoenitates academicae* (Erlangen, 1787), 3:262–63. Goats and other animals were used to suckle syphilitic children in foundling hospitals in the eighteenth century or when there was a shortage of human nurses. Valerie Fildes, *Wet Nursing: A History from Antiquity to the Present* (Oxford: Basil Blackwell, 1988), 147.

58. Mervyn Levy, *The Moons of Paradise: Some Reflections on the Appearance of the Female Breast in Art* (London: Arthur Barker Limited, 1962), 55.

59. William Godwin, *Memoirs of the Author of a Vindication of the Rights of Woman* (London, 1798), 183.

60. Dissatisfaction with wet nursing began in the 1680s; the height of the campaign, however, came in the eighteenth century. See Jane Sharp, *The Midwives Book* (London, 1671), 353, 361–62. See also George Sussman, *Selling Mothers' Milk: The Wet-Nursing Business in France, 1715–1914* (Urbana: University of Illinois Press, 1982); Nancy Senior, "Aspects of Infant Feeding in Eighteenth-century France," *Eighteenth-Century Studies*, 16 (1983): 367; Valerie Fildes, *Breasts, Bottles and Babies: A History of Infant Feeding* (Edinburgh: Edinburgh University Press, 1986); and Mary Sheriff, "Fragonard's Erotic Mothers and the Politics of Reproduction," in Lynn Hunt, ed., *Eroticism and the Body Politic* (Baltimore: The Johns Hopkins University Press, 1991), 14–40.

61. Carl Linnaeus, "Nutrix Noverca," 3:262–63. This work was translated by J.E. Gilibert as "La Nourrice marâtre, ou Dissertation sur les suites funestes du nourrissage mercénaire," in *Les Chef-d'oeuvres de Monsieur de Sauvages* (Lyon, 1770), 2:215–44.

62. See Hunt, *Politics, Culture, and Class in the French Revolution*, esp. part 1; also Warner, *Monuments and Maidens*, chs. 12 and 13; and Madelyn Gutwirth, *The Twilight of the Goddesses*.

63. Cited in Darline Levy, Harriet Applewhite, and Mary Johnson, eds., *Women in Revolutionary Paris, 1789–1795* (Urbana: University of Illinois Press, 1979), 219.

64. In 1905 the *International Code of Botanical Nomenclature* designated Linnaeus's *Species plantarum* of 1753 the starting point for botanical nomenclature. See Frans A. Stafleu, *Linnaeus and the Linnaeans: The Spreading of Their Ideas in Systematic Botany, 1735–1789* (Utrecht: A. Oosthoek's Uitgeversmaatschappy: N.V., 1971), 110. The tenth edition of Linnaeus's *Systema naturae* and Clerck's *Aranei Svecici* together form the starting point of modern zoological nomenclature. See *International Code of Zoological Nomenclature*, ed. W.D. Ride (London: The British Museum, 1985), I.3.

65. Christine de Pizan, *The Book of the City of Ladies* (1405), trans. Earl Jeffrey Richards (New York: Persea, 1982), 70.

66. Marilyn Strathern, "An Awkward Relationship: The Case of Feminism and Anthropology," *Signs* 12 (1987): 276–92; Cheris Kramarae and Dale Spender, *The Knowledge Explosion: Generations of Feminist Scholarship* (New York: Teachers College Press, Columbia University, 1992); Anna Mastroianni, Ruth Faden, and Daniel Federman, eds., *Women and Health Research* (Washington, DC: National Academy Press, 1994), vol. 2; Linda Fedigan, "Is Primatology a Feminist Science?" in Lori Hager ed., *Women in Human Evolution*, (New York: Routledge, 1997); Londa Schiebinger, *How Feminism Has Changed Science* (Cambridge, MA: Harvard University Press, forthcoming).

# Index

Aaron, 166
Abbas, Haly, 278
Abraham, 166
Accademia dei Lincei, 256
Achillini, Alessandro, 57n33
Ackerknecht, Erwin, 309, 310
Acosta, José, 240, 243, 244
Adam, 62, 66-67, 68, 106-113, 180, 249, 286
Aegidius, William (William Gilliszoon), 57n33
Aesculapius, 272
Aesop, 152
Agricola, Georg, 240
Agrippa, Henry Cornelius, 2, 49, 253, 265
Aldrovandi, Ulisse, 240, 243, 245, 251
Aleandro, Girolamo, 166-167, 168
Alexander the Great, 225, 247, 249, 268
Alexandrini, Giulio, 281
Alsted, Johann, 14, 16, 17, 43
Amaltheia, 329
Ambrose, Saint, 130
Ambrosiana, 47
Amo, Anton, 320
Anaxagoras, 62
Anaximander, 70, 71
Ancients, 4, 50, 51, 64, 93, 94, 145, 149, 186, 193, 195, 198, 244, 250, 253, 286, 288
André, Bernard, 73n7
Annas, Julia, 127
Anquetil-Duperon, Abraham-Hyacinthe, 225
Aphrodite, 328
Apostles, 133
Aquilano, Scipio, 70
Arbuthnot, John, 285
Archimedes, 24, 65, 301
Argenterio, Giovanni, 294n63
Aristotle, 1, 4, 15, 17, 35, 37, 38, 42, 46-47, 50, 56n25, 62, 63, 65, 67, 69, 71, 72, 73n9, 86, 88, 92, 94, 95, 98n35, 112, 118, 130, 132, 133, 135, 142n26, 143, 146, 147, 148, 152-153, 155, 191-192, 205, 225, 230, 231, 232, 240, 242, 243, 245, 247, 248, 249-250, 268-269, 284, 319, 321, 322
Aristoxenus, 188
Arnold, Gottfried, 104, 113, 119
Arnold, Matthew, 5

*ars grammatica*, 5
*ars musica*, 6
Arundel, Thomas Howar, earl of, 239, 254
Ascletarion, 271
Ashworth, William, 251
Augustine, Saint, 20, 76n40, 133
Augustus, 267
Avicenna, 49, 50, 57n34, 279

Bacci, Andrea, 281
Bacon, Francis, 1, 3, 7, 8, 9, 13, 14, 41, 42-57, 66, 77n47, 92-93, 163, 216, 225, 230, 231, 239-260, 283, 301
Bacon, Nicholas, 239, 246, 255
Bacon, Roger, 157n12
Baltus, 119
Balzac, Guez de, 163
Barbaro, Ermolao, 57n31
Barberini, Francesco, 167, 183n44
Barbeyrac, Jean, 5, 132-133
Barclay, John, 323
Bardi, Giovanni dei, 188, 189, 192, 196
Baronius, Caesar, 49
Bayle, Peter, 56n26, 104, 113, 114, 115, 116, 118, 228
Bede the Venerable, Saint, 195
Benci, Francesco, 201n34
Bentham, Jeremy, 135, 136
Bentivoglio, Cardinal Guido, 183n44
Bentley, Richard, 164
Berlin Academy of Sciences, 299, 302
Bernal, Martin, 61, 124n51
Bernard, Claude, 304
Berosus, 270
Besson, Jacques, 57n33
Bianchini, 265
*Bildung*, 300, 302, 313n26
Billon, M., 182n22
Bismarck, Otto von, 300, 308
Blackwell, Constance, vii, 4
Blair, Ann, vii, 3
Blum, Rudolf, 52
Blumenbach, Johann Friedrich, 224
Blumenthal, 213
Boccaccio, Giovanni, 143, 189, 200n13
Boccadiferro, Ludovico, 57n33
Bodin, Jean, 3, 7, 9, 29-38, 42, 49, 207

Boerhaave, Hermann, 284, 288, 294n52, 294n61
Boethius, 15, 57n31, 186, 187, 188, 189, 193, 199n2
Böhme, Jacob, 115
Bonaparte, Napoleon, 305
Borrichius, Olaus, 65, 68
Bossuet, Jacques Benigne, 95, 129
Bouchard, Jean-Jacques, 164
Boulanger, Nicole Antoine, 229
Bourdieu, Pierre, 24
Boyle, Robert, 71, 258n12, 319
Brahe, Tycho, 267, 269
Brancusi, Constantin, 139
Brandolini, Raffaele, 189
Broberg, Gunnar, 334n52
Brücke, Ernst, 299, 308
Brucker, Carol Friederic, 82n107
Brucker, Johann Jacob, 4, 6, 7, 8, 19, 20, 21, 61-82, 84, 87-88, 90, 104, 116, 119
Bruno, Giordano, 20, 49, 57n33, 240, 248, 253
Buckingham, George Villiers, 1st duke of, 254, 255
Buddeus, Johann Franz, 4, 19, 63, 71, 86, 88, 89, 90, 92, 96, 98n35, 101n87, 103-125
Budé, Guillaume, 15, 49
Buffon, Georges Louis Le Clerc, 224, 230
Burckhardt, Jacob, 164
Burghley, William Cecil, Lord, 245, 246, 247
Burley, Walter, 64, 76n42
Burnet, Thomas, 68, 70
Burton, John, 287

Cabala, 67, 72, 112, 114, 115
Cahan, David, 312n16
Caius, John, 242-243
Caloni, Jacques, 267-268
Calvin, Jean, 49
Calvo, Fabio, 270
Camillo, Giulio, 47
Campanella, Tommaso, 55n6, 57n33, 72
Capodivacca, Girolamo, 57n34
Cardano, Girolamo, 7, 8, 20, 49, 72, 191, 240, 248, 253, 263-275, 281, 282
Carl Gustav of Sweden, 207, 215
Carl Gustavssohn of Sweden, 208
Casaubon, Isaac, 49, 62, 74n19, 164, 166, 249
Casseretz, Manuel de Costa, 172
Cassiodorus, 15, 57n31
Cassirer, Ernst, 22
Cave, Terence, 33

Cavendish, Margaret (Duchess of Newcastle), 319, 320
Cecil, Robert, 245, 246, 247, 254
Celsus, Aulus Cornelius, 284
Cerbu, Tom, 181n5
Charles X of Sweden, 208
Charles XI of Sweden, 206
Charles XII of Sweden, 208
Charpentier, Jacques, 49, 57n33
Charron, Pierre, 16
Chauvin, Etienne, 15
Chemnitz, Bogislaw, 219n20
Chesne, Joseph du, 49
Cheyne, George, 284, 285
Chladenius, Martin, 94
Christina of Sweden, 214, 221n46
Chrysippus, 134
Cicero, 5, 45, 47, 89, 99n46, 128, 143, 146, 147, 151, 158n19
Clarke, Samuel, 131, 132, 141n18
Clauberg, Johannes, 94
Claudius, 272
Clement of Alexandria, 64, 67, 68, 80n83, 97n14
Cocles, 49
College of Coimbra, 50
Collegium Anthologicum, 205-206
Collegium Gellanium, 205
Columbus, Christopher, 255
commonplace books, 30-32
Comte, Auguste, 322
Condillac, Etienne Bonnot de, 18, 154, 157n16, 230
Condorcet, Marie Jean Antoine Nicholas Caretat, Marquis de, 228, 230-233, 322
Conring, Hermann, 65, 66, 68, 69, 70, 78n63, 207, 284
Constantine, 269
Cook, James, 227
Cooper, Thomas, 15
Cope, Walter, 254
Copenhaver, Brian P., 56n25
Copernicus, Nicolas, 7, 49, 71, 109, 122n23, 247, 261, 262, 263, 264, 268, 269, 305, 321
Cordes, Jean des, 45
Cornaro, Luigi, 281-282, 283-284, 293n29
Costa, Gustavo, 148
Cousin, Victor, 83, 84, 97n8
Crapanzano, Vincent, 224
Crasso, Junio Paulo, 280
Crato, Johannes, 267
Creuzer, Georg Friedrich, 173

Crinesius, Christopher, 169, 170, 183n43
Crinito, Pietro, 272
Croll, Oswald, 49, 253
Crusius, Christian August, 131
Cudworth, 65, 69, 112, 114, 115, 118, 119, 138, 139
Cullen, William, 293n40
Cuvier, Georges, 303

D'Alembert, Jean, 229, 231
Dahlmann, Peter, 205-206
Dante Alighieri, 143, 145, 152, 154
Darwin, Charles, 305
David, 225
Davies, John, 246
Decalog, 136
Dee, John, 243, 244, 248, 253
della Scala family, 165, 167
Delminio, Camillo, 34
Democritus, 95, 319
Derrida, Jacques, 2, 24
Descartes, René, 2, 4, 5, 13, 14, 17, 66, 71, 77n47, 88, 91, 92, 93, 94, 95, 98n35, 138, 143, 146-151, 153, 154, 157n14, 229-230, 231, 319
Descordes, Jean, 42
Deuteronomy, 169
Diana of Ephesus, 329
Dickenson, Edmund, 68
Diderot, Dénis, 18, 21, 97n10, 138, 226, 231
dietetics, 277-295 *passim*.
Dilthey, William, 22, 100n75, 159n31, 224
Diocletian, 271
Dionysius of Halicarnassus, 218n9
Dioscorides, 240, 248
Domitian, 272
Donagan, Alan, 127
Doni, Giovanni Battista, 197-198, 202n60
Doria, Paolo Mattia, 148, 158n22
Doyle, Arthur Conan, 265
Droysen, Johann Gustav, 311
Du Bois-Reymond, Emile, 8, 297-317
Dupleix, Scipion, 130
Dupuy, Christophe, 42, 45, 56n26, 183n44
Dupuy, Claude, 42, 45, 56n26, 182n20
Dupuy, Jacques, 42, 45, 56n26, 169
Dupuy, Pierre, 42, 45, 56n26, 167, 168, 170
Durand, Guillaume, 57n33
Dürer, Albrecht, 264
Duret, Claude, 170
Durkheim, Emile, 224
Eclecticism, 2, 4, 5, 14, 19, 66, 72, 77n47, 78n68, 83-101, 104, 112, 123n41, 199, 217
l'Ecluse, Charles de, 181n5, 243
Edward VI of England, 245
Eleazar, 166
Elizabeth I of England, 239, 245, 246, 247, 254, 255
Elyot, Thomas, 73n7, 281, 293n19
encyclopedia, 1, 3, 7, 14, 16, 18, 21, 22, 41, 42, 43, 44, 47, 48, 53, 54n3, 62, 92, 105, 164, 165, 240, 242, 247, 249, 251-252, 253, 290, 326
Epicurus, 56n25, 62, 88, 147-148, 154
Epiphianius, Saint, 64
*episteme*, 13, 23, 323
Erasmus, Desiderius, 5, 32, 280, 303
Erastratus, 287
Erastus, Thomas, 49
Erlexben, Dorothea, 320
Esperti, Abbé, 144
Essex, Robert Devereux, earl of, 245, 247
Euclid, 70, 81n102
Eudemus, 269
Eudoxus, 269, 270
Eusebius, 68
Euthyphro, 138
Eutropius, 205, 218n9
Ezekiel, 131
Ezra, 169, 177, 178

Fabricius, Johann Albert, 63, 76n37, 118, 245
Falaiseau, 212
Faßmann, David, 103
Fechner, Gustav, 305
Feingold, Mordechai, 243
Ferdinand II of Spain, 264
Ferguson, 224, 226, 229
Feuerlin, Jakob Wilhelm, 88, 92, 100n68, 110, 122n33
Feyerabend, Paul, 22
Fichte, Johann Gottlieb, 83
Ficino, Marsilio, 62, 63, 66, 75n28, 130, 154
fideism, 116, 118
Filelfo, Francesco, 57n31
Findlen, Paula, vii, 7
Florus, 218n9
Fontenay-Mareuil, François du Val, 45
Fontenelle, Bernard le Bovier de, 103
Forliviensis, Jacobus, 57n34
Forster, George, 227
Foucault, Michel, 2, 22, 23-24, 26, 223, 227, 232

Fougerolles, François de, 39n15
Fracastoro, Girolamo, 253
Francke, August Hermann, 103
Franckenstein, Christian Friedrich, 205, 206, 218n9
Franklin, Julian, 28n46
Fraser, J. G., 170-171
Frederick William III of Prussia, 207, 212, 215
Frederick II of Prussia, 210, 306
Frederick William of Hohenzollern, Great Elector of Brandenburg, 207, 212, 213
Freind, John, 277, 285, 287
Freud, Sigmund, 26
Friese, 215
Fuchs, Paul von, 212
Fumerton, Patricia, 253
Fustel de Coulanges, Numa Denis, 224

Gadamer, Hans-Georg, 17, 24, 159n31
Gaffurio, Franchino, 186, 198, 199n2
Gale, T., 63, 64
Galen, 50, 64, 65, 278, 279, 280, 281, 282, 286, 289, 291n11, 294n63, 322
Galiani, Fernando, 225
Galilei, Galileo, 14, 25, 49, 189, 301, 321
Galilei, Vincenzo, 6, 187, 188, 190, 191, 193, 195-198, 199n2
Gallaup-Chasteuil, François de, 182n22, 183n41
Garbo, Dino del, 57n34
Garbo, Tommaso del, 57n34
Garve, Christian, 20, 25
Gassendi, Pierre, 56n25, 57n33, 64, 67, 148, 163, 164, 181n5, 183n40
Gaurico, Pomponio, 272
Gaza, Theodore, 57n31
Geertz, Clifford, 224, 232
Gellius, Aulus, 205
Genebrard, M., 182n22
Genesis, 65, 109, 110, 122n33
Geoffroy, Claude, 325
George of Trebizond, 57n31
Gérando, Joseph Marie de, 224
Gerard, John, 242-245
Gerhard, Ephraim, 88
Gesner, Konrad, 3, 41, 51, 240, 243, 244, 251
Gibbon, Edward, 164, 165
Gierke, Otto, 6
Gilbert, William, 50, 57n33, 243, 253
Gillespie, Michael Allen, 28n46
Giuliani, Veronica, 329
Giustiniani, Vicenzo, 188
Glafey, Adam, 133, 204

Glanvill, Joseph, 319
Glarean, Heinrich, 195
Gloeckner, Hieronymus Georg, 84
Glorious Revolution, 208, 212
Goclenius, 15
Goethe, Johann Wolfgang von, 8, 302-303, 305-307, 314n65, 315n76
Gordian, 272
Gordon, Bernard, 57n34
Gorlaeus, David, 57n33
Goyet, Francis, 32
Gradmann, Christoph, 313n26
Grafton, Anthony, vii, 7, 74n19, 165, 166
Graham, Loren, 28n51, 223
Grant, Edward, 38
Grew, Nehemiah, 325
Grimaldi, 229
Grotius, Hugo de, 98n35, 203, 206, 218n11, 226, 229, 230
Gualdo, Francesco, 183n44
Guarini, Giovanni Battista, 192
Guido of Arezzo, 195, 196
Gundling, Nikolaus Hieronymus, 19, 64, 66, 67, 77n51, 78n63, 79n75, 103-125

Habermas, Jürgen, 24
habitus, 16, 108
Hacking, Ian, 16
Hadrian, 269, 272
Haeckel, Ernst, 306
Hakluyt, Richard, 225
Halle, Jean Nöel, 290
Hallé, Pierre, 45
Haller, Albrecht von, 303, 326
Hallmann, Eduard, 298
Hapsburgs, 219n20
Hardt, Hermann van der, 110, 122n33
Harriot, Thomas, 243, 244, 258n23
Harvey, William, 287
Hayck, Taddäus, 265
Hecuba, 324
Hegel, Georg Wilhelm Friedrich, 16, 21, 25, 83, 84, 87, 94, 97n8, 97n10, 150-151, 159n31, 203, 301, 302
Heidegger, Martin, 17, 23
Heinsius, Daniel, 171
Helmholtz, Hermann, 299, 308, 312n15, 313n26, 314n65
Henry VII of England, 73n7
Herbert, Mary, Countess of Pembroke, 245
Herder, Johann Gottfried, 231
Hermes Trismegistus, 7, 61, 62, 63, 66, 69, 72, 74n19, 82n104, 270, 272

Herodotus, 140n11, 205, 218n9, 225
Hertzberg, Count, 210
Hessen-Rheinfels, Ernst von, 211, 219n21
Heumann, Christoph August, 19-20, 64, 68, 70, 76n42, 77n53, 88, 89, 101n90, 116, 123n43, 123n46
Hierocles, 132
Hildebert the Venerable, 13
Hilliard, Nicholas, 246
Hipparchus, 261, 270, 272
Hippel, Theodor von, 320, 324
Hippocrates, 65, 115, 117, 270, 271, 279, 286, 290, 294n41
*historia atheismi*, 104, 119
*historia literaria*, 3, 52-53, 94, 103, 104
*historia philosophica*, 4, 18, 19
historicism, 21
Hobbes, Thomas, 66, 77n47, 113, 114, 115, 116, 117, 118, 132, 135, 138, 141n18, 203, 206, 218n11, 226, 229, 230, 319
Hoffman, Friedrich, 284, 285, 287, 294n41
Hohenzollern, House of, 300
Holst, Amalia, 320, 324
Holstenius, Lucas, 167, 169
Homer, 70, 145, 154
Horace, 17, 57n36, 90, 143
Horn, Georg, 18, 85
Hornius, Georgius, 63, 64
Houwald, Christoph, 213
Huber, Ulrich, 106
Huet, Pierre Daniel, 17-18
Hugh of Sienna, 57n34
Huizinga, Johan, 164
Humboldt, Wilhelm Freiherr von, 300, 302, 311
Hume, David, 16, 136, 229
humoralism, 278, 282, 283, 290, 332n21
Huppert, George, 28n46
Hurault, Pol, 182n22
Husserl, Edmund, 13, 25, 159n31
Hutchenson, Francis, 226
Huygens, Christian, 319
hygeine, 277-295 *passim*.

Iamblichus, 14, 66, 69
Ideologues, 2
Indians, 113
induction, 43, 66, 72, 248, 301, 303
*ingenium*, 149, 154, 155
Inquisition, 263
Ireneus, Saint, 130
Irwin, 127
Isidore of Seville, 15, 200n13

Isocrates, 140n11
Italic sect, 14-15

Jabir, 268
James I of England, 248, 254, 255-256
Jardine, Nicholas, vii, 3, 8, 56n25, 262
Jerome, 170
Jesuits, 62
Jesus Christ, 69, 131, 133, 179, 180, 263, 271
Johannitius, 278
John of Saxony, 57n33
John XXII, 195
Jonsius, Johann, 20, 62-63
Jove, 151, 152
Joy, Lynn S., 56n25

Kames, Henry Home, 224
Kant, Immanuel, 13, 20, 21, 22, 25, 84, 94, 134-135, 150, 151, 153, 154, 158n29, 203, 223, 224, 227-229, 231-233, 321, 324
Keckermann, Bartholomew, 16, 17, 43
Keill, James, 287
Kelley, Donald, vii, 28n26, 230, 320
Kepler, Johann, 6, 7, 8, 49, 56n25, 189, 261-264, 268, 272-276
Kircher, Athanasius, 64, 65, 66
Kriegsmann, Wilhelm Christoph, 66
Kristeller, Paul Oskar, 48
Kuhn, Thomas, 3, 19, 20, 23, 24, 230

La Boderie, Guy le Fèvre de, 189
La Croix du Maine, François Grude, 47
La Gardie, Magnus de, 206, 209, 213
La Mettrie, Julien Offray de, 326
Lactantius, 48, 67
Laertius, Diogenes, 4, 15, 18, 19, 61, 64, 67, 70, 73n8, 84, 85, 133, 142n23
Lafitau, Joseph François, 5, 9, 226, 228, 231
Lange, Joachim, 106, 110, 111-112, 118
Larroque, Tamizey de, 181n5
Latin West, 50
Laudan, Rachel, 307
Lauremberg, Petrus, 284
Law, William, 129
Le Clerc, Daniel, 65, 66, 68, 69, 70, 80n86, 124n53, 277, 286, 291n6
Le Clerc, Jean, 64, 65, 71, 77n49, 108-109, 112, 115, 116, 118, 119, 122n33, 124n53
Le Gendre, Gilbert-Charles, 16
Le Jay, Gui-Michel, 172, 184n47
Leibniz, Gottfried Wilhelm, 6, 13, 18, 19, 66, 77n47, 87, 95, 120n2, 150, 211, 212

Leopold I, Holy Roman Emperor, 222n67
Lepenies, Wolf, 28n51, 223
Leporin, Christian, 320
Leris, Michel, 230
Leschassier, Jacques, 166
Lessius, Leonard, 282
Lévi-Strauss, Claude, 23
Linacre, Thomas, 280, 287
Lindeman, Erik, 208-209, 213
Linnaeus, Carl, 8, 325-335
Lipsius, Justus, 85
Livy, 218n9
Lloyd, Geoffrey, 71
L'Obel, Mathias de, 245
Loche, Gilles de, 173
Locke, John, 5, 13, 14, 63, 71, 77n49, 108-109, 112, 114, 118, 123n33, 124n50, 124n53, 131, 132, 146-151, 153, 154, 157n16, 229-230
Lombard, Peter, 49
Lucretius, 190
Ludewig, J. P., 210
Ludwig, Karl, 206, 299, 304, 308
Lull, Ramon, 49
Lusitanus, Amatus, 57n34
*lusus naturae*, 252
Luther, Martin, 49, 88, 95, 107, 108, 109, 111, 213, 267, 303

Machiavelli, Niccolo, 259n45
MacIntyre, Alasdair, 127
Mackenzee, 110
Mackenzie, James, 8, 277-278, 284, 285-295
Macrobius, 21
Maddox, Isaac, Lord Bishop of Worcester, 288, 291n5
Magnus, Albertus, 49, 50, 57n33, 248
mammalia, 9, 327-330, 333n46
Mandeville, Bernard, 129
Marianne, 328
Marsham, Sir John, 113
Marx, Karl, 24, 26, 302
Maternus, Firmicus, 272
mathesis, 14, 15, 18, 24, 25, 26
Mauthner, Fritz, 312n16
medical humanism, 280
Mei, Girolamo, 187-188, 192, 195-197, 199n2
Meiners, Christoph, 134
Melanchthon, Philipp, 205
Menestrier, Claude-François, 183n44
Mercuriale, Girolamo, 281
Mesmes, Henri de, 41, 42, 44, 45, 46, 51

Methuen, Charlotte, 103
Michelangelo, 196
Miedis, Bernardinus Gomesius, 57n33
Mikkeli, Heiki, vii, 8
Mill, John Stuart, 128
Miller, Peter, vii, 5-6
Minuti, Theophile, 171-173
Moderns, 4, 49, 50, 93, 94, 95, 145, 149, 154, 195, 198, 199n2, 287, 288
Moffett, Thomas, 243-245, 253
Moller, Joannes, 75n25
Momigliano, Arnaldo, 28n46, 163, 165
Mommsen, Theodor, 301, 308, 313n26
Monardes, Nicolás, 243
Montaigne, Michel de, 3, 9, 16, 29-38
Montanus, Joannes Baptista, 57n34
Montesquieu, Charles Louis de Secondat, Baron de, 224, 231
Montfaucon, Bernard de, 173
Moore, George Edward, 138, 139
Mora, Jose Ferrater, 159n31
moral philosophy, 9, 127-142, 147, 153, 154, 155, 203
Moran, Bruce, 267
More, Henry, 67, 112, 114, 115, 117, 131, 141n15
Moreau, René, 41, 42
Morhof, Georg Daniel, 62, 63, 64, 66, 75n25, 163
Morin, Jean, 166, 167-170, 171, 174, 177, 184n47
Mosca, Felice, 157n11
Moses, 61, 62, 67, 68, 80n92, 81n95, 130, 132, 170, 173, 177, 179, 270
Mosheim, Johann Lorenz, 65, 69, 119
Motherby, G., 285
Moulton, Thomas, 281
Moyer, Ann, viii, 6
Muis, Simon de, 170, 177, 184n47
Müller, Johannes, 298, 300, 302-308, 316n87
Mulsow, Martin, viii

Nägeli, Karl von, 305
natural law, 6-7, 103, 106, 107, 108, 111-115, 116, 119, 121n17, 132, 133, 203-222, 227-229, 230, 322
natural philosophy, 1, 7, 8, 34, 35, 38, 43, 44, 48, 49, 50, 51, 55n11, 56n25, 57n33, 57n34, 57n35, 64, 65, 66, 71, 187, 239-260 *passim.*, 262-263, 264, 266, 280, 298, 302, 303, 305, 306, 308, 310, 319, 322
natural theology, 36, 112, 113, 115, 116, 118
Naudé, Gabriel, 3, 41-57, 263

Necepso, 272
Nelles, Paul, viii, 3
Nepos, Cornelius, 218n9
Nero, 271-272
Newton, Isaac, 61, 118, 231, 321
Nicolini, Fausto, 157n11
Nifo, Agostino, 50, 57n33
Noah, 62, 68, 129, 130, 131
Northumberland, Henry Percy, 9th earl of, 258n23
*novatores*, 17, 49, 50, 51
Nunes, Fernand, 172

O'neil, Onora, 231
Olearius, Gottfried, 64, 85, 91, 92, 98n22
opinion, 16-18, 44
Origen, 64
Orpheus, 69, 70, 247
Osiander, Andreas, 262, 265, 268
Ovid, 156, 253

Pagden, Anthony, viii, 7
Pagel, Walter, 309, 310
Palissy, Bernard, 245
Paracelsus, 49, 50, 65, 240, 248, 252-253, 266
Paré, Ambroise, 245
Parfit, Derek, 136, 137
Paris Academy of Sciences, 256
Paris Polyglot Bible, 166, 169, 184n47
Parlements, 29
Parmenides, 115
Partibus, Jacobus de, 57n34
Pascal, Blaise, 110
Patriarchs, 62
Patrizi, Francesco, 50, 57n33, 62, 63, 75n28, 192, 240
Paullini, Christian Franz, 213
Peiresc, Nicholas-Claude Fabri de, 5-6, 56n25, 163-184
Penny, Thomas, 243
Pereira, Benito, 62
Pergaeus, Apollonius, 65
Perosiris, 272
Persians, 113, 132
Persius, 32
Petit, Samuel, 168-169, 183n39
Petrarch, 143, 156n9
Petreius, Johannes, 264, 274n11
Peurbach, Georg, 264, 274n18
Phaedrus, 218n9
Philo, 64, 68, 76n41
*Physikalische Gesellschaft*, 298
Piaget, Jean, 28n50

Pian del Carpini, Giovanni da, 225
Picinnino, 272
Pico della Mirandola, Giovanni, 49, 57n31, 62, 66, 67, 75n28, 154, 264
Pietism, 105, 106, 111, 112, 119
Pindar, 16
Pinelli, Gian Vincenzo, 45, 56n19, 182n20, 197
Pingree, David, 71
Pintard, René, 55n6, 56n26
Pius II, 329
Pizan, Christine de, 330
plant sexuality, 9, 325-326
Platina, 195
Plato, 13, 15, 16, 62, 63, 69, 72, 74n19, 75n28, 76n40, 77n47, 80n92, 84, 88, 95, 98n35, 114, 115, 116, 117, 118, 119, 124n62, 128, 131, 133, 134, 138, 139, 143, 146, 147, 148, 153, 155, 185, 186, 188, 189, 192, 194, 196, 249, 253, 268
Platter, Thomas, 254
Plautus, 183n39
Pletho, 62, 63, 66, 75n28, 77n49
Pliny, 49, 239, 240, 242, 243, 244, 247, 248, 249-252, 266-267, 325
Plutarch, 36, 64, 70, 280
Pocock, J. G. A., 24, 28n46
Poggio Bracciolini, Gian Francesco, 57n31
Poiret, Pierre, 66, 118
Poliziano, Angelo, 5, 42, 46-47, 53, 57n31, 265
Polybius, 218n9
Polyhistor, Alexander, 15
Pomponazzi, Pietro, 49, 50, 57n33, 73n9
Pope, Alexander, 14
Popelinière, Lancelot Voisin de la, 165
Popper, Karl, 16
Porphyry, 74n19, 268
Porrée, Gilbert de la, 15
Porta, Giovan Battista della, 240
positivism, 2, 21, 48, 308-311, 322
Possevino, Antonio, 3, 41
Postel, Guillaume, 113, 170, 184n49
Potamon, 64, 84, 88
Potter, John, 80n83
Pozzo, Cassiano dal, 56n26, 183n44
*praecognita*, 17
Praxiteles, 139
Pregitzer, 209
*prisca philosophia*, 77n49
*prisca sapientia*, 7, 64, 65, 66, 76n37, 119
*prisca theologia*, 4, 61-64, 67, 80n83
Pritchard, Evans, 224
Proclus, 70, 81n102, 261, 268, 269

Progressive Liberal Party, 299
Protagoras, 128
Protestant Reformation, 303
Psellus, 77n49
Ptolemy, 7, 8, 47, 261, 265, 266, 268, 269, 270, 271, 272, 273
Pufendorf, Esaias, 205
Pufendorf, Samuel, 6, 81n95, 107, 108, 111, 113, 114, 132, 203-222, 227, 229
Purchas, Samuel, 225, 230
Puritanism, 255
pyrrhonism, 52, 53, 56n26, 111, 119, 123n46
Pythagoras, 5, 6, 14-15, 17, 37, 88, 98n35, 114, 121n17, 129-136, 140n11, 142n26, 151, 185, 186, 187, 188, 189, 191, 194, 268, 269, 286

quadrivium, 2, 6, 25, 186, 189, 199
Quintilian, 157n11, 187, 196

Radcliffe-Brown, Alfred Reginald, 224
Raey, Johannes de, 91
Ralegh, Walter, 244, 258n23
Ramazzini, Bernardino, 282
Ramos, Antonio Perez, 230
Ramus, Petrus, 2, 13, 34, 49, 261-262, 267-269, 272
Ramusio, Giovanni, 225, 230
Ranke, Leopold von, 301, 308, 311
Raphael, 155
Rawley, William, 255
Reale, Giovanni, 134
Rechenberg, Adam, 204
Regiomontanus, J., 265, 272, 274n18
Régis, Pierre Sylvan, 17
Reid, Thomas, 128, 141n18
Reimmann, Jakob Friedrich, 110, 116, 119, 122n33, 124n62
Remus, 329
Republic of Letters, 3, 4, 5, 9, 213
Reuchlin, Johann, 49, 67
Rheticus, Georg Joachim, 7, 263-275
Ribier, Jacques, 45
Richelieu, Armand Jean du Plessis, duc de, 184n47
Romanticism, 308, 310
Romulus, 329
Rondelet, Guillaume, 245
Rosenroth, Knorr van, 67
Rosseli, Frater Hannibal, 79n75
Rossi, Paolo, 64
Rötenbeck, Georg Paul, 98n25
Rothschuh, Karl, 316n87

Rousseau, Jean-Jacques, 156, 229, 231, 322
Royal College of Physicians, 287
Royal Society of London, 256, 319, 320
Rudolf II, 273
Rudolf, Hiob, 213
Rudolphi, 302-303
Rufus, Curtius, 218n9

Sahagun, Bernardino de, 225
Sailly, Denys de, 169-170, 178
Samaritan Pentateuch, 6, 165, 166-168, 170-173, 180
Samaritan Targum, 168, 169, 183n41
Samaritans, 164-184
Sanches, Francesco, 2, 49
Sancy, M. de, 180
Santinello, Giovanni, 62, 129
Santorio, Santorio, 286-287, 290
*sapientia*, 4, 48, 62, 65, 66, 68, 93, 108, 147, 158n19, 328, 334n52
Sarpi, Paolo, 158n23
Sarton, George, 309-310, 316n87
Sartre, Jean-Paul, 159n31
Savonarola, Girolamo, 264
Scaliger, Julius Caesar, 6, 49, 164-184, 264
Schelhammer, Günther Christopher, 284
Schiebinger, Londa, viii, 9
Schleiermacher, Friedrich, 93
Schneewind, Jerome B., viii, 4, 5, 121n17
Schneider, Ulrich, viii, 4, 67
scholasticism, 95, 112, 118, 140, 146, 231
Schreiber, Hieronymus, 262
Schurmann, Anna van, 320
Scythian lamb, 252
Sebon, Ramon, 36
*secta Ionica*, 70
sectarianism, 70, 85, 89, 90, 91, 93, 94, 96, 217
Seidler, Michael, viii, 6
Seilern, Johann Friedrich, 213, 219n21
Selden, John, 130-131, 173
Seligmann, Gottlob Friedrich, 98n22, 98n25
Seneca, 44, 105, 108, 122n23, 122n33
Sennert, D., 68
*sensus communis*, 46
Septuagint, 167
seven Sages, 151, 152
Severinus, 49, 50
Sextus Empiricus, 49, 64
sexual difference, 248, 321, 322-330 *passim.*, 332n21
Shaftesbury, Anthony Ashley Cooper, 3rd earl of, 225-226, 227
Siculus, Diodorus, 218n9

Sidgwick, Henry, 135
Sidney, Philip, 246
Sidonius, 57n31
Sigerist, Henry, 309, 310
Simon, Richard, 94
Simplicius, 268-269
Sinclair, John, 277, 290
Sirleto, Cardinal Guglielmo, 45, 56n19
Skinner, Quentin, 24
Sleidan, Johannes, 205
Smellie, William, 326
Smith, Adam, 142n27, 226, 229, 231
Socinians, 108
Socrates, 5, 15, 88, 98n35, 127, 128, 129, 133, 134, 135, 138, 151, 152, 153, 154, 155
Soemmerring, Samuel Thomas von, 323
Solomon, 225, 247, 256
Sophion, 15
Sophism, 128
Souverain, 116, 119
Sparrman, Andrew, 227
Spencer, Edmund, 246
Spencer, John, 81n95, 109, 113, 122n30
Speroni, Sperone, 192, 201n34
Spinoza, Benedict, 68, 94, 112, 113, 114, 115, 117, 148
Srybius, J. J., 63
St. Bartholomew's massacre, 262
Stahl, Georg Ernst, 302
Stanley, Thomas, 18, 64, 70, 81n102, 85, 130, 133, 141n23
Stäudlin, Carl Friedrich, 134, 135
Sterne, Laurence, 163
Steuco, Agostino, 63, 64, 75n28
Stewart, Dugald, 231
Stoicism, 105, 112, 115, 123n41, 134, 135, 147-148, 154
Stolle, Gottlieb, 92
Stott, Rosalie, 293n40
Strauch, Johannes, 205
*studia humanitatis*, 4, 145, 185
Sturm, Johann Christoph, 4, 86, 88, 89
Stuss, Johann Heinrich, 88, 99n41
Suárez, Francisco, 50, 230
Suetonius, 200n13, 205, 218n9, 271
Sybel, Heinrich von, 308

Tacitus, 32, 144, 205, 206, 218n9, 271, 273
Taranta, Valescus, 57n34
Taylor, 127
Taylor, Charles, 111, 230-233
Telesio, Antonio, 50, 57n33
Telesio, Bernardo, 240, 253

Teller, Romanus, 88, 89, 99n46
Temkin, Oswei, 309
*Tentamen Philosophicum*, 298
*Tentamen Physicum*, 298, 300
Thales, 4, 61, 62, 67, 69-71, 81n102, 88, 262, 266
Theophrastus, 248, 325
Thiroux d'Arconville, Marie, 323
Thomas Aquinas, 49, 228, 229
Thomasius, Christian, 4, 19, 64, 65-66, 77n47, 77n50, 77n51, 78n68, 86, 88, 89, 92, 97n10, 103, 105, 106, 107, 111, 112, 118, 119, 209, 210
Thomasius, Jakob, 19, 93, 113, 115, 123n41
Thomson, Richard, 166
Thornton, Robert, 326
Thou, Jacques Auguste de, 45, 184n49
Thrasyllus, 271, 273
Thriverus, H., 57n34
Thubalcain, 65, 68
Thucydides, 218n9
Tiberius, 271, 272
Tigrini, Orazio, 189, 190, 199n2
Titius, Johann Friedrich, 98n25
Toledo, Francisco de, 62
Tomitano, Bernardino, 192
Topsell, Edward, 243
Torrigiani, Torrigiano de, 280
Tortelli, Giovanni, 61
Toulmin, Stephen, 23
Tradescant, John, 245, 254
Treitschke, Heinrich von, 308
Trendelenburg, Friedrich Adolf, 21
trivium, 2, 251, 259n50
Turgot, Anne Robert Jacques, 226
Turner, William, 242-243, 245
Tyard, Pontus de, 189

Vaillant, Sébastien, 325, 333n38
Valla, Giorgio, 186
Valla, Lorenzo, 5
Valle, Pietro della, 167, 168, 169, 172, 174, 175, 176n2, 183n39
Varro, Marcus Terentius, 200n13
Veneto, Francesco Giorgio, 113
Verene, Donald, viii, 5
Vergil, Polydore, 3, 32, 42, 49
Vesalius, 265, 287
Vesuvius, 239
Vico, Giambattista, 5, 9, 25, 133-134, 143-159, 224, 226, 229, 230, 231
Villanova, Arnaldo da, 280
Vincent of Beauvais, 18

Virchow, Rudolf, 8, 297-317
Virgil, 143, 146
Vitruvius, 186
Vives, Juan Luis, 14, 100n71
Vives, Ludovico, 64, 73n7, 93, 154
vivisection, 303-304, 306
Voltaire, François Marie Arouet, 226
Vossius, Gerhard, 14, 18, 64, 85, 86, 91, 93, 97n18, 112
Vossius, Isaac, 97n18
Vulcan, 68

Wachter, Johann Georg, 114
Wagner, Gabriel, 124n51
Wagner, S. G., 222n67
Walch, Johann Georg, 89
Wallis, George, 285
Weigel, Erhard, 216
Weingart, Peter, 223
Weissenborn, L. W., 320, 324
Wesenfeld, Arnold, 86, 88, 89, 90, 91, 92, 93
Whig interpretation of history, 3, 24-25, 127
White, James, 244, 258n23
Wilhelm I of Prussia, 305
William IV of Hesse-Kassel, 267
Williams, Bernard, 127, 128, 135

Willughby, Peregrine Bertie, Lord, 245
*Wissenschaft*, 13, 302
witchcraft, 29, 31, 32, 33, 36, 37, 265
Wolf, Heironymus, 272
Wolff, Caspar Friedrich, 305
Wolff, Christian, 63, 70, 71, 76n35, 81n103, 83, 84, 94, 100n80, 113, 119
Wollstonecraft, Mary, 320, 329
Wooton, Edward, 244

Xenophon, 127, 218n9

Yates, Frances, 34

Zacconi, Lodovico, 188, 194
Zarlino, Gioseffo, 187, 189, 190-197, 199n2
Zedler, Johann Heinrich, 76n33
Zeno, 88, 98n35, 134
Zeus, 329
Zierold, Johann Wilhelm, 112
Zimara, Marcantonio, 57n33
Zimmermann, Johann Jakob, 115
Zöllner, Karl Friedrich, 306, 312n16
Zopf, Johann Heinrich, 88
Zoroaster, 61, 62, 63, 64, 66
Zwinger, Jacob, 43

GENERAL THEOLOGICAL SEMINARY
NEW YORK